*The
Archaeology
of Social
Boundaries*

Smithsonian Series in Archaeological Inquiry

BRUCE D. SMITH AND ROBERT MCC. ADAMS, SERIES EDITORS

The Smithsonian Series in Archaeological Inquiry presents original case studies that address important general research problems and demonstrate the values of particular theoretical and/or methodological approaches. Titles include well-focused, edited collections as well as works by individual authors. The series is open to all subject areas, geographical regions, and theoretical modes.

Advisory Board

Linda Cordell, University of Colorado Museum
Kent V. Flannery, University of Michigan
George C. Frison, University of Wyoming
Olga F. Linares, Smithsonian Tropical Research Institute
David Hurst Thomas, American Museum of Natural History
John E. Yellen, National Science Foundation

The Archaeology of Social Boundaries

EDITED BY MIRIAM T. STARK

SMITHSONIAN INSTITUTION PRESS
WASHINGTON AND LONDON

© 1998 by the Smithsonian Institution.
All rights are reserved.

Copy editor: Robin Gould
Acquisitions-production editor: Robert Lockhart Jr.
Composition of text and art: Blue Heron Typesetters, Inc.

Library of Congress Cataloging-in-Publication Data
The archaeology of social boundaries / edited by Miriam T. Stark.
    p.   cm.
Includes bibliographical references and index.
ISBN 1-56098-779-0 (alk. paper)
    1. Social archaeology.   2. Ethnoarchaeology.   3. Material culture.   4. Pottery, Prehistoric—
Analysis.   5. Architecture, Prehistoric.   6. Human territoriality.   I. Stark, Miriam T.
CC72.4.A74 1998
930.1—dc21   97-40202
                                                                                                                                   CIP

British Library Cataloguing-in-Publication Data is available.

Manufactured in the United States of America.
04 03 02 01 00 99 98   5 4 3 2 1

∞ ♻ The recycled paper used in this publication meets the minimum requirements of the American National Standard for Information Sciences—Permanence of Paper for Printed Library Materials ANSI Z39.48-1984.

For permission to reproduce illustrations appearing in this book, please correspond directly with the owners of the works, as listed in the individual captions. The Smithsonian Institution Press does not retain reproduction rights for these illustrations individually or maintain a file of addresses for photo sources.

Contents

Illustrations and Tables   vii

Contributors   xi

Preface   xvii

1   Technical Choices and Social Boundaries in Material Culture Patterning: An Introduction   1
    MIRIAM T. STARK

2   Social Boundaries, Technical Systems, and the Use of Space and Technology in the Kalahari   12
    ROBERT K. HITCHCOCK AND LAURENCE E. BARTRAM, JR.

3   Material Culture, Social Fields, and Social Boundaries on the Sepik Coast of New Guinea   50
    ROBERT L. WELSCH AND JOHN EDWARD TERRELL

4   Social and Technical Identity in a Clay Crystal Ball   78
    OLIVIER P. GOSSELAIN

5   Scale, Style, and Cultural Variation: Technological Traditions in the Northern Mandara Mountains   107
    SCOTT MACEACHERN

6 The Cultural Origins of Technical Choice: Unraveling Algonquian and Iroquoian Ceramic Traditions in the Northeast   132
ELIZABETH S. CHILTON

7 Technological Patterning and Social Boundaries: Ceramic Variability in Southern New England, A.D. 1000–1675   161
ROBERT G. GOODBY

8 Coursed Adobe Architecture, Style, and Social Boundaries in the American Southwest   183
CATHERINE CAMERON

9 Social Boundaries and Technical Choices in Tonto Basin Prehistory   208
MIRIAM T. STARK, MARK D. ELSON, AND JEFFERY J. CLARK

10 *Habitus*, Techniques, Style: An Integrated Approach to the Social Understanding of Material Culture and Boundaries   232
MICHAEL DIETLER AND INGRID HERBICH

11 Technology, Style, and Social Practices: Archaeological Approaches   264
MICHELLE HEGMON

References Cited   281

Index   341

Illustrations and Tables

Illustrations

2.1. Botswana, showing Kua research area   14
2.2. Tswana land management system   17
2.3. Sociopolitical class system in Botswana   20
2.4. Ranges of Central Kalahari communities   26
3.1. Sepik coast, northern New Guinea   63
3.2. Varieties of trade relationships along the Sepik coast   72
4.1. Location of potters and linguistic groups surveyed in southern Cameroon between 1990 and 1992   84
4.2. Techniques and materials used by south Cameroonian potters for completing the main stages of the manufacturing process   86
4.3. Several variations in manufacturing technique by addition of coils   88
4.4. Granulometric distribution of clays exploited by south Cameroonian potters with respect to processing techniques   90
4.5. Extrapolated distribution of fashioning techniques with respect to linguistic boundaries   93
4.6. Identity of the person from whom potters interviewed learned the craft   95
4.7. Typical stages in learning to fashion a vessel   96
4.8. Mobility with respect to the learning site among the potters interviewed   97
4.9. Distance between the residential site of the potters interviewed and their learning site   97

4.10. Distance between the place of birth of the teacher and her or his residential site at the time of observations   98
4.11. Movements related to learning within the study area   98
5.1. The Mandara Mountains   116
5.2. Ethnic groups in and around the northern Mandara Mountains   117
5.3. A partial listing of territorial lineage groups between Mora and Mayo Ouldemé   118
5.4. Northern Mandara water-carrying pots   120
5.5. Ceramic traditions around Mora   121
6.1. Southern New England and eastern New York, showing the location of key sites in this study   135
6.2. Reconstructions of indigenous architectural forms   140
6.3. Vessel shapes of the Late Woodland period   145
6.4. Primary inclusion type by site   148
6.5. Thermal expansion curve   149
6.6. Temper density by site   150
6.7. Vessel wall thickness   151
6.8. Body sherd curvature   152
6.9. Surface treatment by site   153
6.10. Vessel lots from the Klock site   154
6.11. Vessel lots from the Guida Farm site   155
6.12. Vessel lots from the Pine Hill site   156
7.1. Tribal territories in southeastern New England   164
7.2. Vessel from the Titicut site, Bridgewater, Massachusetts   170
7.3. Castellated rim sherd from Swansea, Massachusetts   172
7.4. Castellated rim sherd from the Titicut site, Bridgewater, Massachusetts   174
7.5. Incised rim sherd from the Titicut site, Bridgewater, Massachusetts   175
8.1. The American Southwest, showing prehistoric culture areas and archaeological regions   184
8.2. Coursed adobe walls at the fourteenth-century site of Arroyo Hondo Pueblo near Santa Fe, New Mexico   189
8.3. Bis sa ani Pueblo, a Chacoan great house constructed during the early twelfth century near Chaco Canyon, New Mexico   193
8.4. Sand Canyon Pueblo, located in the northern San Juan region near Cortez, Colorado   195
8.5. Plan view of the initial occupation of Arroyo Hondo Pueblo, a fourteenth-century site located in the northern Rio Grande region near Santa Fe, New Mexico   196

8.6. Stone masonry typical of the northern San Juan region at Twin Towers, a thirteenth-century structure at Hovenweep National Monument, southeastern Utah   197
8.7. Stone masonry at Rowe Ruin, a fourteenth-century site located near Santa Fe, New Mexico   198
8.8. Artist's reconstruction of a coursed adobe Hohokam compound   200
8.9. Southwestern deserts   202
9.1. The Tonto Basin in central Arizona with surrounding prehistoric culture areas, as traditionally defined   210
9.2. The Roosevelt Community Development Study (RCD) project area and neighboring sites that date to the early Classic period (ca. A.D. 1150–1350)   214
9.3. Example of "houses-in-pits" from the Colonial period (ca. A.D. 750–950). Locus B at the Hedge Apple site (AZ V:5:189/1605)   218
9.4. Example of pithouses from the early Ceramic period in the Tonto Basin (ca. 100 B.C.–A.D. 600). Excavated portion of Locus B at the Eagle Ridge site (AZ V:5:104/1045)   219
9.5. Examples of masonry compound and room block architectural forms from the RCD and Livingston study areas   221
9.6. Room contiguity indexes   222
9.7. Spatial distribution of Group I (room block) and Group II (compound) residential units defined by the room contiguity index   223
9.8. Distribution of Group I and Group II residential units in the eastern Tonto Basin during the early Classic period (ca. A.D. 1150–1350)   224
9.9. Plan of the Meddler Point site (AZ V:5:4/26), Locus A and B   225
9.10. Plan of the Griffin Wash site (AZ V:5:90/96)   226
9.11. Varieties of thirteenth-century corrugated ceramics recovered from Classic period contexts   229
10.1. The Luo region, showing tribal and Luo subtribal boundaries, major markets with pottery, and major active clay sources associated with potter communities   249
10.2. Examples of one pot category (water storage-cooling pot) from six different potter communities selected to schematically illustrate the nature of micro-style differences   251
10.3. Selected examples of water storage-cooling pots produced within a single potter community near Ng'iya market   252
10.4. Distribution by primary consumers of 1,104 pots emanating from Ng'iya market   255

10.5. A Luo homestead *(dala)* with both circular and rectangular versions of houses   257
10.6. An idealized Luo homestead coupled with a kinship diagram   258

Tables

2.1. Range size, population size, and population density of Kalahari hunter-gatherer populations   24
6.1. Late Woodland ceramic traditions in southern New England and New York   144
6.2. Sample data from the attribute analysis of the Pine Hill assemblage   147
6.3. Common ceramic types for southern New England and New York, A.D. 1000–1700   158
7.1. Technological attributes of Narragansett and Wampanoag vessels   168
7.2. Decorative attributes of Narragansett and Wampanoag vessels   169
9.1. Steps in the operational sequence of hand-built ceramic manufacture   215

Contributors

LAURENCE E. BARTRAM, JR. (Ph.D. University of Wisconsin-Madison, 1993) is an assistant professor of anthropology at Franklin and Marshall College. His research interests include the archaeology and ethnoarchaeology of foraging people, paleoanthropology, and zooarchaeology and the social meaning of archaeofaunas. Some of his recent publications include: A Comparison of Kua (Botswana) and Hadza (Tanzania) Bow and Arrow Hunting, in *Projectile Technologies: Archaeological, Experimental, and Ethnoarchaeological Perspectives* (H. Knecht, ed.; Plenum Press, 1997); Perspectives on Skeletal Part Profiles and Utility Curves from Eastern Kalahari Ethnoarchaeology, in *From Bones to Behavior: Ethnoarchaeological and Experimental Contributions to the Interpretation of Faunal Remains* (J. Hudson, ed.; Center for Archaeological Investigations, SIU-Carbondale, 1993); and Variability in Camp Structure and Bone Food Refuse Patterning at Kua San Camps (with E. M. Kroll and H. T. Bunn), in *The Interpretation of Archaeological Spatial Patterning* (E. M. Kroll and T. D. Price, eds.; Plenum Press, 1991).

CATHERINE CAMERON (Ph.D. University of Arizona, 1991) is an assistant professor in the Department of Anthropology at the University of Colorado. Her research interests include ethnoarchaeology, Southwestern archaeology, architectural analysis, lithic technology, and site formation processes. She currently directs a project excavating a Chacoan Great House in southeastern Utah. Some of her recent publications include Observations on the Pueblo House and Household, in *People Who Lived in Big Houses* (G. Coupland and E. Banning, eds.; Prehistory Press, 1996); Migration and the Movement of Southwestern Peoples *(Journal of*

Anthropological Archaeology 14, 1995); Multi-Story Construction in Southwestern Pueblo Architecture, in *Interpreting Southwestern Diversity: Underlying Principles and Overarching Patterns* (J. J. Reid and P. R. Fish, eds.; Arizona State University Anthropological Papers, 1995).

ELIZABETH S. CHILTON (Ph.D. University of Massachusetts, 1996) is an assistant professor in the Department of Anthropology at Harvard University. Her doctoral dissertation examined the relationship between ceramic variation and cultural context in the Northeastern United States during the late prehistoric period. Her publications include Late Woodland Archaeology in the Middle Connecticut Valley: Ceramic Complexity and Cultural Dynamics (*Journal of Middle Atlantic Archaeology* 12, 1996).

JEFFERY J. CLARK (Ph.D. University of Arizona, 1997) is a research archaeologist with the Center for Desert Archaeology in Tucson, Arizona. His doctoral dissertation focused on the Salado horizon in the American Southwest. Currently, Clark is in the process of publishing the results of another large CRM project in the Tonto Basin as part of his work at Desert Archaeology. His research interests include architectural analyses and detection and assessment of migration in the archaeological record. Some of his recent publications include *Early Stages of the Evolution of Mesopotamian Civilization; Soviet Excavations in the Sinjar Plain* (J. Clark and N. Yoffee, eds.; University of Arizona Press, 1993); and Causes and Consequences of Migration in the 13th Century Tonto Basin (with M. Stark and M. Elson) (*Journal of Anthropological Archaeology* 14, 1995).

MICHAEL DIETLER (Ph.D. University of California-Berkeley, 1990) is an associate professor in the Department of Anthropology at the University of Chicago. He has published numerous articles on material culture theory in collaboration with Ingrid Herbich based upon their ethnoarchaeological research in Africa. His other research interests include political economy in Iron Age France. Some of his recent publications include: Our Ancestors the Gauls: Archaeology, Ethnic Nationalism, and the Manipulation of Celtic Identity in Modern Europe (*American Anthropologist* 96, 1994); *Habitus* et reproduction sociale des techniques: L'intelligence du style en archéologie et en ethnoarchéologie, in *De la préhistoire aux missiles balistiques: L'intelligence sociale des techniques* (B. Latour and P. Lemonnier, eds.; La Découverte, 1994); and Ceramics and Ethnic Identity: Ethnoarchaeological Observations on the Distribution of Pottery Styles and the Relationship between the Social Contexts of Production and Consumption, in *Terre cuite et société: La céramique, document technique, économique, culturel* (D. Binder and F. Audouze, eds.; Éditions APDCA, 1994).

MARK D. ELSON (Ph.D. University of Arizona, 1996) is a senior research archaeologist with the Center for Desert Archaeology in Tucson, Arizona. His research interests include prehistoric economic systems, monumental architecture, and social integration. Some of his recent publications include: Tonto Basin Local Systems: Implications for Cultural Affiliation, Migration, and the Salado (with M. Stark and J. Clark), in *Prehistoric Salado Culture of the American Southwest* (J. Dean and T. Lincoln, eds.; Amerind Foundation, 1997) and Causes and Consequences of Migration in the 13th Century Tonto Basin (with M. Stark and J. Clark) (*Journal of Anthropological Archaeology* 14, 1995).

ROBERT G. GOODBY (Ph.D. Brown University, 1994) is an assistant professor in the Department of Anthropology at the University of New Hampshire; he is also a project archaeologist for Victoria Bunker, Inc., and president of the New Hampshire Archeological Society. His research interests include ceramic analysis, ethnohistory, and New England. Some of his recent publications include: Processualism, Post-Processualism, and Cultural Resource Management in New England, in *Cultural Resource Management: Archaeological Research, Preservation Planning, and Public Education in the Northeastern United States* (J. Kerber, ed.; Bergin and Garvey, 1994) and Native American Ceramics from the Rock's Road Site, Seabrook, New Hampshire (*The New Hampshire Archaeologist* 35, 1995).

OLIVIER P. GOSSELAIN (Ph.D. University of Brussels, 1995) is a research assistant at the University Libre de Bruxelles. His research interests include material culture studies, pottery technology, and cultural identity. Some of his recent publications include: The Bonfire of the Enquiries: Pottery Firing Temperature in Archaeology: What For? (*Journal of Archaeological Science* 19, 1992); Technology and Style: Potters and Pottery among Bafia of Cameroon (*Man* 27, 1992); Skimming through Potters' Agenda: An Ethnoarchaeological Study of Clay Selection Strategies in Cameroon, in *Society, Culture and Technology in Africa* (S. T. Childs, ed.; MASCA-University of Pennsylvania, 1995); and The Ceramics and Society Project: An Ethnographic and Experimental Approach to Technological Choices, in *The Aim of Laboratory Analyses of Ceramics in Archaeology* (K.V.H.A.A. Konferense 34, Stockholm, 1995).

MICHELLE HEGMON (Ph.D. University of Michigan, 1990) is an assistant professor in the Department of Anthropology at Arizona State University. Her research interests include social organization, material culture, and the American Southwest. Some of her recent publications include: *The Social Dynamics of Pottery Style in the Early Puebloan Southwest* (Crow Canyon Archaeological Center, 1995) and Archaeological Research on Style (*Annual Review of Anthropology* 21, 1992).

INGRID HERBICH (Ph.D. candidate, University of California-Berkeley) is a researcher at the University of Chicago. Her research interests include style, ceramic production and distribution, and ethnoarchaeology. Some of her recent publications include: Aspects of the Ceramic System of the Luo of Kenya, in *Töpferei- und Keramikforschung 2* (H. Lüdtke and R. Vossen, eds.; Habelt, 1991); and Space, Time, and Symbolic Structure in the Luo Homestead: An Ethnoarchaeological Study of "Settlement Biography" in Africa, in *Actes du XIIe Congrès international des sciences préhistoriques et protohistoriques, Bratislava, Czechoslovakia, September 1–7, 1991*, vol. 1 (J. Pavúk, ed.; Archaeological Institute of the Slovak Academy of Sciences, 1993).

ROBERT K. HITCHCOCK (Ph.D. University of New Mexico, 1982) is an associate professor and chair of the Department of Anthropology at the University of Nebraska. His research interests concentrate on human rights, environment-human interactions, ethnoarchaeology, and the impact of socioeconomic development. Some of his recent publications include: *Organizing to Survive: The Politics of Indigenous Peoples' Human Rights Struggles* (Routledge, 1997); Each According to Need and Fashion: Spear and Arrow Use among San Hunters of the Kalahari (with Peter Bleed), in *Projectile Technologies: Archaeological, Experimental, and Ethnoarchaeological Perspectives* (H. Knecht, ed.; Plenum Press, 1997; and Kua: Farmer/Foragers of the Eastern Kalahari, Botswana (with H. I. D. Vierich), in *Cultural Diversity among Twentieth Century Foragers: An African Perspective* (S. Kent, ed.; Cambridge University Press, 1996).

SCOTT MACEACHERN (Ph.D. University of Calgary, 1990) is an assistant professor in the Department of Sociology and Anthropology at Bowdoin College. He is the director of the Projet Maya-Wandala, and is engaged in archaeological research on state formation processes and state-nonstate relationships in northern Cameroon. Some of his recent publications include: Foreign Countries: The Development of Ethnoarchaeology in Sub-Saharan Africa (*Journal of World Prehistory* 10[3], 1996:243–304); "Symbolic Reservoirs" and Cultural Relations between Ethnic Groups (*African Archaeological Review* 12, 1994:205–224); and Selling the Iron for Their Shackles: Wandala-Montagnard Interactions in Northern Cameroon (*Journal of African History* 34, 1993:247–270).

MIRIAM T. STARK (Ph.D. University of Arizona, 1993) is an assistant professor in the Department of Anthropology at the University of Hawai'i. Her research interests include material culture studies, economic systems, ethnoarchaeology, and ceramic studies in Southeast Asia and in the American Southwest. Some of her recent publications include: Economic Intensification and Ceramic Specialization in the Philippines: A View from Kalinga (*Research in Economic Anthropology* 16, 1995); Pottery Exchange and the Regional System: A Dalupa Case Study, in *Kalinga Ethnoarchaeology* (W. A. Longacre and J. M. Skibo, eds; Smithsonian Institution, 1994); and Re-fitting the Cracked and Broken Facade: A Plea for Empiricism in the Collection and Use of Ethnoarchaeological Data, in *Archaeological Theory: Who Sets the Agenda?* (A. Sherratt and N. Yoffee, eds.; Cambridge University Press, 1993).

JOHN EDWARD TERRELL (Ph.D. Harvard University, 1976) is curator of anthropology and director of the New Guinea Research Program at The Field Museum, Chicago. He is author of numerous papers and books, including: *Prehistory in the Pacific Islands: A Study of Variation in Language, Customs, and Human Biology* (Cambridge University Press, 1986); Trade Networks, Areal Integration, and Diversity along the North Coast of New Guinea (with R. L. Welsch) (*Asian Perspectives* 29, 1990); Language and Culture on the North Coast of New Guinea (with R. L. Welsch and J. A. Nadolski) (*American Anthropologist* 94, 1992); The Dimensions of Social Life in the Pacific: Human Diversity and the Myth of the Primitive Isolate (with T. L. Hunt and C. Gosden) (*Current Anthropology* 36, 1997); and Lapita and the Temporal Geography of Prehistory (*Antiquity* 71, 1997).

ROBERT L. WELSCH (Ph.D. University of Washington, 1982) is currently adjunct curator of anthropology at The Field Museum and visiting professor of anthropology at Dartmouth College. He is a cultural anthropologist who studies the art and material culture of Melanesia and Indonesia. His publications about the Sepik Coast include *An American Anthropologist in Melanesia: A. B. Lewis and the Joseph N. Field South Pacific Expedition, 1909–1913* (University of Hawai'i Press, 1998), as well as Continuity and Change in Economic Relations along the Aitape Coast of Papua New Guinea, 1909–1990 (with J. Terrell) (*Pacific Studies* 14, 1991); Language and Culture on the North Coast of New Guinea (with J. Terrell and J. A. Nadolski) (*American Anthropologist* 94, 1992); Language, Culture, and Data on the North Coast of New Guinea (*Journal of Quantitative Anthropology* 6, 1996); and Collaborative Regional Anthropology in New Guinea: From the New Guinea Micro-Evolution Project to the A. B. Lewis Project and Beyond (*Pacific Studies* 19, 1996: 143–186).

Preface

The study of social boundaries is a perennial archaeological interest, whether one focuses on prehistoric "ethnicity," migration, economics, or state formation. It is also problematic on conceptual and theoretical grounds. Having mined other disciplines for theoretical frameworks and analytical methods to examine ancient social boundaries, most archaeologists are still dissatisfied with results of their studies. New methods are sorely needed in such research.

This volume explores the utility of methods and concepts from two different schools of thought: one European, and one North American. One approach, inspired by the French *techniques et culture* or *technologie* school, explores links between cognition and technical choice, and their reflection in material culture patterning. The other approach, drawing from North Americanist research on "style," examines formal variation as it is expressed in the goods of everyday life. Although these two technologically oriented approaches from opposite sides of the Atlantic *seem* complementary, little synthetic research has been done to date. Chapters in this volume focus on the relationship between technical choices, social boundaries, and material culture patterning.

My involvement in this research grew, initially, out of my ethnoarchaeological and archaeological research at the University of Arizona. I conducted my dissertation research as part of the Kalinga Ethnoarchaeological Project, which Bill Longacre launched in 1973. He spent the 1975–1976 year in Kalinga, and several of us returned for the 1987–1988 academic year in the Pasil Municipality of Kalinga-Apayao Province. My research concerned pottery production and distributional systems in the village of Dalupa, and I focused on part-time pottery specialization as an economic strategy.

One unanticipated outgrowth of my doctoral project lay in documenting processes of technological and stylistic innovation that had occurred between Bill's initial fieldwork in 1975–1976 and the late 1980s. In so doing, Dalupa potters taught us how technological differences (rather than simple stylistic variation) distinguished their pottery from surrounding ceramic traditions. It was the vessel shape, as much as it was aspects of decoration, that made a pot a Pasil pot. In visiting potters from a neighboring region, they reiterated this motif, and demonstrated how variation in key production steps generated their regional style of ceramic jars.

Somewhat later, I participated in the Roosevelt Community Development (hereafter RCD) study in the Tonto Basin (central Arizona) as part of the Bureau of Reclamation's Central Arizona Project. One research goal in the Roosevelt Community Development study was to investigate the formation and shifts in social boundaries and cultural identity.

Most previous studies in the Tonto Basin and its environs had implicitly or explicitly equated "ethnicity" with material culture distributions, particularly decorated ceramics. In applying North Americanist theories of style to our research, we tried, in various ways, to draw contrasts between isochrestic, symbolic, and iconological patterning. This approach proved ineffective, since our primary data base was undecorated (or utilitarian) ceramics, and compositional studies suggested that local potters did not make decorated ceramics until the end of the developmental sequence. Moreover, their utilitarian pottery circulated in restricted networks. How could we study iconological or symbolic aspects of undecorated goods that circulated in very restricted systems? Conventional methods of stylistic analyses proved inadequate to the task at hand.

We wondered whether alternative methods for studying stylistic variation in ceramics might yield more productive results. Inspired by ethnoarchaeological research, we turned to the literature on technology and culture and applied technological approaches to study locally manufactured pottery and vernacular architecture. What we discovered through research on these technical systems were different scales of social boundaries than had previously been detected in the archaeological record of the Tonto Basin.

The results intrigued us, and I decided to organize a symposium for the 1995 meeting of the Society for American Archaeology in Minneapolis, Minnesota, that explored the limits of this technological approach. Most papers in this volume were presented in this symposium. I challenged participants with active interests in social boundaries, migration, and "ethnicity" to study technical choices in the material culture patterning of their research areas. I asked each contributor to consult and incorporate selected examples of the French tradition of *technologie* approach and recent Americanist treatments of style and technological style. Some authors in this volume had already attempted to synthesize European and Anglo-American intellectual traditions in their research on social boundaries. Most, however, were trained in the North Americanist tradition and had limited (or no) exposure to methods of the *technologie* approach.

Authors used several different research strategies—archaeological, ethnohistoric, and ethnoarchaeological—in their case studies. The time period and geographic region varied from one paper to the next. Authors examined a range of media that varied from ceramics and architecture to personal ornaments and site structure. Unifying the various studies were efforts to explore the central theme of this symposium: to find out how, and whether, technical choices (expressed in technological styles and technical systems) mirror social boundaries. Authors drew a variety of conclusions from the use of a technological approach or their data sets. For some, analysis of technical choices identified previously undetected patterns of variation. In every case study, application of this approach produced fresh insights on the relationship between social boundaries and material culture patterning.

This volume represents the time and labor of many generous people who have helped me in all stages of the process. Each chapter contains specific acknowledgments, but I would also like to thank individuals and institutions who have made particularly meaningful contributions to this book's production. Several key sources provided funding for this book: a National Science Foundation grant (BNS 87-10275) to William Longacre supported my Kalinga research. I also received grants from the Arizona-Nevada Academy of Science, the University of Arizona Graduate Development Fund, the Department of Anthropology Educational Fund, the Center for Desert Archaeology (Tucson), and the Conservation Analytical Laboratory (Smithsonian Institution).

I also extend my heartiest thanks to individuals who have given me infrastructural support to produce this book. In this regard, I am especially grateful to Bill Doelle and Desert Archaeology, Inc. for encouraging me to pursue my research goals through various contract projects, for providing funding for meetings and production costs, and for providing an intellectual community. Ronald Bishop and Lambert Van Zelst at the Conservation Analytic Laboratory (Smithsonian Institution) provided a congenial working environment and institutional support during work on this volume. For assistance with drafting, I thank Ronald Beckwith and Ellen Knight (through Desert Archaeology, Inc.); for graphics and editorial assistance, I thank Jo Lynn Gunness (Department of Anthropology, University of Hawai'i); Kanani Paraso and Timothy Rieth (University of Hawai'i) also provided technical assistance throughout the production process. For photographic assistance, I thank Joe Singer (Honolulu). Smithsonian Institution Press editors Daniel Goodwin, Duke Johns, and especially Bob Lockhart helped me through the complicated process of publishing my first book. So too did Ruth

Spiegel and Robin Gould, who have patiently guided me through the production aspect.

Thoughtful discussions with many colleagues enriched my understanding of approaches and ideas used in this volume. Some of these individuals helped me to understand social contexts of production in Kalinga, including Josephine Bommogas, Pia Awing, Bill Longacre, Cristina Tima, Rosalina Busog, and particularly, the Dalupa potters. Other individuals helped me to interpret the material patterning of the Tonto Basin through conversation and critique, including Mark Elson, James Heidke, Jeff Clark, Steve Lekson, and Ben Nelson. Conversations on related topics with Marcia-Anne Dobres, Laura Levi, James Sackett, Michael Schiffer, and Mary Van Buren sharpened my focus and suggested new avenues of inquiry. Carol Kramer and Bill Longacre have faithfully followed my work for more than a decade now; their encouragement and guidance have helped me immeasurably.

I am, of course, most grateful to Joanna Stark and Stanley Stark, who have supported me from my inception; few are blessed with such loving parents. Ongoing dialogue with Jim Bayman, in spoken and written form, helps me find better ways to express my thoughts and to develop new ideas in my research. To all my kind friends, in so many places, I offer my deepest thanks.

1

Technical Choices and Social Boundaries in Material Culture Patterning: An Introduction

MIRIAM T. STARK

Our discipline has recently witnessed a proliferation of theoretical approaches. There are now so many, in fact, that many of us question whether any themes or interests still unite archaeologists into a single field of inquiry. Most of us, however, still share at least one common interest: we study material culture to answer questions about the past. The specific medium of material culture and the particular past (or present) that we study varies widely from one archaeologist to the next. It is also true that many archaeologists study formal variation in material culture through space: in site morphology, in artifact types, in architectural forms, and in raw materials, for example. A primary goal in studying formal variation across space is to identify social groups, whose boundaries are marked by distinctive patterns in the archaeological record. The study of social boundaries is a perennial archaeological interest, whether one focuses specifically on ancient "ethnicity," migration, or economic systems.

Not only do archaeologists study spatial variability, but we also study ancient technology to illuminate processes and activities of past societies. We suggest in this book that relevant methods for doing so may be found in a growing body of French literature on *techniques et culture,* whose distant North American cousin is an "anthropology of technology" (following Pfaffenberger 1992). This resurgence of archaeological interest in material culture and technology involves different trends in the Anglo-American and European traditions of archaeology. Although these two technologically oriented approaches from opposite sides of the Atlantic *seem* complementary, little synthetic research has been done to date (but see Dietler and Herbich 1994a).

To begin remedying this situation, this volume focuses on relationships among technical choices, social boundaries, and material culture patterning. As archaeologists, we continually debate the validity of different conceptual tools for measuring social boundaries and the meanings of the patterns that we find. Archaeologists have often assumed that studies of *style* will provide clues for distinguishing social groups in the archaeological record. A wealth of stylistic studies is now available from most portions of the globe and for almost every period (for recent reviews, see Carr and Neitzel 1995; Conkey and Hastorf 1990; and Hegmon 1992). Various debates have used ethnoarchaeological or archaeological data to argue that style has iconological, emblemic, assertive, symbolic, isochrestic, or even "stylistic" aspects. Style is either active, contested, negotiated, residual, or unconscious. Yet it is distressing to hear one of the debates' most thoughtful representatives intone that, "despite increasing concern with style . . . we are not much closer to agreement on basic questions than we were in the early 1960s" (Plog 1995:369).

This chapter's goal is to provide a theoretical and historical context for studies of social boundaries, technical choices, and material culture patterning contained in this volume. We must rethink fundamental questions concerning how we study material culture patterning and why we seek a unified theory of style (Carr and Neitzel 1995:6–8; Conkey 1990:8–11; Dunnell 1978; Sackett 1990). What is the relationship of style (as it has been variously defined) to social boundaries? Is it possible to detect conditions under which social boundaries and material culture patterning will coincide? Where might we look for guidance in developing more appropriate theory regarding social boundaries and material culture patterning? No ready answers to such questions have yet been found. Perhaps we can begin to resolve such intractable problems associated with conventional approaches by embracing a *technological* approach to artifact variability.

STYLE, SOCIAL BOUNDARIES, AND METHODS

Papers in this volume offer innovative methods and heuristic tools from both European and Americanist archaeology to study social boundaries. The French tradition of *technologie* explores links between cognition and technical choice by examining the process by which variation is created during the manufacturing sequence. The Americanist approach examines formal variation in finished products and uses a number of different techniques to interpret spatial patterning in the archaeological record.

Two basic premises form a working foundation for contributions in this volume: (1) French archaeologists have already developed a useful set of methods for studying formal variation, glossed as the manufacturing sequence or *chaîne opéra-*

toire; and (2) Anglo-American archaeologists have refined various techniques for undertaking spatial analysis and interpreting distributional patterning. Each contributor tries to bridge these divergent perspectives toward a common agenda. Let us first explore briefly the history of the relationship of material culture studies and anthropological archaeology in the Americanist intellectual tradition.

MATERIAL CULTURE AND TECHNOLOGY STUDIES IN THE AMERICANIST TRADITION

As is widely known, material culture studies formed the foundation of early anthropological research and guided the development of Americanist archaeology (Miller 1987:110–112; Pfaffenberger 1992; Stott and Reynolds 1987). Archaeologists have undertaken replicative studies of lithics and ceramics, for example, since at least the eighteenth century (Coles 1973:13–14; Trigger 1989:61). Anthropologists and archaeologists also published detailed descriptions of traditional technologies during the late nineteenth and twentieth centuries (for reviews, see Conkey 1989 and Ingersoll 1987). Many of these studies were undertaken in situations of rapid assimilation and fit into the Boasian tradition of particularistic anthropological research. Most cultural anthropologists lost interest in studies of technology and material culture at some point after 1920, as goals of anthropological research changed. However, archaeologists retained their interest in this topic as it contributed to the construction of culture histories and regional chronologies.

It was not until the "New Archaeology" came to power in the 1960s that material culture studies faded in popularity among anthropological archaeologists. The Binfordian program criticized these "normative" approaches that culture historians routinely employed to classify material culture into culture areas (e.g., Binford 1965). Most New Archaeologists and processual archaeologists insisted that distributional patterning in the archaeological record could not be equated with ethnic (or ethnolinguistic) groups of the past (Cordell and Yannie 1991; Shennan 1989). Studying technological variation was called passé, and reconstructions of ancient technologies lacked merit unless and until they were placed in an explanatory framework. One reason New Archaeologists did not develop a theoretical framework of technology and material culture was that they viewed technology primarily as an "extrasomatic means of adaptation" (Binford 1965).

Throughout this period, most archaeologists equated conventional material culture studies with uncontextualized description that was considered "boring . . . and theoretically limited" (Ingersoll 1987:4). In many areas of the world, regional chronologies that were based on studies of material culture variability were now in place. As a result, processual archaeologists turned their attention to explanation, and material culture studies were duly exiled to the archaeological basement. However, research on material culture by ethnologists, art historians, and other scholars continued during this interim on a wide range of media (e.g., Lechtman and Merrill 1977; Richardson 1974; Schlereth 1982).

The New Archaeologists' strident disavowal of material culture studies lay in their association of such research with narrow description and a lack of an appropriate theoretical framework (e.g., Longacre 1975). Only one area of archaeological research explicitly devoted to material culture studies gained popularity during the 1970s and 1980s. This field has variously been called "Action archaeology" (Kleindeinst and Watson 1956), "experimental archaeology" (Ingersoll and Macdonald 1977) or "middle-range" or "actualistic" research (Binford 1983). By the 1970s, studies of traditional technology—ethnoarchaeological and experimental—were subordinated to research that provided models of spatial variability or tested specific social inferences (see, for example, Gould 1980; Ingersoll and Macdonald 1977; Kramer 1979; Longacre 1981; Schiffer 1976).

New Archaeologists never dispensed entirely with their interest in technology and were keenly interested in studying formal variation, two intrinsic qualities of material culture. If anything, Binford's use of functional and systems theory approaches to understanding formal variation and technology elevated these studies to a higher level. One of the enduring legacies of the New Archaeology lay in its division of material culture variability into discrete realms of technology, function, and style. Technology was defined as raw materials and production steps, while function became associated with utilitarian or instrumental purposes (following Sackett 1990). As noted previously, style was viewed as a kind of residual quality, whose primary function was emblematic, selectively neutral, or even epiphenomenal (also see Dietler and Herbich, this volume).

Increasing numbers of studies were published that focused on style, function, and technology. Archaeologists appreciated the heuristic value of this tripartite scheme, as evidenced in the explosion of applications using Wobst's (1977) information exchange model (see Hegmon 1992 for a review), the growth of Ceramic Ecology (e.g., Arnold 1985), and in research by Darwinian archaeologists (Dunnell 1978). More intrepid researchers sought to reunite these qualities in novel case studies that combined technological research with ethnoarchaeological techniques to study traditional material culture (e.g., Aronson et al. 1994; Childs 1991; Wiessner 1990).

However, many ethnoarchaeologists who engaged in technological studies became frustrated in attempts to implement this processual model. Ethnoarchaeological studies of traditional societies from Africa (e.g., Braithwaite 1982; Hodder 1979b) to Asia (Longacre 1981; Stark 1995b) showed that makers and users of material culture routinely blur the boundaries between technology, function, and style. As research in traditional cultures has demonstrated, style is not simply decoration, and technical choices are not governed simply by environmental pressures. Rather, these behaviors are socially informed actions that reflect a shared understanding of how things are done. As such, material reflections of technical choices are not neutral, and some are surprisingly resistant to change. These studies suggest that we move beyond dichotomizing models that oppose style and function to embrace more holistic understandings of material culture variability.

Some Americanist archaeologists today may be unfamiliar with the history of material culture studies, but most are clearly amenable to technological studies. Indeed, one might say that Americanist archaeologists have come full-circle to a latter-day appreciation of these topics. This recent resurgence of interest in material culture in Americanist archaeology is paralleled by contemporary developments in the British Isles. This trend reflects no fewer than four factors: (1) the growing impact of experimental-ethnoarchaeological studies on archaeological interpretation; (2) the introduction of new analytic instrumentation into our field from the natural and physical sciences; (3) an increased interest in technological studies as sources of inference for both analytical and theoretical research; and (4) an unsatisfactory conceptual framework for studying spatial patterning and formal variation in material culture.

MATERIAL CULTURE AND TECHNOLOGY STUDIES IN THE FRENCH TRADITION

Technology and culture studies have enjoyed an important place in French theoretical archaeology for several decades (Cleuziou et al. 1991:115–118). Where the "revolution" in Anglo-American archaeology demanded new conceptual frameworks and increased methodological rigor, an equally important shift in French archaeology involved the union of ethnology and archaeology. One outgrowth of this merger, with deep roots in Leroi-Gourhan's work, is the development of a methodology for the study of techniques (see, for example, Lemonnier 1986, 1992; Mahias 1993; Sellet 1993; van der Leeuw 1993). To date, these developments are not strongly associated with a particular theoretical school in archaeology, because of a continued emphasis on analytical refinement (e.g., Perles 1987).

One of Leroi-Gourhan's greatest contributions to French archaeology was his contention that we could understand social structures and belief systems of a society through the study of its technology (also see White 1993). Drawing partly on previous work by Mauss (1936), Leroi-Gourhan suggested that human behavior is characterized by *chaînes opératoires,* or deeply embedded operational sequences. These sequences comprise the foundation of a society's technology and are reflected in all manner of material culture, from everyday tools to the organization of space (Leroi-Gourhan 1993:305, 319).

When we view products as outcomes of multiple technical choices made during the manufacturing sequence, then undecorated artifacts—mundane, utilitarian goods—are also an appropriate study. Any technology is a system of behaviors and techniques, from architectural construction technology to cooking practices. Behaviors and techniques are guided by human choices, and most steps in any technological process can be carried out using several alternative approaches. It is this arbitrariness in technological traits that generates variability in material culture patterning. In artifact types, we may describe this totality as *technological style* (Lechtman 1977). We may describe the combination of manufacturing practices that a group uses to make different goods as its *technical system* (Lemonnier 1986, 1992).

In most media used to manufacture traditional technology, technical problems have alternative solutions. Most everyday goods are made through a series of repetitive and mundane activities; their consistency reflects "the way things are always done" in a local tradition (Wiessner 1984:161, 195). Alternatives selected by artisans in their choice of materials and in the form of their products reflect a thoroughly internalized understanding of the manufacturing tradition. They generally pass this knowledge from one generation to the next (Gosselain 1992b; Lechtman 1977:15; Sackett 1986:268–269, 1990:33, 37), and some aspects of the operational sequence are more stable through time than are others. These technical choices, rather than simply raw materials and or design styles, are crucial in determining the outcome of a product. Technological styles thus reflect conscious and unconscious elements of technical choices.

The interdisciplinary nature of technological studies is evident in the French tradition of *technologie,* because a social anthropologist, P. Lemonnier, has been one of its articulate advocates for the English-language audience (see especially Lemonnier 1986, 1992, 1993b). His ethnological background gives him a similar perspective to ethnoarchaeologists: in each approach, one can observe decision-making behavior, the material outcomes of these sequences, and the spatial distribution of each set of technical choices. One goal of cultural technology studies by French archaeologists has been to describe and understand the entire operational sequence (from raw material procurement to finished good) for a particular group of artifacts. Again, we may see parallels between the French and Americanist schools: Schiffer's (1976) behavioral chain has strong affinities with the concept of the operational sequence (e.g., Sellet 1993).

Much of this theoretical research on culture and technology has gone unnoticed by the Americanist archaeological audience, perhaps primarily for linguistic reasons (but see Dietler and Herbich 1989). Fortunately, English-language publications by Lemonnier (1986, 1992) and Leroi-Gourhan (1993, in translation) have recently appeared. So, too, have English-language reviews of the French literature on culture and technology (see especially Dobres and Hoffman 1994; Lemonnier 1993b) and English-language articles by European archaeologists who follow this approach (e.g., Mahias 1993; Roux et al. 1995; Sellet 1993; van der Leeuw 1993). The appearance of this literature has begun to make French archaeological approaches more accessible to Anglo-American archaeologists. White (1993:xviii) muses that, had translations of this French literature been available twenty years earlier, we might have avoided the *style vs. function* debate altogether!

Anglo-American and French intellectual traditions in material culture studies have distinct trajectories, yet there are surprising parallels in the subjects under study and potential compatibility in the conceptual frameworks that each employs. Recent research on both sides of the Atlantic emphasizes the importance of understanding how technical behavior—the manufacture and use of material culture—creates and mediates social relations. Ironically, the theoretical polarization that now pervades Euroamerican archaeology precludes a shared vision of building material culture theory precisely at a time of renewed anthropological interest in the subject. One premise of this volume is that seemingly disparate schools of thought contain elements of a uniquely archaeological theory of material culture. The cross-fertilization of these perspectives in thinking about social boundaries, technical choices, and "style" promises to yield fresh insights.

STRUCTURE OF THE VOLUME

This volume contains ten case studies and one review chapter. It begins with several ethnoarchaeological case studies in which technical choices are examined in conjunction with their resultant distributional patterning. The first case study, by Robert Hitchcock and Laurence Bartram, explores these issues with findings from their long-term research program in the Kalahari desert. The second case study, by Robert Welsch and John Terrell, uses early twentieth-century ethnological collections and contemporary research to study social boundaries on the Sepik coast of New Guinea. Olivier Gosselain describes his ceramic ethnoarchaeological research in Central Africa in the next chapter. Scott MacEachern follows him with a chapter that discusses contemporary notions of ethnicity and material culture in Cameroon and Nigeria.

Four North American archaeological case studies are then presented to test the limits of this method. The first, by Elizabeth Chilton, examines technical variation in Algonquian and Iroquoian ceramic traditions. A second case study, by Robert Goodby, focuses on technological patterning and social boundaries in southern New England during the late prehistoric and contact periods. Catherine Cameron studies architectural variation in three regions of the late prehistoric Southwest to reevaluate conventional explanations of culture change. Miriam Stark, Mark Elson, and Jeffery Clark focus on the archaeology of the Tonto Basin (Arizona) to study social implications of variability in vernacular architecture and in utilitarian pottery.

Two synthetic discussions of analytical methods, social boundaries, and material culture patterning conclude this volume. The first, by Michael Dietler and Ingrid Herbich, reviews French and Anglo-American intellectual traditions. Their critical analysis of the two traditions serves as a foundation for the proposal of an integrated, independent theoretical position, which further incorporates elements of the Bourdieuian practice theory that is not common to either. They illustrate their approach with ethnographic material from Kenyan research. Michelle Hegmon concludes with a thoughtful summary of recent developments in stylistic studies in Americanist archaeology.

CONCLUDING THOUGHTS

Findings in this volume make it clear that we require continued efforts to unite disparate intellectual traditions. It should come as no surprise that these authors, using divergent theoretical approaches to their research, provide equally divergent conclusions. This lack of agreement provides a good starting point for future work, and may be summarized in terms of the following questions:

1. Can social boundaries be identified in the archaeological record?
2. If such boundaries exist, what methods can we use to examine them in the material record?
3. What social processes and what kinds of social groups can we discern by studying discontinuities in the archaeological record?
4. How can we improve our general understanding of the relationship between technical choices and material culture patterning?

(1) Can social boundaries be identified in the archaeological record? We cannot shake our fascination with identifying social groups in the material record, although most archaeologists question the normative assumptions that such a quest requires. Every archaeological study in this volume searches for (and identifies) social boundaries in material culture patterning. Ethnoarchaeological studies pursue parallel research under the assumption that social boundaries are worth seeking.

The nature of social boundaries is clearly complex, and material culture systems are historically situated phenomena. Many previous studies have concluded that the relationship between style and social boundaries is highly contextualized (following Conkey 1990; Hodder 1979b; Lechtman 1977; Lemonnier 1986; Wiessner

1983): groups signal boundaries using different media from one another, and upon occasion, do not signal boundaries at all. This point is best illustrated in chapters that rely on ethnographic fieldwork (Hitchcock and Bartram, Welsch and Terrell, Gosselain, MacEachern, Dietler and Herbich); many of these studies focus on the relationship between contexts of production and contexts of use. Ethnohistorical studies in the volume by Goodby and Chilton use historical and linguistic evidence to support their contextual interpretations. Chapters by Welsch and Terrell and by Goodby remind us that social boundaries reflect a variety of activities that include efforts to maintain, cross, or blur these boundaries. Chapters in this volume demonstrate that the search for social boundaries in material culture patterning is a productive avenue of research.

(2) If such boundaries exist, what methods can we use to examine them in the material record? Most authors generally approve of this new method for studying material culture variability that incorporates technological and stylistic qualities of the manufactured object. Gosselain's study uses the *chaîne opératoire* approach to greatest effect, and finds it well-suited to unwinding the skein of social relations that affect material culture patterning in South Cameroon. Other authors who focus on technological aspects of material culture, like Cameron and Hitchcock and Bartram, find that this approach reveals new patterns of variation. The technological approach requires us to weld back together style, function, and technology into an integrated whole. Variation in technical systems generates more stable and resilient patterning of social boundaries than does "stylistic variation," but we clearly must learn more about the factors that affect technical behavior.

Ethnoarchaeologists who adhere most closely to the *technologie* approach, such as Dietler and Herbich and Gosselain, identify particular steps in the operational sequence that provide the most information on cultural affiliation. Case studies by Chilton and by Stark, Elson, and Clark show the utility of this approach for studying fine-grained boundaries in the archaeological record. Chapters by Dietler and Herbich and by Hegmon emphasize the need to examine both agency (i.e., the actors responsible for manufacturing culture) and structure (i.e., the material patterning that results from these behaviors). They suggest that Bourdieu's theory of practice and his use of the concept of *habitus* can help bridge the gap between competing views of material culture.

(3) What social processes and what kinds of social groups can we discern by studying such distributions of material culture? Selected chapters in the volume bemoan the lack of fit between the types of social boundaries that we may want to see and the nature of material culture patterning (e.g., Goodby, MacEachern): the social boundaries that we identify cannot be equated with modern ethnic boundaries. Some of the most thoughtful researchers on both sides of the Atlantic still equate social boundaries with ethnicity (e.g., Lemonnier 1986, 1992; Sackett 1990).

As MacEachern and Goodby (this volume) so effectively contend, "ethnicity" is a highly contested and problematic modern concept that eludes translation into archaeological terms, particularly in nonstate societies.

Several papers point out the need for more research on the relationship between social boundaries, scale, and material culture patterning. Chapters by Welsch and Terrell and by MacEachern focus most intensively on the nature and scale of the groups within defined social boundaries. Along the north coast of New Guinea, processes that generate large "social fields" involve the exchange of goods and marriage partners in the quest to build alliances between "friends." These social fields are neither culture areas nor ethnic groups, as they cross-cut multiple ethnolinguistic boundaries. In MacEachern's study, material culture marks social boundaries in Africa that are similarly vast, and do not conform to colonial demarcations of ethnic groups or "tribes."

Social boundaries described in this volume demarcate entities at social scales that cultural anthropologists rarely address, and that archaeologists routinely try to understand. These entities are larger than villages, but smaller than regions or culture areas or ethnic groups. Archaeologists in the American Southwest, for example, have described these entities as branches (Colton 1939) or communities (Wills and Leonard 1994). Various studies in this volume introduce social explanations for how this patterning is created. Welsch and Terrell, for example, discuss social linkages that bind together individual villages (and sometimes groups of villages) into larger units described as "communities of culture" or "social fields." Stark, Elson, and Clark suggest that prehistoric populations in the Tonto Basin were linked into "local systems," or well-bounded, small-scale systems whose participants engaged in regular face-to-face interaction. These local systems were forged and maintained through the exchange of goods and spouses and the formation of dyadic relationships, friendships, and political alliances.

(4) How can we improve our general understanding of the relationship between technical choices and material culture patterning? Many of the volume's authors, irrespective of theoretical orientation, call for more research on the nature of social boundaries and on the relationship between technical choices and material culture patterning. Some ethnoarchaeologists like MacEachern call for systematic long-term ethnographic research that examines spatial scales of social boundaries vis-à-vis their material expression. Other authors, like Dietler and Herbich and Hegmon, advocate the adoption of extant theoretical frameworks to facilitate our archaeological research.

One critical difference among authors lies in their willingness to develop integrated models that have cross-cultural utility, the kind that are generally associated with processual archaeology. It is perhaps not surprising that at least two of the archaeological case studies (Goodby, Chilton) emphasize particularistic approaches to understanding social boundaries, while some authors who believe in unifying principles that govern patterning (Dietler and Herbich, Gosselain, and Hegmon) work with ethnoarchaeological material. This topic clearly requires future work in several directions.

The goal of this volume is to develop new approaches for understanding social boundaries in the archaeological record. What is novel about this book is not its use of French analytical concepts such as *chaîne opératoire* nor its focus on social boundaries. Instead, it is the volume's efforts to conjoin these approaches into a new, and hopefully more productive, strategy for understanding material culture patterning. Papers in this volume illustrate that examining technical choices provides insights on the scale and types of social boundaries in a wide diversity of settings and time periods. Like any pioneering effort, applications of this technological approach are clearly more useful for certain types of material culture than others. Through their findings and critiques, case studies in this volume enrich our understanding of how technological behavior generates and reflects social boundaries and contributes to a growing archaeological theory of material culture.

2

Social Boundaries, Technical Systems, and the Use of Space and Technology in the Kalahari

ROBERT K. HITCHCOCK AND
LAURENCE E. BARTRAM, JR.

Understanding how social boundaries are expressed via material culture is a subject of significant concern in archaeology (Conkey and Hastorf 1990; Lechtman and Merrill 1977; Wiessner 1983). One approach to material culture is that of Lemonnier (1986) who addresses the issue of technical systems, or those relationships among tools, techniques, sets of knowledge, and society. A technical system consists of the interplay between techniques, or the effective actions of people, and the material, social, and ideological contexts in which human activities occur. Humans make choices on the basis of what they have been taught, their personal experiences, and their assessments of the costs and benefits of particular actions. Variations in technical choices are determined situationally through processes of decision-making within a context of perceived opportunities and constraints. In some cases, technical systems are utilized by people to mark distinctions in social identities, statuses, and roles (Lechtman 1977). They are also manipulated purposely by individuals seeking particular ends.

In order to understand the relationships between technical systems and social boundaries, it is necessary to examine the structure, content, and dynamics of technological organization and society. Nelson (1991:57) notes that *dynamics* refers to the plans or strategies that guide the technological component of human behavior. As archaeologists, we are interested in the dynamic processes that shape technological choices, human activities, and the formation of the archaeological record.

In this chapter, we present ethnoarchaeological data at several scales collected among foragers, part-time foragers, wage workers, and agropastoralists in southern Africa. We consider the nature of social boundaries and zones of interaction between linguistically and ethnically distinct co-resident populations, and how the contrasting social uses of space and technology among various southern African groups reflects variability in behavior, social structure, and political organization at various levels, ranging from individual artifacts to sites and clusters of sites. While we make no claims of general applicability, we feel that an understanding of how social space is constructed ethnographically can only be helpful in our efforts to determine its meaning(s) in the past.

Understanding social spaces of the past would certainly be easier for archaeologists if the boundary element or "container" was tangible or delimited concretely. In the best of all possible worlds, we would be excavating fences. Sometimes these fences would be nested in complex patterns and would intersect in complex ways, but each would be labeled clearly and would enclose a single, unambiguous, and mutually exclusive social space. The fences would vary in their strength and permeability. The position of these fences would obviously shift stratigraphically, sometimes abruptly so. The spaces contained within them might reflect kinship, economic activities, ethnicity, gender, or political power, and they would each be distinctive.

In the absence of such idealized boundaries, archaeologists have sought to approximate them with a "connect-the-dots" approach, accomplished by mapping carefully chosen artifacts, features, or sites whose positions can be linked arguably to a specific social unit. Line segments drawn between the relevant artifacts or features create the boundaries that we seek to define the limits of social spaces.

Boundaries may be defined in two ways: (1) as the perimeter of an area containing a higher than background density of objects of a certain kind, thought to delimit a particular social or political entity, or (2) as a pattern of boundary markers surrounding a low-density center. Defining social units via material culture is a challenge that is particularly suited to ethnoarchaeology, the study of the material byproducts of contemporary human behavior (Binford 1978a, 1978b; Gould 1990; Yellen 1977). In general terms, ethnoarchaeology contributes to archaeological theory by building inferential bridges between archaeological data and human actions, and by placing archaeologists in provocative situations where new issues for exploration can be revealed. The utility of ethnoarchaeology is that it illuminates the processes by which the archaeological record comes into being.

ARCHAEOLOGICAL AND ETHNOARCHAEOLOGICAL RESEARCH IN SOUTHERN AFRICA

Ethnoarchaeological research among San (Bushmen, Basarwa) in the Kalahari Desert region of southern Africa has been instrumental in developing principles

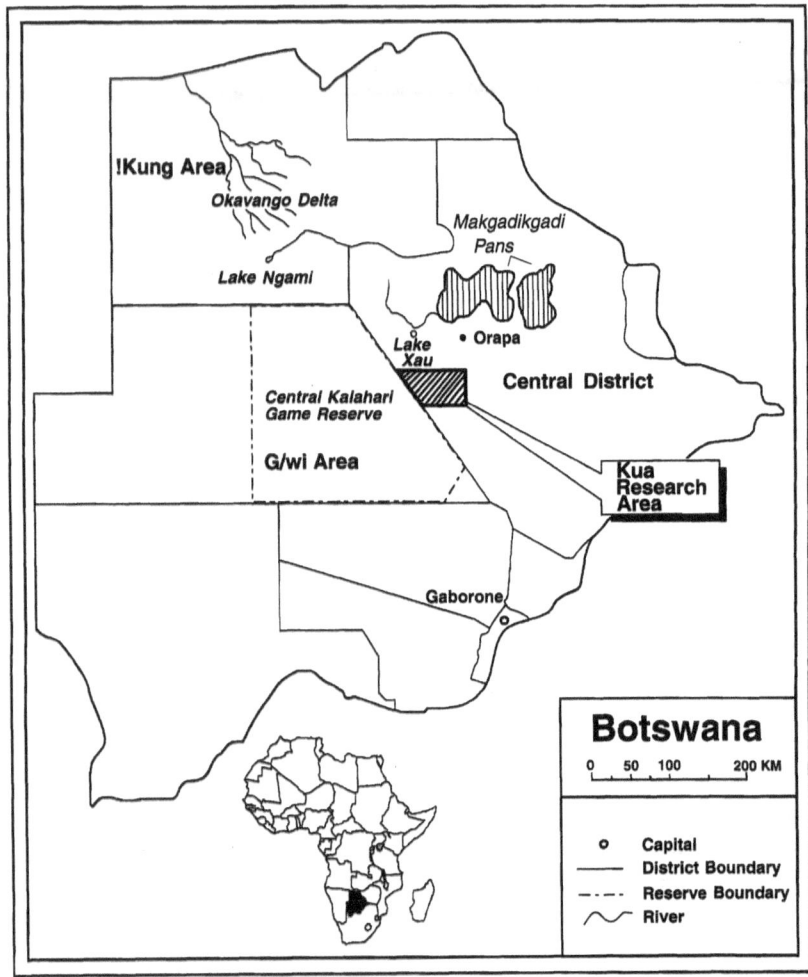

Figure 2.1. Botswana, showing Kua research area.

for the interpretation of site structure and mobility strategies (Hitchcock 1982, 1987; Kent and Vierich 1989; Yellen 1977), faunal remains (Bartram 1993), rock art (Lewis-Williams 1982, 1983), projectile points (Wiessner 1983), and items decorated with glass beads (Wiessner 1984). These analyses suggest that while social information may in fact be contained in material culture, the interpretation of that information is neither easy nor straightforward.

We examine the archaeology and ethnoarchaeology of a diverse mix of populations residing in the east-central and central Kalahari Desert regions of the Republic of Botswana (Fig. 2.1). The region where the study was undertaken can be characterized as semiarid, with rainfall averaging between 400 and 500 mm per annum. Rainfall is highly variable both in space and time; rains tend to occur seasonally, primarily in the period between October–November and April. The vegetation of the eastern and central Kalahari is typical of African savannas and in the study area is dominated by grasses and low shrubs punctuated with "islands" of trees dominated by *Acacia* species (Thomas and Shaw 1991).

Two important geomorphological features in the region are fossil river valleys, which are incised into the Kalahari landscape, and pans, round or elliptical depressions that are up to 2 km in diameter and which in some cases have sand dunes nearby. Water collects in the pans, many of which contain relatively impermeable clays, during the rainy season. These places serve to attract both wild animals and people, who come there to exploit the concentrated and diverse resources.

Archaeological surveys in the area indicate that humans have utilized the eastern and central Kalahari for a substantial period of time, possibly for over a million years. Early, Middle, and Late Stone Age sites have been found in the Makgadikgadi Pans region, just to the north of the study area, and in the hardveld to the east (Cooke 1979; Cooke and Patterson 1960; Hitchcock 1978:93–95; Walker 1995). Middle and Late Stone Age sites were found in the bed of the Letlhakane River and near pans in the east-central Kalahari by members of the University of New Mexico Kalahari Project and land-use planning personnel from the Ngwato (Central District) Land Board. These sites average between 100 and 1,500 sq m in size and consist primarily of scatters of stone tools and debitage. Most of the Stone Age sites are located in areas where surface water was or is available. It is probable, however, that Stone Age foraging groups also ranged away from river courses and pans, exploiting moisture-bearing plants such as melons and roots with the aid of wooden digging sticks.

Iron Age archaeological sites, some of which date back over 1,500 years, have been found in the east-central Kalahari and in the nearby hardveld (Denbow 1983; Hitchcock 1978). These sites vary in size and shape but generally fall into three categories. The smaller sites are found either in open plains areas or on hilltops. They are approximately 1,000–5,000 sq m in size and contain midden deposits made up of dung, faunal remains, and cultural materials. Features on these sites included a kraal (corral) for domestic animals surrounded by a ring of pole and *daga* houses. (*Daga* houses are constructed of a mixture of termite mound earth, cattle dung, and water. This mud is strong and is used for walls and floors of traditional housing.) The next category of sites are found on hilltops and include deeper midden deposits (up to 1.25 m) that cover an area of around 10,000 sq m, and the same kinds of cultural materials and features as the smaller sites along with the foundations of grain bins and in some cases terracing and retaining walls. The largest Iron Age sites, three in number, are located on steep hilltops approximately 100 km apart. These sites average between 80,000 and 100,000 sq m in size, and they contain deep midden deposits of 2 m or more. Features on these sites include stratified house floors, fences, kraals, and trash deposits. One of these sites, Bosutswe, is located in the east-central Kalahari study area near the village of Mmashoro (Denbow 1983). The largest sites are centrally located and are surrounded by smaller Iron Age sites.

Occupation duration of the Iron Age sites varied from the smallest to the largest ones; the smaller sizes appear to have been occupied periodically or semicontinuously from 10 to 25 years while the larger centers were occupied more or less continuously for periods of 150 to 250 years (Denbow 1983:101–102). Cultural materials on these sites included ceramics, iron ore, slag, and metal artifacts, glass, bone, shell, and clay beads, and the remains of both domestic and wild animals. Local Tswana called the middens *marotobola* (trash heaps) because they contained ash, faunal remains, potsherds, and other prehistoric cultural materials (Denbow 1983:84). Many of these sites supported stands of a particular kind of grass, *Cenchrus ciliaris*, which tends to colonize kraals once they are no longer used; this grass apparently can grow in spite of the high concentrations of nitrates and phosphates found in areas where domestic animals have been kept for extended periods.

Contemporary Tswana sites in the eastern Kalahari include cattle posts *(meraka)* consisting of houses, kraals, and a water point, usually a well or borehole (Fig. 2.2). They also included small villages made up of one or several extended family compounds close to agricultural areas known as *masimo* (fields, lands). In addition, the Tswana had extensive villages or towns which consisted of a large number of houses in yards surrounded by fences, public meeting areas *(digkotla)*, and specialized areas such as chiefs' places and cemeteries. These towns were divided into sections or wards in which groups of Tswana or non-Tswana people, who had been incorporated into the Tswana state, resided (Schapera 1938, 1943).

Ethnoarchaeological research efforts were concentrated in the Western Sandveld Region of Central District in a 28,000 sq km area stretching from the southern boundary of the Makgadikgadi Pans south to the village of Lephepe on the border of the Kweneng District, west into the Central Kalahari Game Reserve, and east to the boundary of the Kalahari sand region and the hardveld (see Fig. 2.1). Fieldwork was carried out by Hitchcock in 1975–76, 1977–79, 1980–82, and 1988 and by Bartram from late September 1985 to August 1986 (Bartram 1993; Hitchcock 1978, 1982, 1988). Part of our work consisted of detailed censuses of all of the people in the area along with investigations of subsistence, land use, technology, and employment patterns.

The total population of the area in 1977–78 was slightly over 4,000 people (Hitchcock 1978:217). It consisted of people who belonged to 22 different ethnic groups (Hitchcock 1978:218–222, tables 8.8 and 8.9). About three-quarters of these

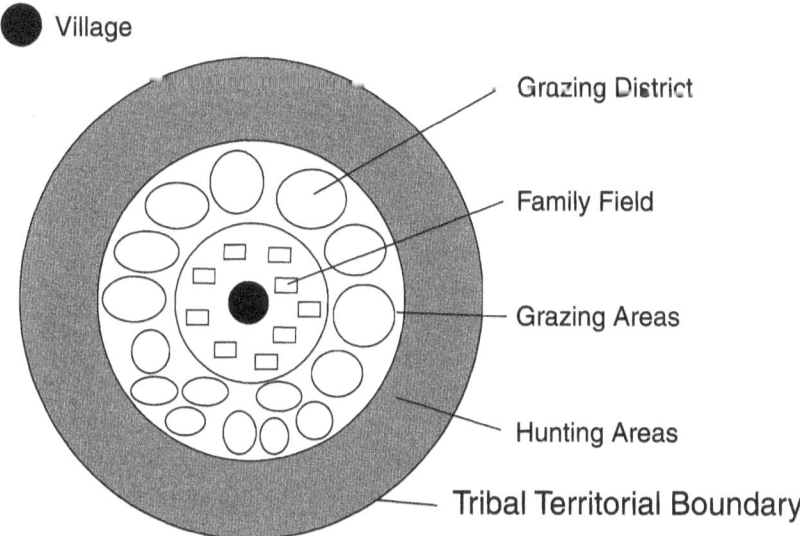

Figure 2.2. Tswana land management system.

people were Basarwa (San, Bushmen). The vast majority of the Basarwa were Northern Kua, while there were also G/wi, G//ana, and Tyua. The latter, who are ethnically and serogenetically distinct from desert-dwelling San, were brought to the Western Sandveld region in the 1950s to work as herders on cattle posts, belonging to the Bamangwato tribal chief's family (Hitchcock 1978, 1982).

A socioeconomic spectrum is evident among the residents of the Western Sandveld region, ranging from sedentary cattle post living to mobile foraging, and from full-time wage laborers to part-time agropastoralists who gain part of their living through the sale of crafts. Kua may be found from one end of this spectrum to the other. In fact, at different times of the year, the same Kua individuals may shift from complete dependence on the agropastoral economy to full-time foraging. Flexibility and mobility, the keys to adaptive success in the frequently hostile habitat of the Kalahari, are conspicuous features of both the subsistence pursuits and settlement patterns of the region's inhabitants.

Kua subsistence technology is portable, efficient, and functional, and in Bleed's (1986) terminology, possesses the design characteristics of a "maintainable system." In contrast to what Bleed (1986) calls a "reliable system," a maintainable system is one suited to the hunting of scattered, ubiquitous game, which are under continual demand. The typical foraging kit carried by Kua adult males consists of a skin carrying bag containing a digging stick, spear, bow, arrows, rope, items used to repair arrows, a poison container (usually made of horn), and sometimes a metal axe. Women usually carry a skin bag or net that contains a digging stick, cord, and sometimes bead-making implements (e.g., a drill and a rubbing stone). Containers used by Kua include wooden mortars, bowls, baskets, and iron pots. The transport of household equipment, personal belongings, food, and children is usually done by people—carrying by hand, over the shoulder, or on top of the head. Some Kua had donkeys or horses they could use as transport aids. These they obtained either as payment for labor on cattle posts or through exchange for goods they manufactured, such as ostrich eggshell bead necklaces and baskets.

Although the Kua are numerically the dominant group in the area that we studied, they have the least political and economic power. The large number of Kua place names encountered reveals their long tenure in the area (Hitchcock 1978). The fact that Kua are often called upon by Tswana herdsmen to serve as guides on hunting expeditions, to perform divination and magic, or to help locate water or delimit the boundaries of a proposed ranch may be because the Kua are regarded as the "indigenous" or aboriginal populations of the Western Sandveld region. Many of the older Kua with whom we conducted interviews affirmed that they were born in the area, and for some of them, this is some seventy years or more ago.

With the expanding numbers of cattle posts and agricultural villages in the area, there is more interethnic interaction today than was the case in the past. This is not to say, however, that the Kua were isolated from other groups; rather, they interacted with them in a variety of ways, exchanging goods with them, working for them as herders and field hands, and, in some cases, entering into long-standing patron-client relations with them (Hitchcock 1978; Vierich and Hitchcock 1996). Even in the 1920s, as Schapera (1930:36) notes, the Kua and other eastern Kalahari San were "all to some extent subject to their Bantu neighbors, often acting as their huntsmen and cattle herders, as well as being required to pay them a regular tribute." Tribute payment consisted of providing Bamangwato and members of other Bantu-speaking agropastoral groups with such as goods as ivory, skins, ostrich feathers, and meat. The Bamangwato profited from the tribute that they obtained, using some of the income generated to expand their cattle herds and to purchase technologically significant items such as guns and wagons. In exchange, the San were able to obtain ammunition, metal tools, ceramics, tobacco, and other goods.

The hunting trade and tribute payment system lasted until the latter part of the nineteenth century when it collapsed, in part as a result of the loss of game animals due to overhunting and the 1896–97 *rinderpest* epidemic. Change also occurred as a consequence of a decision by the Bamangwato chief, Khama I, to renounce the collection of tribute (Hitchcock 1978; Miers and Crowder 1988; Parsons 1973). The nature of the interactions between Kua and other groups shifted, with Kua either working as field hands in exchange for a portion of the crop produced, an arrangement known in Setswana as *majako* (Vierich 1981), or as domestic servants, guides, or ritual specialists such as healers or rainmakers.

It should be kept in mind that the Kua and other Basarwa were key players in the Kalahari agropastoral system. Cattle belonging to Tswana, Kalanga, Herero, and other groups were herded by Kua, who in some cases were able to get a cow a year in exchange for their labor. The vast majority of the cattle posts in the eastern Kalahari were overseen by Kua herders and their families; thus, it is not too farfetched to suggest, as Parsons (1973:96) does, that the Ngwago state existed in part because of the availability of Basarwa labor. Without the livestock management skills of the Kua and the relatively lush grazing of the eastern Kalahari, made accessible through the expansion of water points in the twentieth century, the Bamangwato would not have been able to become as wealthy and powerful as they did.

AGROPASTORAL POPULATIONS IN THE EASTERN KALAHARI

The group with whom the Kua interacted most extensively in the eastern Kalahari from the eighteenth century onwards was the Bamangwato, one of the eight Tswana tribes after which the country of Botswana is named. The Bamangwato are agropastoral people who expanded into the eastern Kalahari and adjacent hardveld areas, incorporating diverse resident groups including the Bakgalagadi, Kalanga, and Bakhurutshe (Hitchcock 1978; Parsons 1973). Because cattle were a highly mobile form of wealth, they could be transferred to individuals in exchange for their allegiance to the individual who owned them. This system, which was known as *mafisa*, had a number of advantages, including spreading one's livestock out, thus reducing risk, and providing holders with domestic animals, which they could use for plowing and milking purposes.

Cattle were the essential resource for the development and maintenance of the Bamangwato *morafe* (tribe). By loaning cattle out, economic and social obligations ensued. Chiefs and other elite tribal members used the livestock loan system to establish alliances and to increase their power base. Under the Bamangwato land tenure system, land was held communally and all tribal members had rights of access to sufficient land for arable, grazing, and residential purposes (Schapera 1943). The tribal chief *(kgosi)* regulated the allocation of land, oversaw the annual cycle of agricultural tasks, and supervised rituals and magico-religious ceremonies, some of which involved sacrificing and consuming cattle.

The Tswana land use system consisted of a centralized town or village *(motse)* where the members of the tribe resided continuously or for most of the year. Surrounding the tribal capital were the lands or agricultural fields where people raised crops. Beyond the agricultural zone were the grazing areas, where cattle posts were established and cattle and small stock (sheep and goat) herds were kept. Many of

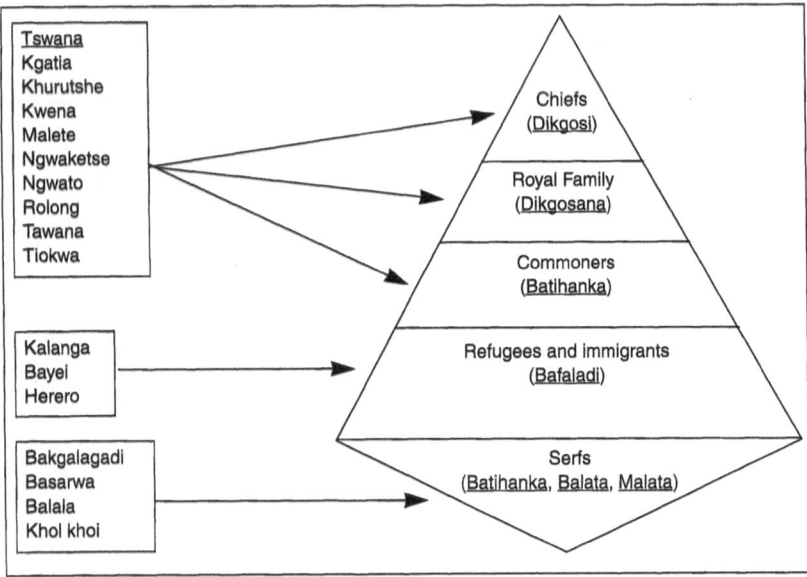

Figure 2.3. Sociopolitical class system in Botswana.

the grazing areas were subdivided into districts *(dinaga)*, which were allocated by chiefs to groups of wards *(dikgotla)*, sections of the tribe that were overseen by headmen with allegiances to the chief. On the peripheries of the tribal territory (the *lefatshe*) were the hunting areas (see Fig. 2.2). This kind of land-use system was overseen by the chief and those whom he appointed to administer it, some of whom were the equivalent of district governors or land managers. Given the Bamangwato tendency to allocate land and water rights only to those people who were defined as members of the tribe, the Basarwa, Bakgalagadi, and other so-called servile groups *(batlhanka)* were in singularly disadvantaged positions, and by the early part of the twentieth century most of them were dispossessed of the land that at one time had been under their sole control (see Fig. 2.3).

Digging wells in grazing districts *(dinaga)* was a crucial factor in bringing about changes in land management and administration patterns. Under Tswana customary law, open surface water was free to be used by anyone who wished. Where water was obtained through the expenditure of capital or labor, as in the case of dam construction or well digging, people were able to keep the water for their personal use and they had *de facto* rights to the grazing areas surrounding the water point (Hitchcock 1980; Schapera 1943).

Changes in water technology initiated early in the twentieth century served to restructure social relations among Kua and other groups. Borehole drilling facilitated expansion of the number of cattle that could be kept. This process, however, led to higher levels of overgrazing and reduction of the resources upon which Kua foragers and poor agropastoralists depended. As a consequence, local people were impoverished, and one of the few options that they had available to them was to enter into institutionalized dependency relationships with cattle owners, whom they worked for in exchange for food, clothing, and other goods. By the 1970s and 1980s, when we carried out our fieldwork, the Kua and their neighbors in the Western Sandveld region were some of the most poverty-stricken people in the country, and they complained bitterly that they no longer led the self-sufficient lives they had as foragers.

LAND USE, LAND TENURE, AND TERRITORIALITY IN THE KALAHARI

The land-use and land-tenure patterns of rural Botswana groups, including Kua, Tyua, Bakgalagadi, Bamwangwato, Kalanga, and Herero in the east-central Kalahari Desert, have a number of features in common with other African land management systems.

Traditionally, African societies managed their land on a communal basis (Schapera 1943). Members of African communities have rights to land, property, and resources insofar as they are members of a specific social group. Land is held in the name of the group. Under these systems of tenure, land cannot be bought or sold, nor can it be pledged as collateral for a loan.

Property in the form of land in Africa consists of what one might describe as a bundle of rights. In many cases, the same piece of land can have a variety of claims on it for various purposes. It is not unusual, therefore, to have complex systems of land and resource rights that are spread widely throughout local communities, and overlapping rights and obligations exist.

Two of the primary factors in land-related matters in Africa are kinship and social alliances. People are allocated land rights on the basis of group membership or, in some cases, through provision by a tribal authority (for example, a chief, a clan elder). An important means of obtaining rights to land in Africa is through marriage. One can also obtain rights to land in Africa through the creation of fictive kinship links. In southern Africa, a person can gain access to land through entering into a patron-client relationship with a person who has recognized land rights. It was not uncommon for individuals to borrow land from a relative, especially land for raising crops.

In some cases, individuals and groups can get land through colonization, the movement into an unutilized area and the establishment of occupancy. Land and resource rights can also be attained through the investment of labor (for example, in clearing a field, constructing a fence, digging a well, planting a tree). There were also cases in the past, and some contemporary contexts, where territorial acquisition occurred through conquest. Land conflicts between individuals and groups did occur in Africa, particularly in those areas where population densities were high.

Land is part and parcel of African sociopolitical systems. It is often perceived as a social and ideological dimension of African society. Local entities in Africa have rights over blocks of land (a band in the case of a foraging society, a lineage or other kind of descent group in the case of a pastoral or agricultural group, such as the Bamangwato). Among the Ju/'hoansi of northwestern Botswana and northeastern Namibia, land is held in the name of a band or local group that consists of several families linked to one another through kinship, marriage, and friendship ties (Lee 1979:334; L. Marshall 1976:157–159). Among the Bamangwato and other Tswana, arable land is held by groups consisting of closely related kin, often a family group made up of several households (a *lekgotlana*), while grazing land is divided among larger segments of Bamangwato society such as wards *(dikgotla)* or tribal sections *(dikgoro)*. Community members who have land rights also have the right to exploit natural resources in those areas, such as wild plant foods and medicines, thatching grass, firewood, and clay for making pottery. They also have the right to exploit wild game that exists in or crosses their land.

In some cases, there are areas that are set aside for the exclusive use of members of the chiefly family, as is the case with grazing and hunting districts among the Bamwangwato (Hitchcock 1978; Schapera 1943). In these situations, the relatives of the chief, the chief himself or herself, or people designated by the chief, have the right not only to use the natural resources but also the human resources there. Local residents of chiefly grazing and hunting areas are expected to work for the chief, serving as guides on hunting trips, laboring in agricultural fields, seeking lost livestock, and providing tribute, known among the Bamangwato as *sehuba* (breast meat).

The Kalahari Desert is by no means an open expanse over which groups move in their never-ending quest for game, plants, grazing, or arable land. Rather, it is a region that is highly diverse ecologically and topographically and which is divided into segments that are recognized by the various groups who reside there as well as by others who use or simply visit the area. Territories of foragers and agropastoralists alike are often named areas with well-known boundaries. These social boundaries are generally respected, and people are supposed to request permission to enter or cross them. People have clearly defined genealogical, social, and ideological ties to these areas, some of which are revealed through the presence of burial areas and other sacred sites.

The San territorial unit is a named area that generally contains sufficient food, water, shade trees, and other resources to sustain a group over the period of a year. Territoriality among San is a subject of considerable discussion and debate (see, for example, Barnard 1992:223–236; Cashdan 1983; Guenther 1981; Heinz 1972, 1979; Hitchcock 1978:242–248; Lee 1979:58–61, 334–339; Marshall 1976:71–79, 184–195; Silberbauer 1981a:99–100, 141–142, 186–187, 191–198; Wiessner 1977:48–59; Wilmsen 1989:51). One of the arguments is over the degree to which groups in these areas have rights of exclusive use of resources within them. It is clear that there is significant variation in territorial boundary protection, with some groups, such as the Tyua of the northeastern Kalahari, marking boundaries carefully and restricting access of nongroup members, and other groups, such as the G/wi and the G//ana in the central Kalahari, generally granting other people permission to enter their territories (Hitchcock 1978, 1982, 1995; Silberbauer 1981a, 1981b; Tanaka 1980). It should be noted, however, that the degree to which boundaries are defined and protected can vary over time, with greater investment in boundary protection when resource densities are low, as is the case, for example, during severe droughts.

Kua and other San were able to gain access to territories of other groups in a number of innovative ways. First, they could marry a person from another group, thus creating an affinal tie that would allow them reciprocal access rights. A second strategy was to establish economic ties through a kind of trade partnership or exchange system. Among the Ju/'hoansi, there is a kind of delayed reciprocity system known as *hxaro*, which was characterized by the transfer of goods across wide areas of the northern Kalahari; the reciprocal exchange relationships facilitated movement not only of goods but also people (Lee 1979; Wiessner 1977). People who shared the same name among the Ju/'hoansi were considered kin and they shared rights to land on the basis of this name relationship (L. Marshall 1976). San inherited land rights from their parents on both sides (Hitchcock 1978; Lee 1979). Individuals were sometimes able to gain access to other groups' land through negotiation with the "owners," those people who had long-standing rights to specific areas. There were also instances in which groups moved into abandoned or uninhabited areas; this occurred, for example, in the central Kalahari region after the smallpox epidemic of 1950–51 (Silberbauer 1981a).

Territoriality had several functions, according to eastern and central Kalahari informants. First, it served to define who had access to resources and who did not. Second, territoriality spaced people out in such a way that there were theoretically fewer conflicts over resources. Third, given that the landscape was divided into specific parcels where people resided for extended periods of time, it was likely that detailed environmental knowledge would be gained from long-term use and monitoring activities. Fourth, territories served a kind of communication function, providing people with information as to the whereabouts of other groups. Finally, one of the useful features of territories was that they enabled groups to conserve their resources, providing them with a means to adjust the numbers of users to the numbers and densities of plants, animals, and other items.

TABLE 2.1
Range Size, Population Size, and Population Density of Kalahari Hunter-Gatherer Populations

Location of Group	Size or Range (sq km)	Size of Population	Population Density*
Northwestern Kalahari			
Dobe=/ai/ai[1]	300–600	9–52	0.43/sq m
Dobe[2]	320	49	0.153
Dobe-/ai/ai[3]	1,000–3,000	—	—
Central Kalahari[4]			
/Xade	906	85	0.094
G!osa	457	21	0.046
Easter Pan	777	50	0.064
Kxaotwe	1,036	64	0.062
Tsxobe	725	70	0.097
Piper Pans	777	53	0.068
Central Kalahari[5]			
G/wi/dom, etc.	2,335	147	0.063
Dantukwe	1,825	41	0.022
Lana	1,637	116	0.071
//oege	2,791	46	0.016
Sibobane	3,415	69	0.020
//Hue	505	45	0.089
Kikao	505	52	0.013
Monatsha	2,615	167	0.064
Metse-amonong	1,825	155	0.085
/o/we	2,673	83	0.031
Molapo-Gyom	4,323	165	0.038
Central Kalahari[6]			
/Kade	4,000	50	0.03–0.058
Western Kalahari[7]			
Ghanzi	30	8–40+	—
Southwestern Kalahari[8]			
N/haite-Hukuntsi-Nwatle	2,200	315	0.143
Pepane-Lehututu-Monong	1,000	247	0.247
Hukuntsi-Tshotswa	1,800	80	0.044
Tshane-Lotlhake	1,600	235	0.138
Kang	1,700	235	0.138
Southern Kalahari, Western Ngwaketse District[9]			
Sekwakwane	160	39	0.238
Mabutsane	400	68	0.170
Kututu	225	59	0.262
Mokgochudi	220	56	0.255
Lokgware Pan	1,600	114	0.071

TABLE 2.1 Continued

Location of Group	Size or Range (sq km)	Size of Population	Population Density*
Eastern Kalahari, Western Sandveld Region (foraging groups)			
Khwee 1	1,100	42	0.038
Khwee 2	1,370	36	0.026
Diphala	940	22	0.023
Ana-O	1,025	29	0.028
Go/o	890	19	0.021
Makokgophane	675	23	0.034
Pulenvane	925	36	0.039
Eastern Kalahari, Western Sandveld Region (part-time hunter-gatherers)			
Man/u	475	57	0.120
Kamokwa	400	39	0.096
Tutu	104	13	0.125
N//aun//au	205	27	0.132
Nyamakatse	186	20	0.108
Mmasogo	314	17	0.079
Kelele	355	30	0.085
Piijinaa	425	44	0.104
Northeastern Kalahari, State Lands Region (north of the Nata River)[10]			
Motomaganyane	195	14	0.072
Gum//gabi	325	88	0.271
Tamashanka	400	32	0.080
Ngwasha	360	50	0.139
Cheberumkwakwa	275	41	0.149
Northeastern Kalahari, Nata River Region (villages along the river)[10]			
Adeshmakae	124	29	0.315
Kwagatsaa	96	27	0.281
Kadisam!kwi	105	36	0.341
Gamtsaa	142	59	0.416
/Garo/gacho	87	21	0.241
Madinoga	90	35	0.389

*calculated as number of persons per square kilometer
[1] Lee 1979:334
[2] Yellen 1977:54
[3] Yellen and Harpending 1972:245
[4] Silberbauer 1972:295, 1981a:193, 1981b:460
[5] Sheller 1977:21, 34
[6] Tanaka 1980:79, 81
[7] Barnard 1979:140
[8] Lawry and Thoma 1978
[9] Childers 1981:30
[10] Hitchcock 1995:189

There are two key concepts of land tenure and resource use in San territorial systems. Local groups have rights over specific territories, which in the Kalahari average between 300 and 5,000 sq km in size (see Table 2.1). They also have rights over the larger area that is viewed as belonging traditionally to the entire ethnic or

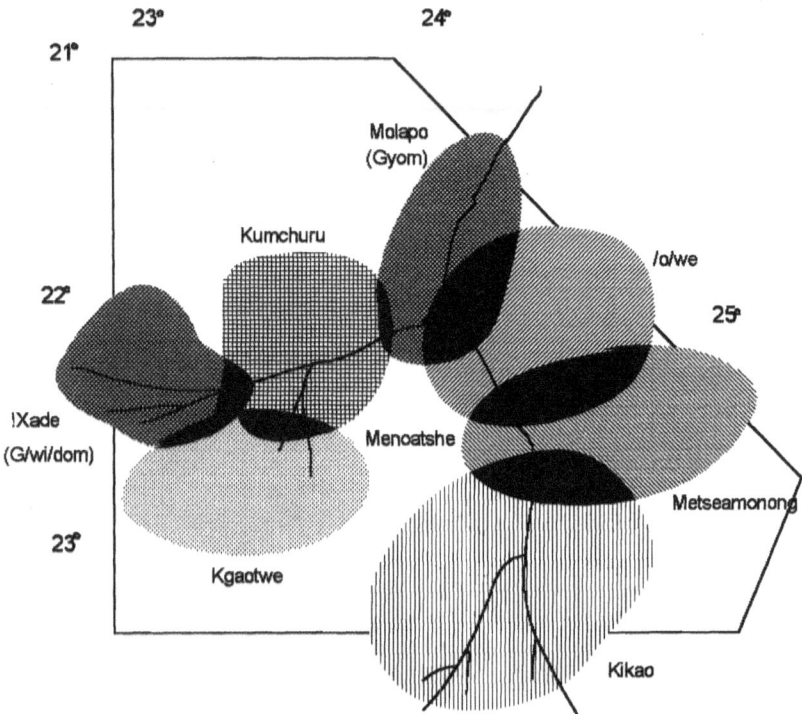

Figure 2.4. Ranges of Central Kalahari communities.

named group. Thus, in the central Kalahari, groups of related G/wi and G//ana have band territories, and they recognize rights to a larger unit called "the great sand face," which belonged traditionally to the indigenous people of the region (see Fig. 2.4). In the eastern Kalahari, the Kua had specific band territories, some of which were arranged in flower petal-like fashion around pans, and they considered the larger region within which the Kua language was spoken to be their own.

San are viewed by some analysts as having strong territorial affiliations (see, for example, Heinz 1972, 1979). It should be noted that it was not only San bands and language groups that recognized and respected traditionally defined group boundaries. Cattle owners in many parts of the Kalahari have also expressed the need to take territoriality and ethnic differences into consideration when deciding upon the placement of water points and hiring workers. It was not uncommon to hear borehole owners say that they had asked permission of local San "territory owners" (*//kaiha*) prior to their decision to drill in particular localities (Hitchcock 1978).

A major change occurred in the eastern Kalahari when new areas were opened to livestock through the digging of wells and the drilling of boreholes. The new water sources meant that cattle could stay in the eastern Kalahari year-round; thus, the system that allowed both the peripheral areas in the east and the sandveld areas to rest for at least part of the year was replaced by one in which cattle stayed in one place continuously. This put greater pressure on grazing and water resources in the area and led to a reduction in the numbers of wildlife and wild plant resources, which were crucial to the adaptive success of foragers.

In the latter part of the nineteenth century, the Bamangwato tribe set up an administrative structure that incorporated diverse groups of people and facilitated control of wide areas (Fig. 2.3). This was done in two ways: by establishing four distinct classes, which correspond roughly to socioeconomic divisions within the society, and by appointing administrators, either district governors or commoners attached in special ways to the chief. The local people, who were termed *bafaladi*, foreigners, were ones who were assimilated peacefully by the Bamangwato or who were incorporated forcibly through conquest. The class (or underclass, as it might be termed) that was essentially dispossessed of land and resource rights in this process was the *bolata*, the serfs or clients, which included San and Bakgalagdi (Schapera 1930:233–234, 1943:43, 260–262).

The paramount chief of the Bamangwato appointed district governors, who were supposed to oversee the various areas in the tribal territory and to provide information about them to the chieftaincy. Like the chief, these overseers had the right to collect tribute from residents of their districts. In some cases, San and Bakgalagdi hunters were provided with guns so that they could hunt elephants, lions, and other animals. Reportedly, the clients manipulated the system to their advantage, shooting twice as many animals as they were told and turning over the required amount of meat, skins, ivory, or rhinoceros horn to their patrons, and selling the rest on the black market (Alec Campbell, pers. comm. 1980).

In the eastern Kalahari there was yet another kind of land division that was of significance. This socially recognized area was the Tswana grazing district. These districts were parceled out not to individuals, as was the case with residential and arable land, but to groups, usually wards or villages linked through kinship and friendship ties (Schapera 1943). The various grazing districts in the Western Sandveld region were divided up into sections, known as *dinaga*. Each *naga*, or sometimes several *dinaga*, had an overseer whose job it was to ensure the proper use of the land and resources in the area. The overseer, known as a *modisa*, the same word for herder, not only obtained rights over the grazing land but also over the people residing on the land. The governor could draw on the services of local people, using them, for example, in corporate hunts. Thus, a new kind of land division was imposed on a landscape already consisting of San territories and ethnically related territorial clusters.

In the eastern Kalahari, as in other areas, the territories of San and Tswana groups were overlapping in some cases and separated from one another in others by areas of what essentially could be described as "no-man's-land." These areas, which generally fell between boundaries, usually were not considered very productive, so people either did not utilize them very often or simply passed through them on the way from one territory to another.

The land tenure picture has become even more complex as a result of environmental, political, and economic changes in the region in the past fifty years. New groups emigrated into the region, including the Herero, who established sizable cattle posts in the 1960s and 1970s. In 1970, the government of Botswana established a series of land boards that replaced the chiefs as land allocation and management authorities. There were sub-land boards set up in parts of the Central District, including ones in Letlhakane and Rakops. These land boards allocated grazing, arable, and residential rights to local people who applied for them. The problem was that members of San communities generally were not granted grazing land, in part because it was assumed that they lacked livestock. In fact, a number of Kua did have cattle, sheep, goats, and donkeys. San who claimed rights to pan areas (e.g., at Lebung in the Western Sandveld region) did not have those rights recognized by the district and subdistrict land boards. As a result, Kua and Tyua found themselves in a position where they had few recognized land rights.

In 1975, the government of Botswana established a land reform and livestock development program, the Tribal Grazing Land Policy (TGLP), which saw blocks of land set aside as commercial ranches being leased out to individuals and small groups. In the eastern Kalahari, a block of 18 ranches was established in the 1980s (Campbell and Main 1991). Some of the residents of this area were required to move elsewhere after the people who got the ranches gained leasehold rights. Some of them went to a nearby communal service center or remote area settlement, Mmaletswai, where they were provided with a borehole, social services, and land for agriculture. The problem was that the size of the area, 64 sq km, was insufficient to sustain people as foragers, so people either had to become wage laborers or they had to migrate out of the area completely. Some Kua and Tyua did this, moving to Serowe, Letlhakane, and other villages to the east and north.

In the latter part of the 1980s and early 1990s, additional veterinary cordon fences were built in the area to control cattle movements. Some of these cordon fences served to restrict movements of wildlife and may well have contributed to the significant decline in game densities over the past two decades (Thomas and Shaw 1991).

An analysis of the social environment of the eastern Kalahari region reveals that the area is complex both in terms of its history and the land-use and tenure systems that exist there. The hunting, gathering, grazing, agricultural, and other economic activities were affected by the changes in land management and tenure systems of the groups that resided in or utilized the region. Social boundaries shifted over time as a result of economic, technological, environmental, social, and political factors. What in the distant past had been a region that contained groups of foragers dispersed relatively widely across the landscape in relatively large territories became an area with more and more social boundaries and smaller land unit sizes, some of them established by San, others by Tswana and other groups who came into the area to work, and still others by the Botswana government. There were changes in social behavior as a result of the shifts in territorial and tenure patterns. Whereas in the past people observed proper etiquette by requesting permission to visit areas where they already had existing social ties, later on, with greater numbers of people and livestock in the region, there was a greater tendency for groups to limit access to their areas. While such a strategy of boundary maintenance was usually accepted at face value by other San, it did not sit well with Tswana groups who had greater political power and whose voices held greater sway in the chiefs' council *(kgotla)* meetings.

By the early 1990s, some of the areas that traditionally had belonged to Kua and other San were either empty, the residents having moved to government settlements, or they contained cattle posts, boreholes, and ranches that were in the hands of other people. The landscape is now divided into Tswana grazing areas, ranches, district council and local government administrative units, and holding pens for livestock near veterinary cordon fences. Boundaries were much less permeable than they had been in the past; in the case of the leasehold cattle ranches, for example, the lessees had the legal right to say whether or not people were allowed to stay or instead had to find alternative places to live. The Kua, who were the original occupants of the Western Sandveld region, and the G/wi and G//ana, who were some of the earliest residents of the central Kalahari, found themselves with few, if any, rights to land. The result was that many of them had moved out of the area in search of more hospitable socioeconomic situations, or they had gone to government-sponsored settlements where they were living in relatively crowded conditions (Hitchcock and Holm 1993; Vierich and Hitchcock 1996). In these settlements, the Kua and other San have a more limited array of economic options, interpersonal conflicts are relatively common, and sizable numbers of people have had to resort to depending on government food and cash-for-work programs for their subsistence and income (Hitchcock 1988; Hitchcock and Holm 1993).

TECHNICAL SYSTEMS AND THE SOCIAL USES OF SPACE

It is clear from the archaeological, ethnoarchaeological, and ethnographic data on the eastern and central Kalahari that social boundaries were observed and respected by a diverse mix of people. At the same time, these boundaries differed in the degree to which they were enforced. There were complex kinds of interactions among both linguistically and ethnically similar and distinct co-resident populations. These interactions ranged from groups and individuals working for other people in exchange for goods, cash, and services to institutionalized patron-client relationships and from sporadic contacts initiated for trading purposes to intermarriage (Hitchcock 1978, 1982; Bartram 1993; Valiente-Noailles 1993; Vierich and Hitchcock 1996).

A common assumption made by archaeologists is that differences in mobility patterns, land use, territoriality, artifact styles, and technological items in the material culture inventory reflect differences in the ethnicity or the ideological orientations of human populations (see, for example, Shanks and Tilley 1987). Structurally oriented and post-processual archaeologists see the archaeological record as being the product of past human thoughts and intents rather than the results of human behavior and natural formation processes. The notion that technical systems can be viewed as "a group of significant choices . . . compatible or incompatible with other choices" and consist of arbitrary and conventionalized assigning of meaning to elements (Lemonnier 1986:172–173) fits squarely into the structuralist perspective. One of the roles of technical systems, according to Lemonnier (1986:173) is to mark difference both between groups and within the same group, demonstrating age, gender, social status, and social, economic, or ideological specialization. Technical systems thus are seen by Lemonnier as signifying systems.

Interviews with Kua, Tyua, Tswana, and other people in the eastern Kalahari indicate that there are differences of opinion over whether or not the styles of artifacts and differences in technical systems are associated with variations in cultural backgrounds of the groups with whom they are associated. For purposes of analysis here, we will look first at the issues of land use and territoriality, and then we will examine the technologies of the peoples in the eastern and central Kalahari. The notions of territoriality and spatial organization of human populations are subjects of significant debate in anthropology (Dyson-Hudson and Smith 1978; Cashdan 1983; Casimir and Rao 1992). In keeping with the broad approach to technical systems, territoriality can be seen not only as control of space within which resources necessary for sustaining life are found, but also as "an interlocking system of sentiments, cognitions, and behaviors that are highly space specific, socially and culturally determined and maintaining, and . . . a class of person-space transactions concerned with issues of setting management, maintenance, legibility, and expression" (R. B. Taylor 1988:6). Territories are seen by people not only in material terms but also in symbolic terms. As one Kua woman put it, our *no* is more than just an area where we find food, firewood, and shade trees, it is the place of our ancestors, and it is the place our children and their children will live."

A Kua man said, "A *no* represents to us the relations we have to trees, bushes, animals, and insects but most of all to each other and to our forefathers." Yet another Kua described his territory as "a sacred place where our fathers and mothers and their fathers and mothers are buried." Clearly, territories represent far more to the Kua than simply areas that people use to obtain food and materials for tool manufacture and construction and where they locate their camps.

The Kua exhibit variations both in space and time in the degree to which they define and defend their boundaries. In the parts of the Western Sandveld that are the most remote, primarily those adjacent to the Central Kalahari Game Reserve, territories are relatively large (roughly 1,000–2,000 sq km in size) and the boundaries ill-defined. In the eastern part of the Western Sandveld, where Kua generally are part-time foragers and small-scale livestock keepers and crop-producers, the territory sizes tend to be smaller (from 100 to 475 sq km) and the boundaries are more precisely defined. In the late dry season or in periods of severe drought, Kua territorial boundaries tend to be marked (with blazes on trees) and reportedly are patrolled in order to ascertain whether outsiders are crossing them. Stress periods result in either greater boundary definition or relaxation of boundaries, the latter being done after sometimes protracted negotiations between the territory's occupants and the people who desire to enter.

The Kua even exhibit what might be described as territorial awareness in the ways in which they locate their camps around a pan. People whose territories are found in the east will establish their residential locations on the east side of a pan. Those groups who have territories to the north of a pan will place their camps on the northern side, and so on around the pan. The result is that a single pan may have as many as five or six groups using it, but each of those groups has a specific "side," as they put it, and hunt and gather in areas radiating out from the pan. It is interesting to note that when individuals come to a new place, they tend to sit on the side of the campfire that reflects the direction from which they came. Thus, according to some Kua informants, it is possible to tell which people around a fire are from which territories simply by looking at seating arrangements around a hearth.

Kua perceive territory in a number of ways. They see it as a kind of resource management device in which access is restricted or granted, depending in part on the state of the resource base and the kinds of socioeconomic relations that they have with other groups. They also see territory as a kind of communication device, one which serves to advertise to people the spatial distributions and rights of various groups. In addition, territory is viewed as a symbol of the social relations among individuals, families, bands, and band clusters or, what Heinz (1979:466–467) refers to as a nexus. Among local groups or bands of Kua there are systems of alliances in which people interact more frequently than they do with other people. While some Kua see these band clusters as being what one might call marriage pools, there were cases where people from those clusters married members of groups from other clusters. The band alliance systems or nexuses consisted of a person's consanguineal and affinal kin as well as those people with whom that person traded, carried out reciprocal visits, and had bonds of friendship. At an even higher organizational level was the language group, or all those people who spoke the Kua language *(Kua dam)* (Bartram 1993:43).

As mentioned earlier, the Kua in the east-central Kalahari were not the only San in the region. On the western fringes of the study area there were G/wi and G//ana, in the southern zone were Tsassi and some Eastern /Hoa, to the east there were the Tshua, to the north the Shua and the Tshhua, and to the northwest there were the Teti (Barnard 1992; Cashdan 1979, 1984; Hitchcock 1978, 1982).

One group which had been brought into the Western Sandveld region in the early 1950s was the Tyua, those who some refer to as being Northern Khoe or Tshwa (see, for example, Barnard 1992:117–131). The Tyua differed significantly from the Kua, though they both spoke Central Bush or Tshu-Khwe languages and had histories of hunting and gathering. In the regions where they normally reside, generally along rivers and around seasonal pans to the east and north of the Makgadikgadi Pans, the Tyua are part-time foragers who also engage in fishing, pastoralism, agriculture, and trade with other groups (Cashdan 1979; Hitchcock 1982, 1995). Sometimes called River Bushmen, the Tyua were notable in part for the fact that they had totemic exogamous groups (Barnard 1992:125–126; Cashdan 1979:41–44; Hitchcock 1982:136–137). The totems, which were inherited through the male line, may have served to facilitate social linkages with other groups, thus functioning as a kind of fictive kinship tie, not unlike the name relationship of the Ju/'hoansi (L. Marshall 1976:238–240).

The Tyua were also noted by observers for the degree to which they were sedentary and overtly territorial. Territorial boundaries were marked, and there were cases of individual point resources such as baobab trees *(Adansonia digitata)* reportedly belonging to specific families, which passed the rights to them down from one generation to the next. Conflicts resulting from perceived infringements on Tyua groups' territories were by no means uncommon, according to some of our informants. Tyua groups along the Nata, Semowane, and Lepasha rivers not only marked their territories by blazing trees and piling up rocks along their boundaries, but they also patrolled those boundaries, sometimes on a weekly or even a daily basis. In order for people to gain access to a Tyua territory, they had to approach the *//kaiha,* either a headman or headwoman, and seek permission. There were reports to the effect that sometimes the negotiations for territorial access lasted several days or even a week or more, and entry was often refused in spite of the extensive discussions.

Technical systems include not only social boundary maintenance and resource management systems but also mobility strategies. People in the central and eastern Kalahari position themselves on the landscape, depending on a number of factors, including the distributions of natural resources and other groups. Mobility strategies are organizational responses to the structural properties of the natural and social environment. In the Kalahari Desert, people make adjustments in movement patterns, group size and composition, resource exploitation strategies, and technology depending on a whole series of variables, including season, resource types and density, and types of transport facilities available (e.g., donkeys). Social factors play significant roles in mobility decisions. People sometimes choose to move to a new location to visit relatives or to carry out trade and exchange (Wiessner 1977). People also move if they feel that the place they are occupying is too crowded and individuals are beginning to get on each other's nerves (Lee 1979). Among the Kua, the death of an individual will often lead to a group abandoning a site and moving elsewhere. Kua also move if illness strikes or if a whole series of mishaps occur, such as people having what they consider to be too many run-ins with large predators (such as lions).

In order to analyze the mobility strategies of foraging and agropastoral populations and the factors that condition them, it is necessary at the outset to draw a distinction between two types of mobility: residential and logistical (Binford 1980; R. L. Kelly 1995:117). Residential mobility can be defined as the movement of an entire group, both producers and dependents, from one point to another on the landscape. Logistical mobility, on the other hand, is the movement to and from a residential location by an individual or task-oriented group for purposes of obtaining matter, energy, or information. Logistical trips consist of either daily forays or expeditions of several days or weeks. The logistical component of a settlement system is related to the organization of production of a society as well as to the distribution of critical resources in the environment, only some of which consist of food items. A Tyua man told one of us (Hitchcock) that the crucial resources to the Tyua were, in order of significance, meat, salt, water, and palm trees. The latter, he said, were important because they provided palm leaves used in the manufacture of baskets, which they traded with Kalanga and other groups for grain, pots, metal tools, and tobacco.

The determinants of mobility strategies in the central and eastern Kalahari tend to vary over time, both on an annual and a longer-term basis. The primary determinant of mobility in the region for the Kua, G/wi, and G//ana is not surface water but rather plants. Water is available in pools for at most two to three months each year; the rest of the time foraging populations must satisfy their moisture requirements by eating water-bearing plants or squeezing juices from the rumens of animals (Silberbauer 1981a, 1981b; Tanaka 1980). In the latter part of the rainy season and into the dry season the Kua depend heavily on melons (e.g., *Citrullus vulgaris* or *Citrullus naudinianus*). During this period they tend to be aggregated in groups ranging from 14 to 42 in size. Mobility is relatively high, with groups making from four to eight moves. In the hot-dry season (August–October), on the other hand, the Kua are dispersed into family groups that live primarily on roots and tubers, especially *bi: (Coccinia rehmannii)*. In this season, Kua family groups usually stayed in a relatively small area, in part because travel was difficult and moisture-bearing plants were limited.

It is interesting to consider the criteria used by eastern Kalahari foragers in determining whether or not to move camp. Some people say that the critical variable is how far they have to walk; thus, there may be an upper threshold on daily foraging trip distance, evidently around 20–25 km, that, when it is reached, serves to convince group members that it is time to move. Other people claimed that it was not the distance traveled but rather the yields that were important. Average daily gathering yields in the eastern Kalahari were between 18 and 25 kg of plant matter (including dirt, husks, and inedible portions). It was noted by some individuals that once the yields dropped to around 5 to 7 kg per day that it was time to locate and move to a new foraging area. A number of Kua informants said that they decided to move after reaching the point where they were gathering most of the day without having the opportunity to sit in the shade and consume a portion of their pickings while they were still out in the field. It is clear that residential shifts are based in part on people's judgments as to the relative costs and benefits of foraging in an area.

The role of information in this system of aggregation and dispersal is important. Prior to the break-up of bands, the localities to be occupied by the various family units are surveyed by individuals or small groups. The resources are assessed carefully and their relative advantages and disadvantages subsequently become the subjects of exhaustive discussion. An advantage of having a territorially based land-use system was that theoretically everyone knew where each group was going to be living. This kind of system, therefore, served as a means of ensuring subsistence security and reduced the threat of competition for scarce resources. Monitoring and information dissemination are crucial in decisions about mobility and land-use strategies.

A crucial determinant of residential shifts in the Kalahari is local resource depletion. The intensity of resource use in an area depends on a number of variables, including the number of consumers (group size), group composition (number of producers), the type and quantity of resources, the efficiency of a group's technology, and the length of occupation of the area. In the eastern Kalahari, there is a correlation between large group size and frequency of residential moves. The frequency of moves varied from once every six to eight weeks in the middle of the wet season to once every six or seven days in the dry season. Larger groups had greater food requirements and thus tended to deplete an area more quickly than might otherwise be the case.

During the rainy season, especially in the period between March and May, daily foraging trip distances tend to be lower than in the dry season, in part because of the relatively high availability of food plants, particularly melons, beans, and berries. In the dry season, on the other hand, trip distance tends to be somewhat higher and energy expenditure greater because there is more reliance on roots and tubers. In the late dry season, when groups in the eastern and central Kalahari are widely dispersed into family units across the landscape, foraging distances again retract because of the difficult conditions. This is the period of greatest stress for Kua, G/wi, and other San populations, and families were observed lying in the shade for long periods during the day in an attempt to conserve moisture and energy. At this time of the year, gathering was usually done only in the early morning and late afternoon in areas relatively close to camp.

In the past, Kua hunter-gatherer groups moved several times from one residential location to another during the course of a year. Those Kua who lived on cattle posts had a slightly different pattern. Kua families tended to remain stationary on cattle posts, moving out primarily to take cattle for grazing or to take meat to the owner of a domestic animal that had died. Long-distance hunting trips were taken by groups of males, sometimes with donkeys and horses, during the rainy season. The numbers of annual moves were also influenced by political decisions. In the eastern Kalahari, Kua groups moved more frequently in 1975–76 than they did in 1978–79, after the declaration of hoof-and-mouth (foot-and-mouth) disease in the area. Individuals were not allowed to take their domestic animals with them when they moved from one area to another as a result of Botswana government restrictions, and as a result they tended to move less frequently.

The Tyua, who were moved into the eastern Kalahari area in the 1950s by cattle owners from the chiefly family of the Bamangwato Tribe, exhibited a different pattern of land use from the Kua. The Tyua on the cattle posts tended to be more sedentary than their Kua neighbors. One reason given for this greater residential stability was that the Tyua herders and their families obtained a greater proportion of their subsistence and income from the cattle owners for whom they worked. Most Tyua cattle post laborers were given bags of maize meal, tobacco, clothing, and other goods in exchange for their labor; they were also paid higher wages in general than the Kua, receiving between 30 and 60 *pula* per month, or approximately $10–$20 (Campbell and Main 1991:45; Hitchcock 1978:314). Tyua herders had the right to drink the milk of the cattle that they were caring for, and they used oxen for purposes of pulling plows and sledges, transport facilities used for carrying firewood, drums of water, and other goods.

The mobility strategies of the Tyua could be characterized as tethered ones in which the cattle post served as the focal points. Tyua groups in the eastern Kalahari stayed close to the cattle posts year-round, though sometimes groups of adult males and boys would go off on expeditions to hunt. Tyua females tended to gather wild plants much less frequently than did the Kua women, in part, they noted, because they got much of their food in the form of milk, grains, and domestic animal meat, but also because they were unfamiliar with the local plants and could not locate roots and bulbs as easily as their Kua neighbors.

A major factor affecting residential mobility in the eastern Kalahari in the 1970s and 1980s was the presence of other groups in the habitat. Nearly every pan had one or more groups which used the place as a dry season fallback point. Some groups began digging wells on the edges of pans so that they could extend the time that they stayed there and so they could water livestock that they had acquired through purchase, loan, or payment for herding work. Kua groups began to aggregate around pans, and eventually some pans, such as Khwee, in the western part of the region, were occupied by as many as five or six different groups, each of them located in camps that were distributed around the pan.

In the 1950s, the government of what was then the Bechuanaland Protectorate underwrote the costs of borehole drilling in the eastern Kalahari in order to provide water for cattle affected by drought (Hitchcock 1978). The expansion in boreholes led to an increase in human and livestock densities in the region. Tyua long-distance hunting and gathering trips had to be planned in such a way as to include areas that had not experienced the devastating effects of overgrazing by livestock. The logistical trips were thus of longer duration since people had to go further west into the parts of the eastern Kalahari that did not have boreholes and cattle posts. Sometimes Tyua logistical groups went into the Central Kalahari Game Reserve to hunt, often with dogs and donkeys, although this was considered illegal according to government faunal conservation legislation (Hitchcock 1988). Tyua men were sometimes arrested for engaging in subsistence hunting, resulting in a loss of crucial labor to Tyua families on the cattle posts. One of the responses to this situation was for Tyua to undertake long-distance trading trips instead of hunting trips, going into towns such as Serowe and Letlhakane to exchange baskets and wooden items, such as spoons that they had carved, for cash, which they then used to purchase goods at local shops.

The data on mobility patterns of the Tyua indicate that their foraging trip distances tend to be much greater than is the case among mobile Kua foraging groups. While the numbers of moves per annum may be lower, they tend to cover larger areas in the course of their expeditions. Long-distance foraging trips were generally undertaken expressly for purposes of obtaining resources in large amounts for storage. The stored food, in turn, meant that fewer foraging trips had to be made.

It is useful to compare logistical hunting trips among Tyua groups with those of the Kua. In the eastern Kalahari most Kua hunting trips were undertaken by a pair of hunters who went out between 3 and 30 km from their camps. Tyua expedition hunts tended to cover much greater distances, sometimes well over 150 km, and they usually included sizable numbers of people, sometimes as many as 20 or 30. Moreover, the Tyua expedition hunts were organized carefully, often with the assistance of a *dzimba,* a man who was particularly skilled in hunting and who was often called upon to take the lead and to make decisions about where to hunt and how to deploy the hunters. These cooperative hunting activities were somewhat reminiscent of the communal hunts organized by Tswana chiefs in which the spoils of the hunt were divided among the participants and those who assisted in the butchering and carrying of the meat (Schapera 1943:258). As was the case with chiefly hunts among the Tswana, an extra portion of the meat obtained in Tyua expedition hunts was set aside for the hunt leader, the *dzimba.* The bones, skin, horns, and other animal parts were processed and made into karosses (skin blankets), carrying bags, and other household items.

One of the observations made by the Tyua, Kua, Bamanagwato, and other people with whom we spoke was that the Tyua living on cattle posts in the Western Sandveld were, for all intents and purposes, nonterritorial. They did not mark territories in the eastern Kalahari, nor did they refuse permission to other people who wished to gather or hunt in the vicinity of the cattle posts where they lived. When asked why they did not maintain social boundaries in the east-central Kalahari whereas they did in the areas in the northeastern Kalahari from which they came, they replied that the eastern Kalahari was not their home; rather, it was the home of the Kua, G/wi, and G//ana. "This is not our land," they noted, "so we have no right to put boundaries on it." It is apparent that the degree to which people maintain boundaries is not simply a product of who they are. Territorial behavior is not innate in some groups and not others; rather, it is a complex set of behavior patterns that relate to structural properties of the social and natural environment and to the meanings which people assign to land, resources, and symbolically significant as well as nonsignificant space. Landscapes are not just regions over which people move; they contain places and things that hold tremendous ideological significance to individuals, groups, and societies. An elderly Kua woman noted that the eastern Kalahari was a "sacred landscape," since it contained the graves of her ancestors, the remains of camps where they had lived, and "special places" where ceremonies such as initiation rites were held.

TECHNOLOGICAL ORGANIZATION

Technology has been a central focus of archaeological research since the inception of the discipline. Recent work on technology has focused on the study of technological organization, or the investigation of "the selection and integration of strategies for making, using, transporting, and discarding tools and the materials needed for their manufacture and maintenance" (Nelson 1991:57). In order to understand technological organization, one must examine the economic, environmental, social, political, and ideological variables that affect those strategies.

Technology can be viewed as an important part of a technical system. Traditionally, archaeologists have assumed that the most common artifacts found in archaeological contexts represented the most important items in a group's technology. Binford (1978a:452; 1978b:338–344), however, has pointed out that societies may differ in the efficiency of their technologies. The Kua and other San have relatively simple but workable technologies; important items of technology are curated or maintained in anticipation of future use. Examples include hunting technology (e.g., bows, arrows, quivers, spears) and items used to manufacture other technological items (e.g., bone needles, awls, adzes).

Among the Kua, moves are made to places that not only contain food, water, and water-bearing plants, but also ones that support specialized resources. These resources might be stone that is used for certain kinds of tools. Some areas contain host plants for larvae that provide poison for hunting arrows. The roots of the *seroka* shrub *(Commiphora pyracanthoides)*, for instance, support the larvae of an insect *(Diamphidia nigro-ornata)*, the entrails of which are squeezed onto the spot just below the arrow point. This poison is virulent and can kill an antelope or other game animal in a relatively short time, depending on the quality of the poison, the place on the body where the animal is hit, and its physical condition.

Some items are purposely not left at residential locations, an example being items and materials used in the manufacture of arrow poison. The technological items that are used to grind up the poison, which consist of small slabs of rock or flat pieces of wood, crushing stones or small wooden pestles, sticks used to apply the poison to the foreshaft below the tip of the arrow, and horn or bone containers in which poison is curated, are all either maintained in protected places where they cannot be reached by small children or they are cached in areas away from the residential location.

Occasionally, items in good condition will be left at a camp or an activity area purposely. This is done by people who anticipate that they will return to the place at some point in the future and will make use of the item. This is done, for example, with nutcracking stones, which are left in mongongo *(Ricinodendron rautanenii)* groves. Other examples include arrow shaft straighteners and stones that are used to sharpen arrow points that are deposited in hunting blinds. These items have been termed "site furniture" by Binford (1978b:339).

Patterns of technological organization contribute to within-system differences. Variability in technology that is seen archaeologically thus may not reflect differences in the cultures or thought processes of the groups that produced the materials and byproducts but rather the environmental, social, and economic contexts of behavior and problem-solving strategies of individuals within a single group or society.

Formation processes of the archaeological record will vary, depending on the degree to which items in the technology are curated. In the case of a curated as opposed to an expedient type of technology, the byproducts of procurement, processing, and consumption will be those most likely to be found in the context of use, thus providing us with important clues to site function. In an analysis of site contents in Kua and Tyua residential locations, it was noted that there was indeed a greater likelihood of worn out items being left at places occupied for extended periods of time (Hitchcock 1982:371). There was another factor at play, however, which had not been anticipated. Tools were sometimes left in abandoned camps and compounds simply because people did not consider them worth carrying at the time they departed. Later on, when a digging stick wore out or a plow blade broke, they returned to the abandoned sites to retrieve the items that had been left there. In this sense, items were not so much discarded as cached for later use. Approximately 20 percent of the transient camps and residential sites that we visited in the eastern Kalahari contained serviceable tools. Given these organizational characteristics, it may be possible to differentiate between site types on the basis of the kinds of remains found on them.

Draper (1975:101) has suggested that a major difference between mobile and sedentary groups is that those groups that are less mobile (more sedentary) will have a richer material inventory. She suggests that the reason for the expansion in the number of items in the technology is that sedentary groups tend to have more tasks to carry out, many of them associated with processing of goods for storage and the preparation of food. Sahlins (1972:11–12) suggests that mobile hunter-gatherer groups would be tied down if they had substantial numbers of material items. As he puts it, "Mobility and property are in contradiction" (Sahlins 1972:12). In his view, mobility reduction would allow people to accumulate more goods since it would no longer be necessary to carry them from one place to another (Sahlins 1972:11–12).

An analysis of the contents of mobile and sedentary groups' sites in the eastern Kalahari, however, tends to cast doubt on Sahlins's arguments about a simple relationship between mobility and technology. Mobile groups sometimes cache items at various points on the landscape, thus reducing transport costs. Mobility reduction does not "allow" the accumulation of goods so much as it requires an expansion in the numbers and types of technological items, particularly those associated with processing, maintenance, and storage of food, tools, and other goods.

In order to analyze the technology of human groups, it is useful to draw a distinction between different types of items, one type being "tools" and the other "facilities." Wagner (1960:92) defines implements as "movable material objects especially designed for the application of energy in precise and controlled ways for particular mechanical tasks." He goes on to point out that tools are "objects used to transmit motion" (Wagner 1960:94). Those tools that transmit motion directly he calls implements, while those used to transmit motion indirectly he calls missiles. Tools, then, would include knives used to cut meat, while missiles include arrows that are shot with the aid of a bow, or spears thrown either by hand or with the aid of a spear-thrower (an atlatl). Those items Wagner calls facilities, on the other hand, are defined as artificial objects whose function is to contain or restrict motion (Wagner 1960:94, 97). Facilities can also be divided into types, one example is containers.

Containers used by the Kua, Tyua, Bamangwato, and other groups in the eastern Kalahari include baskets and pots. Facilities also include houses, which are used to protect people from the elements on the one hand and to hold items of value (e.g., skin sleeping mats) on the other. Another example of a container found in Herero pastoral sites are skin bags in which sour milk (known to the Bamangwato as *madila*) is kept. Other kinds of facilities include "bases" such as trails and "barriers" such as fences and walls.

As part of the ethnoarchaeological investigations conducted in the eastern and central Kalahari, we investigated the contents of sites, recording the tools, facilities, and debris such as faunal remains and the by-products of tool manufacture. We compiled material inventories at various types of sites, ranging from residential sites to special purpose sites (e.g., butchering sites and ambush hunting localities) (see Bartram 1993; Hitchcock 1982:373–380, table 27). The structure and organization of Kua mobile foraging technology in the eastern Kalahari is such that many of the items they possess are implements. These tools are often multipurpose; a digging stick, for example, can double as a kind of hammer or as a spoon. Kua also have some facilities, examples are small wooden mortars and horn containers for arrow poison. The numbers of items in the Kua material inventories ranged from 18 to 75. There tended to be far more items on residential sites than special purpose ones; residential sites had an average of 35 different items on them, whereas special purpose sites averaged around 5.

The material inventories of Tyua, Bamangwato, and Herero groups tend to be much more complex than those of mobile foragers, consisting of substantial numbers of facilities ranging from large beer pots to corrals (kraals) for keeping livestock. The sites that were occupied or used by part-time foragers, pastoralists, agropastoralists, and groups who depended primarily on agriculture tended to have sizable numbers of material items, ranging from 25 to 167. The average number of items on the more complex sites was 120. The long-term residential sites where people not only kept livestock but also engaged in agriculture, trading, wage labor, and occasional foraging were the richest ones in terms of material cultural inventory. In these sites there were often items that were associated with the processing and maintenance of goods. The facilities on these sites ranged from substantial housing, with walls and floors made of mud and cattle dung and thatched roofs, to pole and stick fences surrounding the houses and associated activity areas. In some long-term residential sites associated with agriculture and pastoralism there were threshing floors, large storage containers for grain and other goods, watering troughs for livestock and chickens, metal drums for holding water, baskets of various sizes and shapes, and specialized equipment such as branding irons and bridles.

Many of the facilities of sedentary groups are used in the production, processing, and preparation of food. Plows were found in the yards of residential compounds close to agricultural fields; large sticks were used to beat grain such as sorghum and millet prior to winnowing it in specialized baskets; and large wooden mortars and pestles were crucial in the processing of maize, sorghum, and millet. The grain is mixed with water in large metal pots and cooked, being stirred occasionally with large wooden spoons or specialized porridge stirrers consisting of a carved wooden stick in which bent pieces of wire are inserted at one end. Many of the agropastoral sites had items that were used to maintain the sites, such as brooms and rakes. A major difference between sites of mobile and sedentary groups was that the latter often contained large metal trunks in which people kept clothing, blankets, and other valuable items. There was a whole range of specialized implements and facilities in sedentary sites, including metal stands for iron pots, files, funnels, pliers, calabash containers for seeds to be used in crop planting, steel traps for capturing predators, and wooden, metal, and plastic plates and bowls. In addition, there were items used in personal decoration, for example, headbands, earrings, bracelets, and necklaces, some of them made of metal, others of glass, and still others of ivory. In two sites there were ancient leather shields that were used, according to informants, in battle.

It is interesting to note that the items in the technology that are considered utilitarian have some of the most distinctive decorations on them. This is true, for example, of gourds and calabashes, which are decorated with various designs, some of which are carved into the surface and are sometimes made more obvious by rubbing charcoal and ash into them. Some of the calabashes were used for milk while others served as bowls. When asked about the decorations on the gourds and calabashes, informants replied that the designs served to identify the users. A number of people noted that the decorations varied according to the age of the person who used them. Other people noted that the variability in decorations on containers had to do with the fact that there was more than one family using a cooking area, and they wanted to ensure that individuals were given the appropriate container. An examination of the calabashes in different kinds of households (nuclear, extended, polygynous) indicated that there was greater variability in design elements in those households that had people from several different kinship groups residing together. In polygynous Herero and Bamangwato households, there tended to be greater variation in designs on gourds and calabashes. This was particularly true where there were three or more co-wives living together. The design differences on calabashes in the eastern Kalahari appear to reflect relationships between kinship and resource allocation. This situation resembles that of the Il Chamus in Kenya examined by Osborn (1996), who exhibit significant variation in designs on calabashes belonging to co-wives in polygynous households who are attempting to ensure adequate distribution of milk to their children, especially in times of food scarcity and competition for access to resources.

Another area of variability in eastern Kalahari technical systems relates to the brands on cattle. Herero and Bamangwato herders maintain detailed information on their cattle, often giving them names, differentiating them according to their age, color, horn shape, breed, and other criteria, and going to great lengths to care for them. Brands are put on cattle not only to mark ownership, but there are different brands that are used on cattle from the same herd. When asked why this was the case, people replied that some of the cattle had been pledged to various children who would later inherit them. They were still managed as a unit and kept in the kraal of the father. Thus, the various brands served to differentiate those cattle that would eventually belong to specific kin. Rather than being a mark of family or community holdings, the brands indicated individualized ownership, with the holdership remaining in the hands of the senior family member. It should be noted that brands have become more specific to families with the commercialization of the livestock industry and the expansion in the numbers of herds in the Western Sandveld as a result of rising numbers of water points, enhanced veterinary care, and more intensive livestock education programs.

A surficial assessment of the technological items employed by people in the central and eastern Kalahari might indicate that there is uniformity in materials, manufacturing and maintenance methods and techniques, purposes of specific items, and styles of artifacts. A more detailed examination of the technology carried out at various scales, from individual artifacts to activity areas, sites, clusters of sites, and entire regions, indicates that there is variability both in the organization of the technology and in the various items that make up the technological inventories of individuals, families, communities, age and gender-based groups, and totemic groups.

It is interesting to note that although Kua sometimes draw distinctions between different types of items according to which groups made them, they were observed making and using the full array of types of items. While some distinctive patterns were recognized by people, these patterns were usually not attributed to being a product of ethnic differences. Some technological items were, however, viewed as containing information or signals that distinguished social variation. These items include the following: (1) arrows, (2) decorated ostrich eggs, (3) ostrich eggshell beads, (4) small leather bags on which glass or ostrich eggshell beads were sewn, (5) ceramics, (6) baskets, (7) houses, (8) rock art, and (9) tattoos. Some attributes of technology were seen to reflect personal or individualistic attributes, one example being tattoos, while at the same time they were said to convey information on other aspects of social organization. Facial and body decorations are relatively easily identifiable and are sometimes remarked on as being associated with life-history events. Some tattoos on the sides of the head near the eyes of adult males, for example, were done at the time of their first kill and were said to be helpful in "helping us see wild animals better." Marks on the shoulders of and chests of people were often associated with efforts by shamans or traditional healers to cure people or to prevent them from becoming ill from disease. Still other marks were done ostensibly to protect people from witchcraft. Thus, tattoos represented a number of different events, processes, and phenomena and generally were not attributable to ethnic identity.

The analysis of projectile technology in Africa has revealed variation that can be ascribed to function, style, material type, prey, season, and cover type (Wiessner 1983; Larick 1985; Hitchcock and Bleed 1997). The arrow tips sometimes consist of sharpened pieces of bone, quill, wood, or, in most contemporary contexts, metal. The material used for most metal arrow points by Kua, G/wi, and G//ana today is fence wire, which is heated, beaten, and then filed into the desired shape. Both Kua and G/wi arrow points are triangular in shape. There are two major differences between G/wi and Kua arrow tips that both groups recognize: (1) the G/wi arrow tips are large while Kua arrow tips are small; and (2) the G/wi arrow tips have longer barbs than do the Kua arrow tips. Kua said that differences between arrow tip types, including the craftsmanship and the sharpness, were due in part to "hunting experience," which they said that G/wi were noted for. They also said that G/wi abilities to make good arrows were greater than theirs because the G/wi used them more. When asked why this was the case, the Kua informants said that the people in the Central Kalahari Game Reserve like the G/wi had the legal right to hunt with bows and arrows, something that was not true for them since they lived in what was considered a grazing area.

Arrow points, link shafts, and main shafts of arrows are found only rarely in Kua residential locations. This was due to the fact that people purposely cleaned up the places where they manufactured, maintained, and repaired their projectiles because of the potential danger of people stepping on them and dying from the poison. They also said that they hid hunting weapons like bows and arrows because they feared that they would be arrested by government game scouts or police officers if they had them in their possession.

Arrows were exchanged by hunters, sometimes over extensive distances; as a result, it was not uncommon for individuals to have in their possession items that they did not make themselves. These arrows sometimes varied in materials, types of link shafts, and size and shape of projectile points. When asked about the causes of variability in projectile technologies, informants noted that the differences were due in part to the types of prey people were hunting, the season during which they were used, the time when hunting occurred (day or night), and the availability of certain materials. They rarely ascribed the differences to the social identity of the makers, except when they were shown G/wi arrows. The G/wi, according to the Tswana, were "dangerous people" because they were so accurate in their arrow shooting. The perceived social boundary between the G/wi and the Tswana was that the former were "fierce foragers" who possessed "the secret of arrow poison." When pressed about this issue, the Tswana said that the real reason the G/wi were so dangerous was "because they turned themselves into lions."

Tswana informants noted that there were differences between groups both in terms of the kinds of weapons that they preferred and in the styles or forms of projectiles that they used. Kua and G/wi, they said, preferred to use bows and arrows. The Tswana and Herero, on the other hand, were "spear-users." It is interesting to note that the Tyua, who were moved into the eastern Kalahari region, were, like the Tswana, "spear users." They were also like the Tswana because they engaged in long-distance expedition hunts involving sizable numbers of participants. The fact that the Tyua were also sedentary, herded livestock, had relatively rich material inventories, spoke Setswana as well as a click language, and were incorporated into Tswana age regiments *(mephato)* served to indicate to some Kua that they were, in some ways, Batswana (people of the larger Tswana tribal polity). Yet because they were considered Basarwa (Bushmen) by the Bamangwato and by themselves, they worked as herders in exchange for food and milk, were sometimes involved in institutionalized patron-client relationships *(bolata)* with high status Bamangwato, and were treated as though they had few, if any rights (e.g., they were not allowed to speak in *kgotla,* the chief's council place) meant that they were, for all intents and purposes, Basarwa. As a result, the Kua and G/wi treated the Tyua like they did each other, allowing them to cross their territorial boundaries if they sought permission in socially appropriate ways, trading goods with them, and sharing food with them when they were under stress. The Kua and G/wi considered the

Tyua to be their "uncles" because they shared with them similarities in language and history.

From an archaeological standpoint, however, the sites of the Tyua more closely resembled those of the Bamangwato, Herero, and other Bantu-speaking groups in the eastern Kalahari. The residential sites contained the full array of agropastoral, herder, or sedentary agriculturalist materials: metal tools, ceramics, the remains of domestic fauna, substantial mud houses, kraals, and dumps. In addition, Tyua faunal procurement strategies included spear hunting at night from blinds. The reason that they did this, they said, was not because they traditionally were ambush hunters, but rather because such a strategy ensured that they would not be arrested by game scouts, who generally were seen as being reluctant to patrol hunting areas at night for fear of large predators such as lions. The presence of large numbers of hunting blinds in a region, therefore, might be attributed as much to the social and political context of a region as it could to the frequency of night hunting or the types of prey being sought. When asked why Kua were bow and arrow hunters and Tyua were spear hunters, members of both groups replied that the difference was not due to inherited ancestral traditions, but rather the kinds of animals they were hunting or the type of vegetation cover they were dealing with. When they were shown spear and dart points from the archaeological record in areas currently occupied by G/wi, G//ana, and Kua, local people there said that "obviously there has been a change in hunting strategies." They did not believe that the spear and arrow points were the products of Tyua, Tshua, or Tswana peoples. Technology, they said, is not a reflection of people's ideas or identities; it is a means of solving problems like getting enough food to eat.

Yet another significant topic in the study of the social dimensions of technical systems relates to the manufacture, use, and exchange of ostrich eggshell items in the Kalahari. The manufacture of beads from the eggshells of ostriches *(Struthio camelus)* is a common activity among San populations (Lee 1979:l00, 232; Silberbauer 1981a:216, 221, 227). Once the eggs are collected in the field and the contents consumed, women (and sometimes men) would break the shells into small uniform square pieces. These small pieces were then placed on a kind of anvil (usually of stone, but sometimes a log) and were drilled with a tool made of a straight twig or branch of *Grewia flava,* which had a piece of wire inserted into the end. Next, the square pieces were strung on a piece of twine, string, or cord (or, occasionally, sinew) and the incomplete "necklace" was rubbed with a stone to round off the sharp edges of the beads. The roughly shaped beads were then rubbed with a finer stone in order to make them uniform in size and to give them a rounded edge. The beads were sometimes sewn onto clothing using intricate designs. They were also traded widely across the Kalahari as part of regional exchange systems, which are known as *hxaro* exchange among the Ju/'hoansi (L. Marshall 1976:304–306; Wiessner 1977, 1984). These exchange ties have changed over time, with some of them being intensified to promote alliances among various groups, and others disappearing as a result of regional conflicts, as was the case in Angola during the wars there.

The most highly valued ornaments exchanged among some San groups are ostrich-eggshell bead necklaces and headbands. Women sometimes wore as many as a dozen ostrich eggshell-bead necklaces, though the average was five. The bead necklaces are important heirlooms, and they reportedly are handed down over as many as seven generations. When older bead necklaces were inspected by individuals, they sometimes remarked on the identity of the makers, pointing out that they were Kua from certain areas of the eastern Kalahari. They said they could tell this by the thickness of the beads and the types of marks they had on them. One of the changes noted by some informants was that the sizes of ostrich-eggshell beads increased as groups became more sedentary and involved in agropastoral activities. Nowadays, with the rapidly expanding commercialization of southern African crafts, including glass beads, the numbers and quality of necklaces worn by individuals is on the decline. There is also less certainty about their makers since the contemporary beads are more uniform than the older ones. Overall, the similarities in bead styles over large areas can be viewed as partly a result of the wide-ranging exchange systems and of the social significance attached to the beads and those who make them. A contemporary problem is that people who are found to have ostrich-eggshell beads in their possession can be arrested since it is now against the law to collect ostrich eggshells without a license. As a consequence, people are hiding their beadwork and the tools that they use to make the beads. It is apparent that there is significant variation in the factors affecting ostrich-eggshell procurement, processing, maintenance, and exchange. Some of these factors reflect changes in material conditions as well as social alliance formation and information dissemination.

Finally, an important area of concern for ethnoarchaeologists is the determination of criteria that can serve to differentiate between bone accumulations that are the result of human activities and those that have come about because of the action of natural agents. There are several natural agents that have been observed accumulating bones in caves or in dens in various parts of the eastern Kalahari and adjacent areas. These agents include hyenas, leopards, and porcupines. Hyena dens, particularly those of brown hyena *(Hyaena brunnea)*, were noted to have large numbers of bones in and around them, some of them as large as skulls of gemsbok. Durability of bones (bone density) was a crucial factor in the survival potential of bones.

Some of the other agents affecting bone patterns in eastern Kalahari sites besides cows, predators, and porcupines included birds, notably vultures and crows, and dogs. It is interesting to note that the Kua have rules about dogs eating warthog bones. According to John Yellen (pers. comm. 1976) Ju/'hoansi do not allow dogs to eat warthog bones as they feel they are dangerous to the dogs' health. Both the Kua and the Ju/'hoansi bury the bones of warthogs so the dogs cannot get at them. Another reason that bones of wild animals were buried by Kua was that they were afraid that game scouts from the Department of Wildlife and National Parks or the police would arrest them for illegal hunting. The conclusion can be drawn that a number of variables affect site structure and content, including occupation duration, numbers of people and animals residing in or coming to the site, types of activities conducted at the site, and post-occupation processes, both cultural and noncultural.

CONCLUSIONS

It can be seen that the factors conditioning variability in technical systems are multiple and complex. Human populations in the eastern and central Kalahari must cope with a wide range of environmental, demographic, social, economic, and political conditions, many of which affect decision-making and technical choices. Environmental monitoring and the sharing of goods and information are strategies employed by Kalahari populations in an attempt to reduce the degree of unpredictability and risk that they face. Adjustments are made in group size and composition, mobility strategies, subsistence systems, and technology in order to adapt to complex environmental and social conditions that tend to vary both spatially and over time.

A major conclusion to be reached based on the analyses presented here is that technical systems do not reflect people's thoughts but rather the material conditions that they face. Social boundaries are maintained in different ways and at different scales. People in the eastern Kalahari observe, demarcate, and protect boundaries under certain conditions, especially in places where there are high human and livestock population densities or resources are scarce. They also allow people to cross those boundaries if they see benefits to such decisions, one example being Kua allowing Tyua to cross territorial boundaries for purposes of hunting, knowing that they will receive some of the meat in exchange, while the risk of arrest is borne by the Tyua.

Variability in the use of space and technology in the Kalahari can be viewed as what Adams (1996:253) sees as "a multilevel phenomenon." The variation in technical systems is demonstrated from a variety of sources, including archaeology, ethnohistory, and ethnoarchaeology. Yet, in the same way that formal variation among artifact types or attributes may express a variety of underlying phenomena, variation in the use of space and technology at different scales may reflect a variety of causes. These causes range from something as mechanical and mundane as the desired location of one's house in the shade or near water, to something much more complicated, such as differentiating between property of the children of co-wives in polygynous households. Sorting these out in specific instances represents a classic equifinality problem for archaeologists.

The archaeological expression of social boundaries in material culture is obviously complex and potentially multivariate. It is patterned at several scales, and dynamic in space and time. The variability seen in technical systems in the eastern Kalahari is only rarely attributable to the ethnic backgrounds or social identities of different peoples. The causes of the variation range from the degree to which groups are mobile or sedentary to the kinds of resources that exist and from technological organization to the political and legal contexts in which people are operating. A major change over time in the eastern Kalahari is the extent to which people had *de jure* rights over land, meaning that they can legally require other people to leave their areas. Changes in land-tenure systems in southern Africa have ensured that small numbers of relatively wealthy people have been able to benefit at the expense of the less-well-off majority. This is especially true in the eastern Kalahari, where high-status families from the dominant tribe in the region established water points and cattle posts, leading to the displacement and impoverishment of indigenous foragers and part-time hunter-gatherers who were forced off the lands that they had occupied for generations. Clearly, the Kalahari never was, and is not now, a borderless world, either from a material or a socipolitical standpoint.

If we are to build theory about technical systems and human adaptations, we must be careful not to take assertions about human tendencies or thought processes for granted. People do not always attempt to signal their social identity with technology, nor do they necessarily make arbitrary technical choices. Natural selection does not operate on the thoughts, ideas, and goals of humans; rather, it operates on choices once they have been made. Our purpose as anthropologists is to explain the nature of past and present human systems. This can only be done if we begin to isolate the various material conditions under which evolutionary changes take place.

ACKNOWLEDGMENTS

An earlier version of this paper was presented in a symposium entitled "Social Boundaries, Technical Choices, and Material Culture Patterning," held at the 60th annual meetings of the Society for American Archaeology in Minneapolis, Minnesota, May 3–7, 1995. We especially wish to thank Miriam Stark, Alan Barnard, Andrew Smith, and Alec Campbell for their help and suggestions on ways to improve this paper. Support of the research reported here was provided by the U.S. National Science Foundation, the Fulbright Program, the Remote Area Development Program of the government of Botswana, the U.S. Agency for International Development, and the Norwegian Ministry of Development Cooperation.

3

Material Culture, Social Fields, and Social Boundaries on the Sepik Coast of New Guinea

ROBERT L. WELSCH AND
JOHN EDWARD TERRELL

The papers in this volume are about material culture and what objects can tell us about the social formations that produce them. Anthropologists exploring the social construction of material culture emphasize that the roles and cultural meanings of things in human life have many valences and are far from simple (Appadurai 1986a; Kopytoff 1986). However, as Malinowski helped pioneer in New Guinea early in this century, studying material culture is one way to map social formations, and vice-versa (Malinowski 1920a, 1920b, 1921, 1922, 1934). More recently, anthropologists have examined the movement of material culture in various parts of New Guinea, including the Papuan Gulf (Barton 1910; Dutton 1982; Groves 1972; Williams 1932–33), the Vitiaz Strait (Fernstrum 1990; Harding 1967), Geelvink Bay (Held 1947, 1957), the Bird's Head (Elmberg 1966; Miedema 1994), the Sepik coast (Barlow 1985; Lipset 1985; Tiesler 1969–70; Welsch and Terrell 1991), and through even more intensive studies of the *kula* (e.g., Damon and Wagner 1989; Leach and Leach 1983). This volume continues this respected tradition.

In each of these New Guinea examples, social processes involving the exchange of objects have led to a distribution of material culture that is considerably larger and more extensive than the villages and face-to-face communities in which the anthropologists and ethnographers of our time have characteristically worked (see also MacEachern in this volume). Each case involves communities speaking several different languages. In none of these cases does the observable distribution of material culture coincide with the boundaries of a single ethnic or ethnolinguistic group; nor does a shared material culture delineate a single society, culture, or culture area. But if in these cases material culture does not map identifiable ethnic groups, cultures, or culture areas as anthropologists have conventionally understood these groupings, what social processes and what kinds of social groupings can we discern by studying such distributions of material culture? Rather than reanalyzing the most familiar of these New Guinea case studies, we consider a case that has been less well described, the Sepik coast. In this region, we have been able to observe these social processes and we can describe the distribution of several classes of material culture. Our firsthand observations allow us to ask: What kinds of social phenomena are being mapped by particular distributions of material culture and what social groupings are being created, maintained, or delineated by exchanges involving material culture?

To understand the kinds of social groupings engendered by these exchanges, we find it useful to think of people as living in "social fields," drawing primarily upon the sports analogy of the playing field. Sporting events usually take place within bounded spaces and by the same token when social fields are bounded—whether intentionally, accidentally, or statistically—we refer to bounded social fields as "social formations." The social formations anthropologists have traditionally described in places like New Guinea are those usually called villages, communities, societies, cultures, tribes, and ethnolinguistic groups. Scholars have generally assumed that all of these various kinds of social formation are relatively small, socially bounded groups having more or less homogeneous cultures and languages. But while social formations corresponding to villages and ethnolinguistic groups are present along the Sepik coast, the distribution of material culture does not map easily onto these small, local social formations. Instead, the exchange of objects along this coast and the geographic distributions of different kinds of objects have led us to study social formations that are much more extensive than those anthropologists have usually investigated.

This paper examines social fields on the Sepik coast both from their material culture frozen in time as museum collections and on the ground during three seasons of fieldwork in 1990, in 1993–1994, and in 1996.[1] We illustrate how people on this coast historically operated within extensive social networks that often included peoples speaking more than a dozen different languages. Much, but by no means all, of the content of social relations on the Sepik coast has involved gifts of foodstuffs, raw materials, and local crafts, including objects such as earthenware pots, bows, arrows, string bags, baskets, ornaments, canoes, and paddles. The question we raise is: To what extent are these social networks reflected and expressed in the material culture found in different villages along the coast? This question has its corollary in two others: What kinds of social formations do distributions of material culture mark off? How much of the structure of these social relations can we identify from the material culture record alone?

After outlining what we mean by the term "social field," we begin our analysis with a short description of how the social networks we have observed along the Sepik coast have operated for the past century and how these networks can be interpreted as social fields. Then we turn to an analysis of how certain kinds of products are regionally distributed, how these products have been exchanged historically, and what their geographic distributions look like along the coast. Finally, we consider the extent to which the location of production centers and the distribution of products delineate social boundaries. Of most concern is how the movement of certain products within particular social fields creates bounded social formations that are much larger than the kinds of social groupings usually studied by anthropologists and archaeologists.

SOCIAL FIELDS AS PLAYING FIELDS

A social field is a social network. It may be thought of as a web of social, economic, and political relations. Over the past two or three decades the discipline of anthropology has embraced the concept of the "social field" or the "social landscape" (see, for example, Barth 1969a; Gosden 1989; Watson 1977, 1983; Wolf 1982). It is now commonplace, even customary, for cultural anthropologists (and even a few archaeologists) to refer to social fields (or similar metaphors) when describing ethnographic (or prehistoric) communities. There can be little doubt that nearly all people live in social fields. People are not only involved with their neighbors (Terrell 1993; see also Gregory and Urry 1985; Lomnitz-Adler 1991), but more often than not they are involved with expansive regional and global systems. But it is not enough to acknowledge that people maneuver within social fields beyond their families and villages. What is needed are case studies that document the particulars of how social fields affect the phenomenal world of people, places, and things that make up the world people inhabit.

A social field is a field of interactions and social influences, both intended and unintended, whether perceived by other parties or not. In the Massim *kula* ring, for example, the sudden release of a large number of *kula* valuables on one island following a mortuary exchange affects the timing and intensity of *kula* activities on many other islands several steps removed from these mortuary activities (Wagner 1989). Similarly, in a process Watson (1977) has called the "Jones effect," the introduction into some New Guinea highlands communities of a new crop, such as the introduction of sweet potatoes that are easier to grow as pig fodder than taro, compelled nearby communities to adopt the new crop. In a somewhat different way, Welsch (1994) has suggested that across the Fly-Digul plain, marriage with women from villages with different languages and cultures may have compelled the husbands and their extended families to participate in a variety of exotic social and economic activities, most importantly bride-price payments, pig feasts, and certain kinds of men's cult ritual.

In these three examples, individuals find themselves affected by social actions that extend far beyond their face-to-face communities. In each case, fluctuation in the availability of valuables, the introduction of a new crop, or participation in new marriage practices changes the range of possible actions for every participant, sometimes temporarily but sometimes permanently.

We also find it useful to think of social fields as playing fields on which and within which people play the game of living. The metaphor of the playing field fits comfortably enough with several of the best studied examples of social fields in New Guinea: Malinowski's description of Trobriand Islanders strategizing over *kula* valuables (Malinowski 1922; see also Damon and Wagner 1989; Leach and Leach 1983); Strathern's Melpa leaders attempting to manipulate, cajole, and influence their friends, peers, and competitors in the *moka* (Strathern 1971, 1979; see also Meggitt 1974); and Watson's descriptions of Eastern Highlanders operating within "crowded fields" that compel individuals and communities, for the sake of survival, to accept changes introduced by their neighbors (the "Jones effect") (Watson 1970, 1977, 1981, 1983). And, of course, most ethnographic descriptions of warfare, headhunting, and raiding in New Guinea have invoked competitive images of similar social fields.

The metaphor of social fields as playing fields, however, conjures up sports analogies that we only partly wish to invoke. Virtually all human activity can be understood at least partly in terms of game theory, in which individuals strategize to achieve particular ends. In many games people are divided into separate teams that compete against one another. Membership in one team necessarily prevents a person from playing on or with another team. It is much more useful to conceptualize social fields on the Sepik coast as comprising a very large playing field, with many groups of players, many different languages, and several different local environments, each with its own economic possibilities. This metaphor of a large playing field, with its distinctive local landmarks, local characteristics, and arrays of friendly and (sometimes) not-so-friendly players is apt, because as in most sporting events, there is an agreed-upon set of basic organizing principles. In sporting events we call these organizing principles "rules"; along the Sepik coast one is more likely to hear individuals discuss these principles as their "expectations" about how other people ought to behave.

When discussing the Sepik coast, however, it is not useful to emphasize competition among players, since most interactions are cooperative and there are no groups or parties against which anyone is systematically competing. Everyone is staunchly proud of their ancestors' abilities in war. Such fierce posturing, however, largely seems to have been a defensive measure that probably prevented hostilities from erupting. Warfare and fighting between groups were reported before the German administration extended its control over the region in the first decade of the twentieth century (e.g., Rodatz 1908, 1909). With a few exceptions, these hostilities appear to have been brief and temporary. Fighting would break out, for example, following some insult or slight, but peaceful relations (or at least polite avoidance) returned after a few skirmishes. Age-old antagonisms between villages are not characteristic of the Sepik coast, nor for the most part does one find "traditional" or "customary" enemies. But distant places or villages, where one had no active social ties, always posed a danger and were assiduously avoided.[2]

THE SEPIK COAST AS A PLAYING FIELD

At the beginning of the dry season in the days before outboard motors and dinghies, the people who lived on the small islands off what is now Aitape set off in outrigger canoes to visit friends along the coast. Calling at one of the coastal villages where they had reliable and well-established friendship relations, the islanders gave gifts of smoked fish, pots, ornaments, and other items to their friends. After enjoying the local hospitality for a few days, they then set off again to visit friends in other communities further down the coast. After visiting with these friends and distributing gifts brought for them, the islanders set off once more to visit yet other friends. Thus they moved along the coast until all of their cargo was dispersed. When all of their gifts had been given away they prepared for their voyage home, receiving gifts of sago, yams and tubers, tobacco, betel nut, baskets or black palm bows, wooden bowls, plumes, and other items that their friends chose to give them. After loading up at the most distant village, they retraced their path calling in once more on the friends they had recently visited. When their outrigger canoes were full they went home, unloaded their cargo, and set off again for the mainland to continue receiving gifts from their various friends in different places.[3]

While islanders may have asked for certain items from their mainland friends, it seems that individuals usually had little control over what gifts they might receive above and beyond sago, which, for islanders, was always the most important goal of the season. For mainlanders, fish, earthenware pots, and shell rings were the most desired objects, but they, too, could not count on getting any particular mix of items from any particular friend in any particular voyaging season. Over the course of several years, everyone eventually got what they desired from their friends, but there was never any bargaining, any haggling, or any discussion of the value, quality, or price of any item.

During each dry season islanders would call at a slightly different set of communities. In this way the islanders would not overstay their welcome or—perhaps more importantly—they would not overtax the sago resources of their friends. Periodic breaks also gave their friends time to ready especially desired "extra" goods a season or two later. Every participant had goals for their interactions with different friends along the coast, and each knew that he (or sometimes she)[4] could maximize his long-term benefits by being a "good friend" to all of his active partners. Central to being a good friend on the Sepik coast is the display of generosity, at times almost excessive generosity.

Despite the fact that each community has its own term for "friend"—there are about a dozen different vernacular terms—virtually the entire coast shares the same basic ideas about what a friend should do and how a friend should behave. By following the same customs and practices that everyone else follows for being a "good friend," everyone on the coast plays on the same "playing field" with the same rules. Individuals from different communities may have specialized functions—as providers of particular goods—much as goalies and centers have different functions on the soccer field. But everyone has the same expectations about decorum and generosity, which ensures that everyone will achieve most of their personal goals in every interaction, either by obtaining valuable products or by accumulating the good will that will provide valuable products in the future.

Although we have emphasized gifts of objects as the content of friendships, it is important to note that people obtain a great deal of satisfaction from the socializing, joking, and feelings of connectedness that come from visiting long-term "friends of the family." Good friends are expected to participate in one another's social activities—attending funerals, initiations, important ritual events, and the like. Especially good friends may be expected to help amass the foods that will be needed for the various feasts that accompany these events. In the old days, friends were also expected to assist in times of defense or revenge against another village. While such alliances do not seem to have been activated very often, the threat that members of one lineage could call on a host of friends as allies undoubtedly served to obviate the need for such actual alliances.

THE ORGANIZATION OF PRODUCTION AND EXCHANGE

Face-to-face social life in villages along the Sepik coast was (and continues to be) organized around small extended families and exogamous lineages living patrivirilocally in hamlets or villages. These agnatic groupings of kinsmen, together with their wives and daughters, were the central corporate groups within which most people have historically (and up to the present time) lived their day-to-day lives.

Each individual had many cousins and affines in other lineages to whom he or she was tied through varied bonds and obligations of kinship, but the most important family relations were with those people who resided on the same lineage lands, especially those people in the same household who worked together, cooked together, and ate together. When a lineage was small the entire lineage might have lived and worked cooperatively, but if a lineage was large, containing many grandchildren and great grandchildren of a single ancestor, it was usually only those with the same father or grandfather who regularly acted as a corporate group. In such cases the entire lineage tended to act as a corporate group only occasionally, especially during funerals and other ceremonial occasions. But men from different households within a lineage could be called upon to assist their fellow lineage members.

Production of items intended for exchange with friends in other communities generally centered on individual households and to a somewhat lesser extent on the extended agnatic families we have called lineages. These family groups were the corporate groups that made pots, sago, smoked fish, shell rings, baskets, bows, canoes, and other products. Within these groups men and women cooperated toward common ends. Men from pottery-making communities, such as Tumleo Island, helped their wives, mothers, and sisters gather clay, while the women actually made the pots. In sago-producing communities, men pounded the sago pith while their womenfolk washed the sago. In fishing communities, men tended to catch larger fish while women were generally responsible for drying them over a fire. In the same way, men generally helped their wives harvest giant clam shells *(tridacna)*, while women actually used the bamboo borers to make the shell rings. In mainland villages, men planted yams, while women tended and weeded them; later men harvested the yams, and women carried them back to the village. In all of these examples and many others besides, production was a cooperative effort within households and lineages that were the central family units along the coast. Individual women or men might sometimes use some of their production for specific personal ends, but in the aggregate, production was primarily for the benefit of the entire family group, whether that group was a single household, two or three households, or a small agnatic lineage. All members of the family could expect to benefit from the family's cooperative production.

The groups of islanders that sailed along the coast visiting friends were generally members of the same family groups. For the most part, they were a set of agnatically related men, plus a few other kinsmen from the same community. They had a single common goal: to obtain all the sago and yams their family would need for the long wet season. To achieve this end each man would have readied his supply of pots, fish, and shell rings that would be needed as gifts to particular friends in specific villages. Beyond this goal, everyone on the canoe might have different personal objectives and might have brought along a variety of smaller gifts to add to the main products they would give away. Any particular voyage along the coast thus combined cooperative and individual interests in complex ways, but the corporate goal of amassing food for the wet season was always paramount over any individual's private interests. If relatives from outside the local lineage had been invited along, they would also have their own objectives for the voyage, but these goals would not have competed with those of the owner of the canoe, whose family had sponsored the voyage. Exchanges were generally organized as transactions between individuals or between two sets of siblings, each group having a single common interest as members of the same family.

A small group of agnatic lineages usually constituted a named hamlet or village. When tensions built up between particular individuals, one lineage or group of lineages might move away from the others to establish their own hamlet. Over time, they might return to the original village site, the rest of the village might join them on the new site, or all of the original lineages might establish a village on a different site altogether. As the threat of warfare declined during the colonial and postcolonial era, the number of small hamlets forming around a single lineage has increased dramatically on the mainland along the coast both to the east and west of Aitape town. We presume that such shifts of residence along with some movement of personnel among and between lineages have long been a feature of life along the Sepik coast.

Historically a cluster of from one to ten hamlets or villages comprised a named "community." Each community had its own identity and character that differed from that of other named communities.[5] At the time of European colonial expansions into the region in the late nineteenth century, there were more than 60 communities along 700 km of coast between Humboldt Bay (now Jayapura) and Astrolabe Bay (now Madang); a similar number of additional communities were situated in the bush adjacent to these coastal communities. In most cases, each named community acted as a separate and distinct polity and had its own rituals, economy, leadership, and ways of doing things. In some communities there were noble families from which the leader of the community, village, or hamlet was chosen in an hereditary or semi-hereditary fashion. In others, leadership was more egalitarian, emphasizing achieved leadership, big men, elders, ritual leaders, and the like.

Ties within each community tended to be organized around hereditary ties of kinship and marriage. But outside the community, nearly every individual and family was linked to many other communities through hereditary ties of friendship. Thus, each adult had an extensive network of social relations along the coast that linked him or her to as few as 5 or 6 or as many as 25 other communities. It is important to note, however, that while people in every village had friends in many other communities, these were not relationships between polities but were organized as relations between individuals or between individual families.[6]

In the nineteenth century, there was no lingua franca that allowed easy communication along the coast. Communication between friends in different communities often required that one or another friend be bilingual in the other's language. As adolescents most men would have spent some months as guests of their father's friends, allowing them a chance to learn the rudiments of the friend's language as well as giving them a chance to get to know their father's friends more intimately.[7]

Thus, Sepik coast (personal) social networks existed despite major differences in language, despite significant differences in political organization, despite differences in environment, and despite economic specialization. Friendship ties involved exchanges of foodstuffs, raw materials, and local crafts. Foodstuffs included sago, fish, yams, bananas, tubers, greens, pork, and game, as well as betel nuts and tobacco. Raw materials included wood for houses and canoes, sago leaf for thatching, bamboo for building materials, shells for ornaments, ochers and clay for pigments, cane for arrows, palm wood for bows, and bone, stone, and shell for tools. Local crafts traditionally included a wide variety of items that circulated in fairly large numbers—earthenware pots, string bags, bows, arrows, canoes, and paddles—and a host of other ornaments, household utensils, and tools that were given away less routinely, but which nevertheless (in the aggregate) made up a significant part of the regional economy.[8]

Central to this regional economy was an uneven distribution of basic resources and local centers for the production of specialized products. Although sago swamps were found in most communities along the coast as well as on the larger islands (especially Walis, Mushu, and Kairiru), sago was absent on all of the smaller islands (Tumleo, Ali, Seleo, Angel, Tarawai, Karesau, and Yuo), and this resource shortage continues to motivate interaction (Tiesler 1969–70; Welsch 1998; Welsch and Terrell 1991). Not surprisingly, these islanders got sago, which was their staple, from sago-rich friends in mainland communities or on Walis and Mushu. In exchange for sago, they gave smoked fish, earthenware pots, shell rings, canoes, and other products. Each island produced its own specialized goods. These specialties were exchanged among their friends on other islands so that each individual or family could build up an inventory of pots, smoked fish, and both large and small shell rings (all island products), and this inventory of island products desired by their mainland friends was used for gifts when they visited the mainland hoping to bring back sago, yams, and other mainland products.

People on the mainland produced a number of specialized products as well. Lagoon villages around Sissano made black palm bows that were much sought after in many parts of the coast. People in the Murik Lakes area (especially communities at Murik, Kirau, and Mendam) were noted for their soft baskets—the so-called Murik baskets. Arapesh- and Boiken-speaking communities east of Matapau—Sawom, Balam, But, Boiken, and their neighbors as far as Wewak—made round wooden bowls that were also highly prized. Bush communities in the hinterland behind most of the coastal shore made an abundance of string bags and arrows that were prized by their coastal friends.

Differences in the environment and the distribution of resources played an important role in determining which communities produced which products. But there also seems to have been a kind of tacit agreement among people that each community or district should have its own specialized good. Even in the case of string bags, bows and arrows, and wooden bowls, all of which were made nearly everywhere, there was general agreement along the coast that nonspecialist makers could not achieve the same high quality as those made by specialist producers.[9] In such instances, it is likely that specialist products were seen as emblematic of the identity of people from a particular village, community, or district, whether or not every inhabitant possessed better-than-average skills in the specialty.

Although we write of these relationships historically in the past tense, readers should note that these friendship relations are still active today. During our fieldwork along the coast, we routinely discussed the content and workings of these social networks both in the past and in the present with informants from many different places. We often observed these friendship relations firsthand. Although friendship continues to be a very important aspect of intervillage relationships, since the advent of markets and the cash economy, friendship is no longer the only social mechanism that moves objects around from one village to another. It is easier to understand how the institution of friendship has shaped the distribution of products on the coast from their historical context rather than from contemporary social patterns.

THE LIMITS OF THIS SEPIK COAST PLAYING FIELD

What emerges from these several aspects of friendship is a vast array of crisscrossing, overlapping, and intersecting personal networks spread out along an even larger playing field that is the Sepik coast. Every participant knows that every one of his friends has other friends farther on along the coast or even deeper into the hinterland. Yet nobody knows the limits of this playing field, nor does anyone care if there are, in fact, limits to this vast field of intersecting social relations. Everyone knows only the limits of his own personal networks within this field. Whether the playing field extends another few kilometers or a few hundred makes no difference to the players, any more than the distance to the extreme edges of a stadium are of consequence to a goalie or fullback in a soccer game. It is enough to know (in detail) only the part of the playing field in which one maneuvers.

While players on the field may not perceive the full extent of these social networks, as outside observers we can glimpse the combined effects of these personal networks by observing the movement and distribution of objects that have traveled along these pathways. What emerges is not a bounded culture area in which all the players are the same. We find instead a (largely) unbounded system of friendship relations in which everyone shares effectively the same ideals and expectations about how friends should behave.

Thus the playing field we describe is not a "culture area," as the term has traditionally been used (Kroeber 1939; Wissler 1923), since there are differences in economic patterns, rituals, local politics and leadership, and the like. But all of these interlocking communities share a common notion of hereditary friendship and common expectations about how friends ought to behave. Individuals also know which local products are desired elsewhere and—despite the absence of haggling—most people share a sense of the value structure of the raw materials and objects that flow through their social and economic environment. Everyone needs to know not only how to behave but also the relative values of different things if they are to express their close and abiding friendship through generosity.

MATERIAL CULTURE DISTRIBUTIONS ALONG THE SEPIK COAST

There were (and still are) differences along this coast in the distributions of specific items of material culture. The most obvious distinction found in the material culture of the Sepik coast is that some objects were made by specialists, while others were made by nearly everyone. Another distinction of importance is that some items have a distribution that is both distinctive and limited to a discrete part of the coast; such items tend not to have been exchanged very much, and, where found, they usually reflect the local technology, local style, and local designs or decorative motifs. We find it most informative to distinguish three classes of objects that incorporate these differences. The first two classes comprise items frequently exchanged and are either (I) specialized products made in only a small number of communities, or (II) products that are made nearly everywhere. In both cases, the objects are found in the assemblages or tool kits of most communities. The third class (III) consists of objects that seem to be only infrequently exchanged; these items tend to have distinctive distributions along the coast. Although space does not permit a detailed analysis of any of these groups of objects, summary descriptions of their distributions will illustrate the extent to which material culture reflects the extensive social networks that have been at work on the Sepik coast.

Class I: Specialist Products

The best (and most complex) example of a specialized product is earthenware pottery. Other specialized products include shell rings, outrigger canoes, wooden bowls, black palm bows, and Murik baskets. Each of these products has a limited set of producing communities, some supplying local demand, others supplying large sections of the coast. Typically these objects were made in fairly large quantities but in only a small number of communities; they were made specifically as exchange goods to attract (as return gifts) the various things their makers desired from friends.

EARTHENWARE POTS   Earthenware presents the most complicated pattern of specialized production on the Sepik coast. Partly because of its distributional complexity and partly because pottery is of such interest to archaeologists, we will give it somewhat greater attention than the other object types discussed below.

There are several different pottery industries, each associated with a distinct producing center. Although there are three (possibly four) different technologies in use, pots or other forms of earthenware are present in the tool kits or assemblages of every community along the coast. Everyone has access to earthenware, either because they make pots themselves or because they get them from friends.

Earthenware comes in a variety of shapes or forms: (a) cooking pots used for cooking over a fire, (b) large pots used to make sago pudding with boiling water, (c) large storage jars for water or sago flour, (d) "frying pans" or cooking shells used for frying sago pancakes, and (e) ceramic ornaments for decorating men's cult houses. Of these shapes and forms only (d) and (e) have a distinctive distribution; the others are found everywhere along the coast.[10]

From A. B. Lewis's observations in 1909–1910 (see Welsch 1998) and from our own fieldwork in 1990 and 1993–1994, it appears that there were three traditional techniques for making pots.[11]

The first consists of a paddle-and-anvil technique, which was used at four main sites along the coast: Humboldt Bay (Kayu Batu), Vanimo Village, Tumleo Island (together with its recent mainland settlements at Raihu Camp and Yakoi), and finally, an industry found in a cluster of three villages east of Wewak (Kaiep, Turubu, and Samap [production ceased]).

The second is a spiral-coiling technique in which the potter rolls out coils of clay about 1–1.5 cm in diameter, building up the sides as the coil spirals around a base. We observed this method at Serra (Puindu) in 1993; A. B. Lewis saw it used in Sissano in 1909, although it ceased as an industry there after the Second World War. Neither we nor Lewis observed the industry at Ramo or Sumo, but some pots from these communities appear to be made by spiral coiling, while others may be made with the slab-building technique. Finally, at Walis Island we collected one pot that was clearly made by spiral coiling; this pot seems to have been made at either Kombio or in Abelam in the Torricelli and Prince Alexander mountains to the south.[12]

A third (slab-building) technique is similar to that found in the spiral coiled industry but uses slabs of clay 3–5 cm wide rather than coils 1 cm in diameter. This method is used in several scattered locations. The best-known site is at Leitere on the coast, but the technique is also used at Goiniri, where uniquely among all of the communities in our study area, potting is practiced by men rather than women. Some pots made in three or four villages south of the Torricelli crest above Amsuku are routinely exchanged into the study area, and these appear to be a mix of pots made with spiral-coiled and slab-building techniques. It is possible that the Ramo and Sumo industry also used this slab-building technique, but definitive data are currently lacking.

Pottery-making is in decline in all of these pottery centers. It is essentially moribund everywhere except at Tumleo and Kaiep-Turubu. Traditionally, these two centers supplied the majority of pots used on the coast from Serra to the Murik Lakes, where a very different inland industry takes over. West of Serra, the two centers at Vanimo and Kayu Batu dominated in the production of pottery. The slab-building industry at Leitere—like the spiral-coiled industries at Ramo, Sumo, Sissano, and Serra—seems to have served relatively local needs. (Figure 3.1 shows the location of these diverse industries in the study area).[13]

What are we to make of the distribution of these three different pottery industries? Technologically, there may be either two or three distinct traditions, since it is possible that the slab and coiled techniques have their origins in a single-coiled tradition, although this is by no means certain. At the same time, there may be a fourth tradition represented at Goiniri, where male potters produce a very bulky, crude product (when compared with all other pottery in the study area).

What is most striking is that none of these three or four pottery traditions is associated exclusively with a single language family. The paddle-and-anvil technique is found in three Austronesian communities (Humboldt Bay, Tumleo, and Kaiep-Turubu), but it is also prominent in Vanimo Village, where an unrelated tonal language in the Non-Austronesian Sko family is spoken.

The spiral-coiled and slab-building industries are more or less clustered in two areas (1) around Leitere, Ramo, Sumo, Sissano, and Serra, and (2) well inland across the Torricelli crest, except for Goiniri, where it has expanded to the north. These

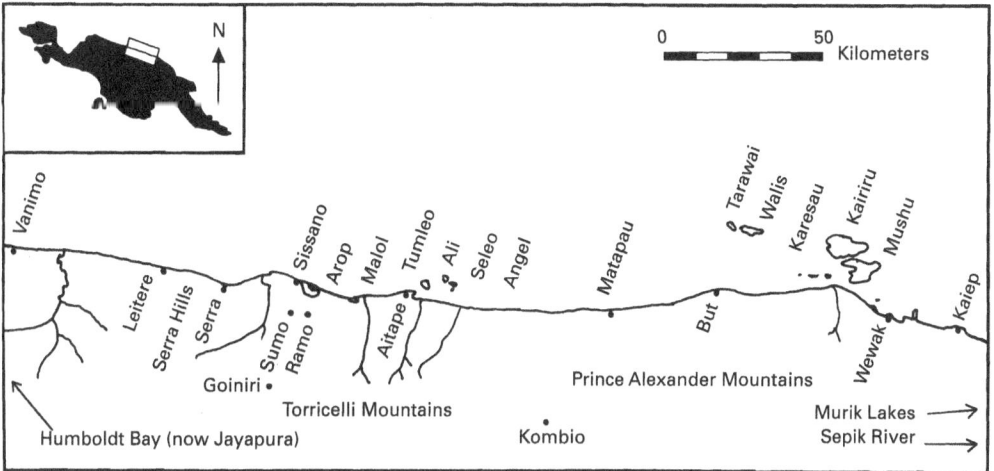

Figure 3.1. Sepik coast, northern New Guinea.

spiral-coiled and slab-building industries may have had a common origin, but they were not restricted to a single language family, since Leitere and Ramo-Sumo speak two related but different Sko languages and Sissano and Serra seem to speak dialects of the same Austronesian language; Goiniri and the trans-Torricelli communities speak unrelated languages. We may see these industries as having diffused from one source (possibly Leitere), but it is important to note that, unlike the paddle-and-anvil industries, these were all extremely local industries; their pots have rarely traveled beyond the communities adjacent to those of their makers.

In contrast to this pattern, Tumleo and Kaiep-Turubu pots saturated the entire coastline, and pots from one or the other source can even now be found in virtually every house along the coast between Wewak and Aitape. Formerly, the same conditions held between Leitere and Tanah Merah Bay, where the majority of pots were from either Vanimo or Kayu Batu.

SHELL RINGS AND OUTRIGGER CANOES Unlike pottery, which was made in mainland and island communities, shell rings and outrigger canoes were principally island products. Outrigger canoes used to be a major product made for export at Ali Island, where they were made for many friends along the coast (Woichom 1979; see also Parkinson 1900, 1979). Only one village (But) along the entire coast from Aitape to Wewak had canoe-building technology and they probably exported some outrigger canoes to their friends along the coast. Smaller numbers of canoes were made at Tumleo, Seleo, and Angel (Erdweg 1902; Welsch 1998), but they were apparently a significant part of the economy at Tarawai and especially at

Karesau and Yuo (for sources see Tiesler 1969–70). The islanders' need for canoes is obvious and they were the main suppliers for most of the mainland villages that desired them. Canoes seem also to have been an important product at the mainland communities of Humboldt Bay, Tanah Merah Bay, Sko, and Vanimo in the west, probably because, geographically speaking, there were no offshore islands, where such craft could become a specialized product (Frankel 1978; Galis 1955). Note that outrigger canoe production included a number of Austronesian-speaking communities, but it also included several areas (But, Tarawai, Vanimo, Sko, and Tanah Merah Bay) where Non-Austronesian languages belonging to three unrelated language families were spoken.[14]

Large outrigger canoes were typically decorated with the local clan motifs of their makers, even if the canoe was made for a friend elsewhere. Overseas friends typically would not have known the meaning of these decorative elements, but after a time, everyone up and down the coast recognized the canoe by these designs.

In historic times shell rings had a very restricted set of producing communities. Shell rings were of three types: (a) large rings, 6–20 cm in diameter, bored with bamboo tubes from giant clam shells *(tridacna)*, (b) small shell rings (a few centimeters in diameter) ground down on a grindstone from the end of a *conus* shell, and (c) a thin *trochus* shell ring resembling the *lailai* of the Vitiaz Strait. All three kinds of rings seem to have been made in two areas: at Ali, Seleo, and Angel islands near Aitape and at Tarawai Island to the east. These four islands supplied the needs of the entire coast.

Communities west of the Serra Hills (Leitere to Humboldt Bay) seem not to have used shell rings in any systematic way; this part of the study area practiced marriage by bride-price payments that involved large glass beads, and as a result, most men in this area were far more interested in beads than shell rings. Not so the mainlanders from Serra to the Murik Lakes, where sister exchange is the dominant form of marriage and shell rings are a sign of wealth and prestige. Early accounts do not explain why these mainlanders worked so hard to get shell rings. It probably had to do with the fact that south of the Torricelli and Prince Alexander ranges there are many groups who practiced bride-price and traditionally used shell rings as the standard unit of value. These systems seem to have been connected to the middle Sepik and Yuat rivers, which were in turn connected to the central highlands. The constant demand for rings from the interior may have provided a constant siphoning off of these rings, which coastal peoples seem to have used exclusively as ornaments in their own communities. In any event, shell rings were ubiquitous along the coast east of Serra, despite their restricted areas of production.

BLACK PALM BOWS AND MURIK BASKETS   Black palm bows were made principally in a cluster of contiguous villages west of Aitape: Malol, Arop,

Warapu, Sissano, and Serra. These bows are known to have been exchanged down the coast to the east. In return, villagers liked to receive Murik baskets, soft woven baskets—often quite large—made of a sedge that grows in the lagoons at the mouth of the Sepik River. These baskets were traditionally made in the several villages around the Murik Lakes (Murik, Kirau, Mendam, and probably a few at Kaup, see Barlow 1985) as well as in two or three villages to the east (Kopar, Watam, and perhaps a few villages further east, see Welsch 1998). Production for gift exchange across the Sepik River centered on the Murik Lakes, as these villagers would not allow their friends from the east to interact directly with their friends from the west, positioning themselves as middlemen between the two areas. Like black palm bows, Murik baskets used to be extremely common as far as Serra and could be found in every community, despite the small number of villages that produced them.

Decorative designs on both black palm bows and Murik baskets are highly varied and reflect local family and lineage designs. People outside their centers of production would not have known the significance of these designs. At least in the Murik Lakes area, baskets made as gifts for friends rarely had important and valuable clan designs on them, as these were reserved for local uses (K. Barlow, pers. comm. 1991).

Class II: Products Made Everywhere

The second group of objects had a very different pattern of production. These products used to be made in many places on the coast, especially in the hinterland bush villages. The two examples considered here are string bags and arrows, both of which are still found in every community west of the Murik Lakes.

Lewis reported in 1909 that certain objects from the interior were found in all villages, especially "arrows and netted bags, which are the specialties of the 'bush' villages, and which they trade to the coast natives for such things as salt, shell ornaments, etc." (Lewis specimen list in Welsch 1998:119–120).

In 1909 Lewis reported that string bags:

> come almost entirely from the interior, tho [sic] a few are made in certain coast villages, and even on the islands. Two groups may be distinguished, the boundary line being about the region of [Aitape]. East of this the bags have an even and more open weave than to the west, where the weave is closer and ribbed. The western bags also average much smaller in size. None of the bags is ornamented with shells as further east, but there is a great variety in color pattern. This design, and the character of the weave of the narrow band around the mouth of the bag, indicate the local place

of origin. A native who is familiar with these styles can tell what village a
bag comes from by looking at it. (Welsch 1998:119–120)

In the years since Lewis's research in the region, netted string bags seem to have
become a much more common product in the coastal villages and on the islands
at Aitape. Bags with a tighter weave are now much more common east of Aitape
than early in the century, suggesting a gradual diffusion of this technique to the
east. These two kinds of bags represent significantly different techniques of the
sort discussed by Lemonnier (1986). The evidence suggests that these techniques
were not associated with particular ethnic or linguistic configurations, nor were
they associated with the same set of communities early in this century as they are
today. In Lewis's day, each village and probably every lineage had its own special
design or pattern. These motifs show some clustering of design elements along
several stretches of the coast. For the most part, however, the meaning of each distinctive motif is only known locally. Recipients of string bags would, of course,
know who had given it, but lacking this information, few would likely have
known how to read the distinctive motifs. These designs were probably used by recipients simply as identifying marks.[15]

Considerably less is known about the distribution of distinctive arrow designs.
Virtually every man along the coast made arrows and those arrows in museum collections display a dazzling array of decorative variations, shapes, and functional
differences. Very little of this variation can be linked to particular sections of the
study area, although it appears that individuals, families, lineages, and villages may
have had their own decorative motifs. What is clear about the distribution of arrows historically is that arrows were routinely given away to friends, and few
friends would have been able to attach any meaning to the decorative motifs. Like
string bags, arrows were found everywhere, and arrows from particular bush villages were distributed over a wide area.

Class III: Products Rarely Exchanged

A third class of objects had fairly widespread distributions, but unlike the other
two classes, these seem to have been exchanged only infrequently. This class includes a much larger set of items, such as bone daggers, headrests (pillows), penis
gourds, spears, spear throwers, hair baskets, and sleeping bags. Each of these items
had a regionally restricted distribution, and none of these objects had the same
distribution as the others. In all these examples, stylistic features seem to delineate
sections of the coast that consist of a contiguous cluster of communities. Cassowary bone daggers and headrests were the most widespread of these object types

(see Terrell and Welsch 1990b).[16] Headrests are ubiquitous along the coast, although all the headrests west of the Serra Hills are much longer and thinner than those to the east. This distribution coincides with a partial break in social networks along the coast, which, as suggested above, is directly related to different marriage patterns (sister exchange to the east, bride-price to the west). Daggers were found on both sides of this line. There are, however, different sets of typical design elements on the daggers made east and west of this cultural divide. For daggers—and probably for headrests as well—although there may be specific meanings attached to certain decorative elements or to designs as a whole, informants in many villages insist that these designs are principally important because they allow owners to identify their daggers readily.

Penis gourds represent the object with the most restricted distribution along the coast, being found in Sko, Wutung, Yako, Warimo, and Vanimo, all communities that speak languages in the Sko family.[17] Although Moore and Romney (1994, 1995) contend this is a significant distribution because it is restricted to a single family, penis gourds are by no means restricted to Sko-speaking areas (see, for example, Austen 1947; Riesenfeld 1946; see also Foy 1902). Penis gourds are found over an extensive area and they are the most distinctive piece of material culture in the interior of Irian Jaya. Their distribution extends as far south of the central cordillera as Ningerum, Muyu, Yonggom, and Awin (Welsch 1994).

Spears, spear throwers, hair baskets, and sleeping bags comprise a set of objects distributed primarily in the eastern part of the study area. None of these objects is currently used to any significant extent anywhere along the coast, although all were regularly observed by Lewis in 1910. The best information available about their traditional distribution comes from early collections at The Field Museum (see Welsch et al. 1992:574-575, nos. 8, 9, 26, 39). These museum collections suggest that spears were primarily used along the stretch of coast that extends from Wewak to Malala, with a few found as far west as the islands at Aitape where they had been given by friends to the east of them. Spear throwers have a more restricted distribution, from around the mouth of the Sepik to Potsdamhafen. Hair baskets were distributed from around Wewak to Potsdamhafen. All three object types were also found in the Schouten Islands of Wogeo, Koil, and Kadowar. Sleeping bags, which principally served as family-sized mosquito nets, were made of the sedges used to make Murik baskets. These have a very narrow distribution in the Murik Lakes, along the lower Sepik River, and in a few villages on the coast to the east of the Sepik. As we have suggested elsewhere (Welsch and Terrell 1994), this distribution may be environmentally determined, probably as a response to larger mosquito populations, although the natural occurrence of the sedges may also be limited to this area. What is most striking about these four object types is that no two of them have the same pattern of distribution. Stylistic variation abounds but such variations generally seem to serve as personal identifying marks, although in the makers, they no doubt have local symbolic or lineage associations.

SUMMARY

Different kinds of objects produced in different places on the Sepik coast of New Guinea traveled different social pathways in many directions. We have found that these pathways rarely correspond to identifiable or deliberately constructed "ethnicities." What is most striking about these social networks is that they existed despite extraordinary differences in language. More than 60 mutually unintelligible languages belonging to perhaps 24 different language families and at least 6 unrelated phyla are spoken along this stretch of coast, with another 20 or so languages spoken in the hinterland adjacent to coastal communities. In most communities everyone spoke a single language or mother tongue. But in several communities (both historically and today) two languages were spoken, usually where a smaller group or community was assimilating into a larger one. There were also one or two communities where two languages seem to have been of roughly equal importance. In all but a few cases, the same language—or very closely related dialects—are spoken in more than one community. Nearly every community had some neighbors who spoke the same or a closely related language as well as other neighbors who spoke completely unrelated languages.[18]

Nonetheless, the aggregate of the many social pathways on the coast shows that there is a common pool of material culture and a common set of expectations about friendship, about how to interact with visitors from afar, and about how to maintain stable and enduring relations with people from other places. Inherited friendship patterns map out a vast "community of culture" on this coast that cannot be parsed by language or other ready markers of corporate group membership or encompassing identities.

When we look at the distribution of different kinds of objects along the coast, we find a complex array of object types, styles, and manufacturing sites. As we have suggested elsewhere (Welsch et al. 1992:571; Welsch 1996a), the distributions of different kinds of material culture reflect different technological choices and different cultural behavior in at least some of the ways suggested by Lemonnier (1986). In this respect, Class III (and to some extent Class II) items are the most interesting. For example, using either bows and arrows or spears (with or without a spear thrower) implies different adaptations to the task of hunting.[19] When these weapons are used in warfare, they lead to different kinds of behavior in the midst of battle. Insofar as one technology offers better efficiency, these technological adaptations are not decisions made by individuals independently. Decisions like these are made under pressure from the broader social field in which people live (Watson 1977). Other technological choices, such as whether men should wear a loin cloth, a penis gourd, or go naked—because they would appear to be of no selective advantage—offer more scope for individual freedom. Yet we generally find on the coast that neither kind of technological choice is associated with particular local ethnicities. Technical choices of both sorts structure a much broader multi-ethnic configuration (what we have called the coastal "community of culture") and contribute to minute individuating differences at the level of the lineage, the hamlet, or the village (what may be referred to as "cultural diacritics," see also Watson 1990).

Similarly, pots made with the paddle-and-anvil technique seem to be more durable and of better quality than those made by spiral-coiling or with the slab-building technique. The slab-built and spiral-coiled pots we collected and observed were considerably less durable and more likely to break when transported than the paddle-and-anvil pots from Tumleo and Kaiep. This technological difference may help explain the wide distribution of Tumleo and Kaiep pottery and the limited distribution of pots from the other centers of manufacture. The decorative (stylistic) variation of Tumleo and Kaiep pots may have (or may have once had) local meaning in the villages where these items are made. But outside of these producing villages such variation has little or no significance to users, except insofar as such variety allows people to identify their own possessions. Indeed, few potters on Tumleo today, for example, can interpret the designs used by other potters on the island and potters are generally unable to explain the motifs on pots in museum collections made a century ago.

ANALYSIS

From this summary of how material culture is distributed along the coast, it is clear that nearly every item has a somewhat different distribution. Individually, these objects and the composite assemblages they belong to do not distinguish any particular language grouping. In fact, it appears that no single object type is truly specific to a particular language or language family. Nor does material culture on the Sepik coast divide the region into ethnic groups or allow us to identify communities, polities, or clans in any systematic way, despite the fact that there is an abundance of stylistic variation that undoubtedly has significance to their makers and owners. Put somewhat differently, the analysis or mapping of objects and decorative styles does not divide up the coast into bounded, social units of any significance that we can discern.

Mapping material culture on the Sepik coast can, however, tell us about the occurrence of resources (or products) and their distribution through various kinds of exchange. Communities that use the same kinds of objects along the coast may be said to belong to the same "resource field." They may or may not, however, also belong to the same social formations. Our informants have made it clear that they are aware of people living outside of or beyond the field of their own social networks. They sometimes obtain objects made by such people, usually through intermediaries, but these "other people" are marginal to their face-to-face friendships.

Four examples from along the Sepik coast illustrate how the Sepik coast can belong to a single "resource field" but different "social fields." These include: (1) the silent trade between the Murik Lakes villages and their hinterland neighbors, (2) Murik control of interaction between communities east and west of the Sepik mouth, (3) the social discontinuity between Serra and Leitere, and (4) the partial social discontinuity between the bush villages and the coast.

(1) *Murik silent trade.* The most striking of all social "nonrelations" occurs in situations known as "silent trade." Although this kind of trade was less common in New Guinea than some early commentators assumed, social relations between the Murik Lakes people and their inland neighbors appear to have been based on a silent system of barter (see Barlow 1985). Murik people deposited their smoked fish at an established place for trade and left. Then, bush people came to take the fish and left an appropriate amount of sago. They then left and the Murik returned to take their sago home. During the entire transaction, it was not necessary for either party to see who they were dealing with. The Murik and their inland neighbors belong to different social fields and in this sense two different social formations, but because they use the same fish and sago, they inhabit the same resource fields.

(2) *Murik coastal middleman trade.* Murik people live at the mouth of the Sepik River. They were critical players in controlling and consciously manipulating social relations to the east and west of them along the coast (see Lipset 1985). But unlike their relations (or lack thereof) with inland folk, their relations with other communities along the coast were warm and cordial on both sides of the Sepik as well as with the islands off the coast. But the Murik were also shrewd traders and knew the value of monopolizing middleman trade whenever the opportunity arose and they kept their friends and partners on the west from dealing with people on the east, and vice-versa. People around Wewak have told us they would have liked to develop friendships with people east of the Sepik, but the Murik prevented them from doing so: if they ventured east of the Sepik on their own, their Murik friends become angry and (at least for a time) cut them off from future exchanges. So when they wanted products from places east of the river, they asked their Murik friends to get what they need for them, or they might occasionally accompany their Murik friends on a visit to villages in the east.

For people east and west of the Sepik, therefore, the river marks the boundary between two social fields or social formations. But as before, both sides of the Murik share the same pool of material culture. Indeed, it is in Murik interests to maintain a high demand for products across this social boundary of their own making. Thus, both sides of the Sepik belong to the same "resource field" even though they belong to different social networks. Only the Murik themselves participate in both social fields.[20]

(3) *The social discontinuity between Serra and Leitere.* Further west along the coast beyond Serra is a rugged patch of terrain known as the Serra Hills and beyond them lies the lagoon community of Leitere. The Serra Hills effectively prevent easy communication by foot between these two adjacent communities and they are known to have been an impassable barrier for at least one early anthropologist in 1909 (Neuhauss 1911). Tiesler (1969–70) has suggested that even the rugged coastline, where the hills jut out into the water forming bluffs would have served as a barrier to communication by sea, since the shoreline offers few safe beaches in the event of a sudden storm. Having visited both sides of the Serra Hills, it is clear to us that the hills themselves do not separate people in Serra from those at Leitere for environmental or geographic reasons. In the dry season, canoes can easily pass from one side to the other without any greater difficulty than elsewhere along the coast. Rather, it is the different social worlds that these two communities inhabit that partly insulates these two communities from one another. In Leitere and villages to the west, marriage involves a bride-price payment using glass beads as the medium of exchange; in Serra and villages to the east, marriage is characteristically through sister exchange, with no exchange of bride wealth. People west of the hills were preoccupied with glass beads that they used in most transactions, including bride-price payments. They had very little interest in people living east of the Serra Hills, who had access to shell rings but no supplies of glass beads. People in the east had little interest in glass beads and generally looked eastward for their friends, while people in Leitere and the eastern villages sought ties in the west. Yet despite these economic differences, our data show that despite these cultural differences, some friendships extend across this divide, even though the density of these ties is far weaker than those ties in any other direction. In this instance, people in Serra and Leitere inhabit different social formations and participate in different social fields. It is not clear to what extent they may be said to inhabit the same "resource field" since both have some contacts with the other and exchange some items, but we suspect these transactions are fairly limited. Unlike the coastal transactions through the Murik, there is no middleman community between the two and in most respects it would seem that here the boundary between these two social formations coincides with the boundary between two resource fields.

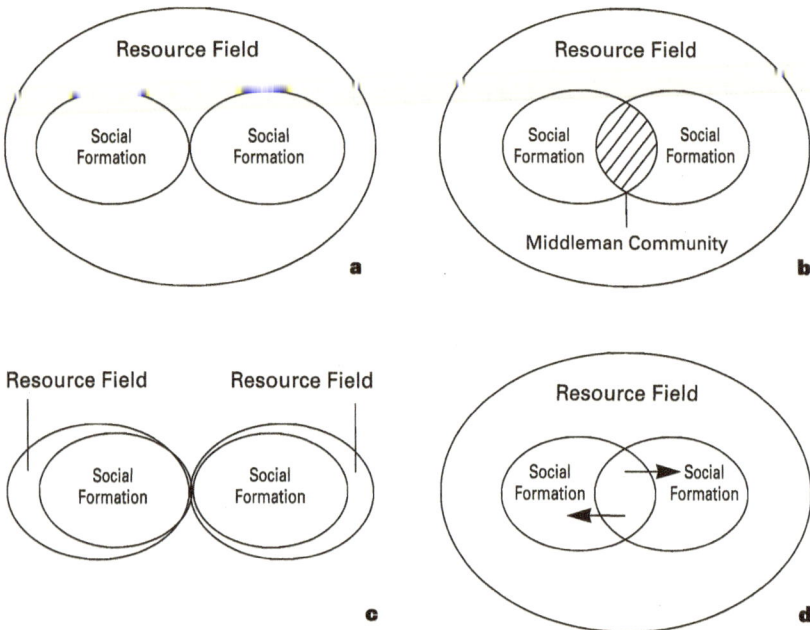

Figure 3.2. Varieties of trade relationships along the Sepik coast: (a) silent trade between communities in two social formations; (b) middleman community monopolizes trade between two social formations; (c) minimal contact across a social and economic boundary; (d) permeable social boundaries between social formations.

(4) *Relations between bush villages and the coast around Aitape.* For the most part, people in the hinterland have relatively limited social relations with people on the coast. If bush people have coastal friends, they visit only the handful of coastal villages that lie nearest their inland hamlets. Informants in both areas—coast and bush—report limited and tenuous relationships between the two areas. Some individuals in both areas flatly deny that they ever had ties or friendships between the two areas and everyone in the Aitape area denies that there was ever a system of silent trade as in Murik. Yet in 1909 A. B. Lewis collected a sizable number of objects in coastal and island villages that were made in the bush and these could only have gotten to where Lewis collected them if there were some kind of transaction occurring between the coast and the bush. What informants are telling us when they minimize the importance of coast-bush relations may be that, as in Leitere and Serra, the two sets of communities inhabit quite different social worlds. On the coast, relationships between communities are organized around hereditary friendship, while in the bush they are oriented around extended net-

works of kinsmen. In fact coastal people actively discourage marriages between friends; bush people seem to prefer creating relations through marriage alliance, in the kinds of ways anthropologists have discussed for the past century. But unlike the social and resource discontinuity that occurs between Serra and Leitere, here along the coast-bush frontier, social relations were limited but the flow of products seems to have been fairly brisk. Nowadays, a number of prominent residents of bush villages have adopted the coastal pattern of relating to people beyond their hamlets, to the extent that over the past fifty years they have developed extensive friendship networks along the lines of those long observed on the coast. Here, belonging to the same resource field does not require belonging to the same social formation. And yet some individuals seem to be moving the social boundaries that formerly existed in this area by incorporating themselves into both the coastal and the inland social formations.

Each of these four situations produces somewhat different kinds of social and economic boundaries between different communities in this part of New Guinea. Each generates a different sort of social formation boundary, which may or may not coincide with that of a resource field. We illustrate their differences graphically in Figure 3.2. These four conditions do not exhaust the possibilities of how different social formations may "meet" one another or the extent to which different social formations coincide with different resource fields. Nevertheless, these diagrams illustrate how boundaries can exist between social formations without necessarily producing significant discontinuities in the distribution of material culture.

DISCUSSION

As Margaret Mead (1938) pointed out for the Mountain Arapesh, who lie on the periphery of our study area, the distinctive cultural and linguistic features in this region of New Guinea that distinguish any particular community from its neighbors appear to be inherently dynamic and unstable, as are their social boundaries. What persists over time are the kinds of larger social formations that we have described here: formations that extend over 200 km or more along the coast, not the local communities that participate in these large social formations.

It seems unlikely that archaeologists would be able to disentangle the complex ways that people on the Sepik coast have related to one another from analysis of the material culture alone. Ethnographers have reified the "village" as the basic building block of culture and society, and as MacEachern (this volume) notes, ethnographers have encouraged archaeologists to find such small, bounded communities in the past, as well. Along with Willey and Phillips (1958:2), we believe that "archaeology is anthropology or it is nothing." But like MacEachern, we do not believe that local ethnography is the only kind of anthropology to be emulated. Anthropology certainly includes ethnographic questions about the practices and social relations in particular villages and communities. But anthropology also includes regional and processual questions about how these communities are linked to their neighbors. Moreover, the kinds of data most often available to archaeologists and students of material culture lend themselves more directly to regional, comparative, and processual questions than they do to questions about local ethnographic ones.

ACKNOWLEDGMENTS

The research upon which this chapter is based was supported by National Science Foundation Grants No. BNS-8819618 and No. DBS-9120301, and National Endowment for the Humanities Grant No. RO-22203-91. Helpful comments about this analysis were provided by Hoyt Alverson, Kirk Endicott, Debbie Hodges, Miriam Stark, Sarah Welsch, and two anonymous reviewers. We are indebted to the many people in West and East Sepik provinces who have shared with us their life experiences and understandings of friendship. We gratefully acknowledge the assistance of these institutions and individuals. Any shortcomings are, of course, our own.

NOTES

1. Welsch and Terrell were on the Sepik coast in April and May 1990 for brief visits to Aitape, Ali, Tumleo, Malol, Sissano, Warapu, Pultalul, Vanimo, and Wutung. Welsch returned to the coast from April 1993 to February 1994, visiting more than ninety villages in Jayapura, Vanimo, Aitape, and Wewak districts. He conducted about 130 interviews with older men about their friendship networks along the coast. Terrell joined him in Papua New Guinea from July to November 1993 to conduct an archaeological survey of the coast. During this field season Terrell also assisted Welsch with his interviews about these extensive social networks. Welsch and Terrell were joined for a total of nearly five months by Wilfred P. Oltomo, senior technical officer at the Papua New Guinea National Museum and Art Gallery, who also assisted with this research. In September and October 1996, Terrell and Welsch returned to Aitape to conduct archaeological excavations, during which time further insights about friendship relations in the region were made.
2. Only in the case of Warapu and Sissano was there ongoing hostility when the Germans "pacified" the region. It is possible that colonial efforts to prevent further hostilities actually

prevented these communities from resolving their differences, which have persisted off and on for almost a century.

3. This description is overly simplified. Islanders might make several trips back to their home island during a season, as their canoe gets loaded with heavy sago and yams. If they have asked one friend for a large quantity of sago, they might feel it more polite to appear less greedy by unloading sago from another friend at their home before returning to their first friend.

4. For convenience, we refer to the individuals who visited friends as if they were male, even though occasionally women were part of a group visiting friends along the coast. Both men and women had friends of both sexes scattered along the coast, but before pacification few women traveled very far from home. Today, with increased mobility on highways and in dinghies, women actually get around a great deal—usually accompanying their husbands or brothers. In the past, women seem to have traveled much less than today, and they never traveled alone. For an early view see Erdweg (1902); Lipset (1985) suggests a similar but more varied pattern in more recent times around Murik and we heard such accounts around Aitape.

5. These named polities, which we call communities, include named local units such as Sissano, Warapu, Arop, Malol, Lemieng (formerly Walman), Paup, Yakamul, Ulau, Suain, and Matapau. These communities generally correspond to the units the German colonial administration called *Landschaft*.

6. For a fuller discussion of these networks see discussions in Welsch (1996b), Welsch and Terrell (1994). A somewhat earlier interpretation of these networks can be found in Terrell and Welsch (1990a, 1990b) and Welsch and Terrell (1991). Barlow (1985), Lipset (1985), and Barlow, Bolton, and Lipset (1986) offer slightly different views of these networks, concentrating their attention on areas well to the east of Aitape.

7. Since the Second World War, Pidgin English has become a well established lingua franca for the entire region; it is universally understood by men, women, old, and young alike. But Pidgin is a twentieth-century phenomenon, only introduced during the German period, particularly after the opening of a government station at Aitape in 1906. By 1910, for example, when Field Museum curator A. B. Lewis visited the region, each village had one or two Pidgin speakers, but the language was by no means a widely known or useful lingua franca (Welsch 1998).

8. Transactions involving foodstuffs and raw materials are still extremely important in most places. With the advent of stores and manufactured goods, certain of these products are no longer made: large canoes are rarely made today on the coast; shell rings are now made only on Tarawai. The production of earthenware pottery, although still an important element of the regional economy, seems to have declined to less than half the production observed by A. B. Lewis in 1909. Nevertheless, many local products are still exchanged with great enthusiasm.

9. Such claims seem to be made, despite the fact that specialist-produced items are not always as well made or as serviceable as homemade products.

10. Although earthenware frying pans were traditionally found on the islands and in the lagoon villages to the west of Aitape (Malol, Arop, Sissano, etc.), the islanders made little use of

them. Frying pans were critical to the diet of everyone in the lagoon villages. Frying pans were also used in Leitere and possibly other villages to the west. Ornaments for the men's cult houses are known for certain only on the islands at Aitape.

11. For this and subsequent discussion, we deal primarily with areas west of the Sepik River. These are areas we have visited and thus we have a better grasp of variation in material culture than in the area east of the Sepik.
12. Now in the National Museum and Art Gallery, Port Moresby.
13. For descriptions of earthenware pottery at Humboldt Bay see Galis (1955) and May and Tuckson (1982:301); at Vanimo see Broadhurst (1975), May and Tuckson (1982:317–325), Welsch (1998, book 2); at Leitere see May and Tuckson (1982:316–317); at Sissano, Serra (Sarai), Ramo, and Sumo see May and Tuckson (1982:316), Welsch (1998, book 2); at Goiniri see May and Tuckson (1982:298); at Kombio and Abelam see Allen (1977), May (1977), May and Tuckson (1982:275, 281, 292ff). For discussion of the better-known tradition at Tumleo see Erdweg (1902), May and Tuckson (1982:308–315), Parkinson (1900, 1979), Tuckson (1977), Tuckson and May (1975), and Welsch (1998, book 2). Discussion of Kaiep pottery can be found in Egloff (1977:72–73), May and Tuckson (1973; 1982:302–307), and Tuckson and May (1975).
14. Tiesler (1969–70) situates the origin and development of these intricate exchange networks in the arrival of Austronesian-speaking immigrants from what is now Indonesia. These people, he claims, introduced outrigger canoes to the region; better transportation allowed wider networks and more regular interaction. The historic distribution of canoe-making technology does not, however, support this theory, because there were as many Non-Austronesian as Austronesian centers producing outrigger canoes. The technology seems to be organized around environmental variables: islanders had a profound need of the technology and either invented or borrowed it. Along this 500 km stretch of coast, the outrigger canoe is clearly not correlated with Austronesian-speaking communities.
15. Discussion of the distribution of motifs is derived from an analysis by Larry Saviers, who used this material for a senior honors project at Northwestern University. Wilfred Oltomo, senior technical officer at the National Museum and Art Gallery assisted both in Aitape District and at The Field Museum in making these observations about the modern distribution of string bags along the coast.
16. General information about either of these object types in early accounts is limited. Lewis (Welsch 1998) makes some comments about objects he saw and collected, see also Erdweg (1902), Graebner (1927), and Parkinson (1900, 1979). More recently Newton (1989) compared stylistic variation in cassowary bone daggers. For a general sense of how these objects were distributed geographically, see Welsch, Terrell, and Nadolski (1992, table 2). In analyzing stylistic variation on headrests, Welsch was assisted at The Field Museum by James Coplan. Both Coplan and Abigail Mack assisted with analyses of bone daggers.
17. Information from Leitere is ambiguous. Informants gave contradictory information about traditional dress in this village. Field Museum collections do not enlighten us as to whether gourds were worn here or not. At Humboldt Bay men wore no clothing whatsoever (Galis 1955), a fact that led Dutch observers to call the region Papua Talanjang, meaning "the naked Papuans."

18. At Small Mushu and Worak villages on Mushu Island, Papuan speakers of Coastal Boiken and Austronesian speakers of Kairiru language appear to form a single community with considerable intermarriage. In precolonial times the situation at Dallmannhafen was probably quite similar, although the subsequent growth and development of Wewak town around this community has essentially obliterated earlier social networks and community structures that would have been apparent to early visitors. After one or two families from Austronesian-speaking Ali Island intermarried and settled at Matapau, originally an Arapesh-speaking (Papuan) village, Ali and Arapesh are of nearly equal importance today. In contrast to these mixed communities, there are others where smaller languages are being swamped by larger ones, such as the moribund Nori (Papuan) at Serra (Austronesian). Something similar seems to be happening at Onei, where the original Papuan language is in decline as young people routinely use Austronesian Serra as their language of preference. Language changes are also in progress at Leitere, Ningera, and possibly west of Vanimo.
19. Early in the century Fritz Graebner (1909) had noted the peculiar distribution of the bow and arrow, as opposed to the spear, in Oceania.
20. Of more than 100 informants west of the Sepik, almost none have any direct knowledge of people and villages east of the river, and Lissant Bolton (pers. comm. 1990) reports that villages east of the river are largely oriented to the east. Here again, participation in the same resource field does not necessarily imply participation in the same social formation.

4

Social and Technical Identity in a Clay Crystal Ball

OLIVIER P. GOSSELAIN

As is apparent in the chapters that constitute this volume, as well as in a series of seminal publications (i.e., Childs and Killick 1993; Dietler and Herbich 1994b; Dobres and Hoffman 1994; Ingold 1988, 1990; Lechtman 1977; Lechtman and Steinberg 1979; Lemonnier 1986, 1992, 1993b; Maret 1980; Pfaffenberger 1988, 1992), our conception of both traditional and modern technologies has changed considerably during the last two decades. Previously ignored or approached solely from a determinist and evolutionist point of view (but see Mauss 1935, 1941; Merrill 1968; or Haudricourt 1987), technical behaviors are now perceived as full social productions. They are shown to constitute culturally grounded systems in which the choice of actors, raw materials, tools, and processing modes does not merely relate to natural pressures, but also to symbolic, religious, economic, and political ones. This is not to say that technical actions stem from social rather than material necessities, but simply that the notions of "meaning" and "function" are practically inseparable when transforming raw materials or satisfying biological needs. Furthermore, technical behaviors do not appear to be randomly adopted: they result from particular learning processes and can thus be viewed as "socially acquired dispositions" (Dietler and Herbich 1994a:465).

These are essentially simple ideas, as described by an anonymous reviewer, and there would be little need to discuss them again were it not that many archaeologists remain oblivious of such anthropological issues. The field of pottery studies especially stands as one bastion in which the Binfordian conception of culture as an "extrasomatic means of adaptation" still prevails. A crude mix of old school ethnoarchaeology, ceramic engineering, and whimsical experiments, it revolves mainly around one basic assumption: the manufacture and subsequent use of vessels is governed by so many ecological and physical constraints that technical behaviors are better explained as adaptive strategies rather than as social (or cultural) choices. Considerable efforts have therefore been devoted to characterize raw materials and finished products, and to relate the observed variations to the mechanical and environmental problems supposedly faced by potters. Although presented as major theoretical and methodological advances (e.g., O'Brien et al. 1994; Rice 1996; Schiffer et al. 1994), such studies are typically built upon preconceptions, unsophisticated arguments, and, quite often, a poor understanding of pottery technology and contexts of use. Moreover, they reinforce the old distinction between style and function ("culture" vs. "nature"), thus depriving archaeologists of a valuable tool for approaching cultural identity and boundaries.

In this chapter, I challenge the views expressed by ceramic ecologists and behaviorial archaeologists by evaluating tenets of their approach using ethnographic research among traditional societies. My first aim is to show how the combined use of ethnography, archaeometry, and experimental archaeology allows a better understanding of pottery *chaînes opératoires* and leads to the identification of other kinds of constraints than the ones usually attributed to pottery-making (see also Gosselain and Livingstone Smith 1995; Gosselain et al. 1996). In the second part of the article, I explore the relationships between technical choices and social identity among a series of south Cameroonian populations and evaluate the potential offered by the reconstruction of manufacturing processes for assessing cultural boundaries in prehistory. Although my case is geographically- and time-restricted, it shows how a systematic comparison of technical styles can assist archaeologists in the study of such critical processes as migratory movements, population mixing, or language diffusion.

But before considering these exciting topics, we must identify the main limitations of the "Ceramic Ecology" approach and clarify the elements needed to build a more culturally oriented approach to pottery techniques.

PROBLEMS WITH CURRENT CERAMIC RESEARCH

Initially developed in an attempt to incorporate laboratory analyses into pottery studies and to relate raw materials and manufacturing techniques to the potter's social and natural environment (Matson 1939, 1965; Shepard 1936), the approaches which have come to be known as "Ceramic Ecology" and "Function and Use" together became one of the main subfields of ceramic analysis, but also one of the most stagnant. During the last two decades, an impressive number of books and articles have been devoted to these approaches, which endlessly hark back to the same basic assumptions.

First, potters have to cope with the potentially negative effects of their environment: specific chemical and physical properties of the available raw materials call for specific processing techniques (Arnold 1971, 1985:20–32; Klemptner and Johnson 1986; Rice 1987; Rye 1976; Stimmell et al. 1982); fuel shortage promotes or restricts the use of particular firing techniques (Arnold 1985:214; Rice 1987:174; Rye and Evans 1976:165; Sinopoli 1991:33; Stimmell and Stromberg 1986:244); poor climatic conditions lead potters to practice their craft on a seasonal basis or to modify their techniques, tools, and facilities to adjust to poor weather (Allen and Zubrow 1989; Arnold 1975, 1985; Kolb 1989; Rice 1987:315–316).

Second, the transformation of clay into a finished product is a very hazardous process: "pottery manufacture is fraught with danger: accidental breakage can occur at just about any point in the production sequence, and quite often does" (Kaiser 1989:1). Since different stages of the manufacturing process are closely connected, a choice made at one level is liable to condition the choices made at the next levels (Rice 1984; Schiffer and Skibo 1987).

Third (and this is by far the most widely held opinion), potters must produce vessels whose mechanical performances are adapted to their intended functions. The general idea is that different contexts of use induce different—and potentially detrimental—kinds of stress (e.g., thermal shock, compression, abrasion) or require specific physical properties (e.g., thermal conductivity, flexural strength, waterproofing) in order to avoid vessel failure or to enhance functional fitness. Potters are forced to choose suitable raw materials and to process them in an appropriate way (Braun 1983; Bronitsky 1986; Ericson et al. 1972; Rice 1987:226–232; Rye 1976; Schiffer and Skibo 1987; Schiffer et al. 1994; Steponaitis 1983, 1984; Young and Stone 1990).

Confronted with such a number of constraints, potters are left with little room for expressing their identity or for meeting nontechnical and nonfunctional concerns. All they can do is to find materials and processing techniques that fit the Laws of Nature, and hence, to follow predetermined paths. This, in turn, greatly facilitates the work of ceramic ecologists, for identifying the constraints that underlie certain technical choices is sufficient for reconstructing the potters' environment as well as social (functional) needs. For example, the adoption of kilns or ovens reveals a desire to maximize production output under poor climatic conditions (Arnold 1985:213–218), and a change in vessel wall thickness and temper size proceeds from a change in dietary practices and a subsequent need to increase heat conductivity and resistance to thermal shock (Braun 1983). Or, the use of a processing technique that does not improve resistance to thermal shock but allows a decrease of vessel weight is indicative of a mobile population, in which the portability of the product is more important than thermal performance (Schiffer and Skibo 1987; Skibo et al. 1989b).

While probably appealing to those archaeologists who still dream of explaining human behaviors in universal terms or who want to process their data in an "elegant" and (apparently) unambiguous manner, these ceramic ecology models of interpretation seem more than dubious. With respect to the supposed relationship between vessel performance characteristics and contexts of use, for example, it is striking to see that most arguments do not proceed from an examination of archaeological or ethnographic evidence. Instead, arguments rely blindly on concepts and theories from ceramic engineering and experiments made under completely artificial conditions (see especially Bronitsky 1986; Schiffer and Skibo 1987; Steponaitis 1983). But while we all agree that a coarse, low-fired clay is less appropriate than a silicon carbide for making space shuttle tiles or motor parts, its efficiency regarding the cooking, handling, or preserving of food does not necessarily need to be questioned. Since we know virtually nothing about stresses that affect vessels in traditional contexts of use,[1] notions such as "mechanical" or "functional" fitness are, so far, practically meaningless for interpreting ceramic manufacture and use in traditional or past societies.

Several scholars have already criticized this tendency to ignore the cultural dimension of technical behaviors and to rely on unicausal explanations when discussing technological change (i.e., Feinman 1989; Miller 1985; Plog 1980a; Spriggs and Miller 1979; van der Leeuw et al. 1991; Woods 1986). Yet materialism continues to gain ground among ceramic specialists, and one witnesses today an alarming resurgence of the Darwinian evolutionary approach, in which technical diversity is mainly explained in terms of "adaptedness," "selective pressure," and "phenotypic convergence" (Neff 1992, 1993; Nieman 1995; O'Brien and Holland 1992; O'Brien et al. 1994).

Besides being scientifically questionable,[2] this anthropological version of Darwinism refers to another outstanding problem in archaeology: the dichotomy between "style" and "function" (Binford 1965; Dunnell 1978). Evolutionary and processual archaeologists assert that technical behaviors constitute passive and culturally meaningless responses to environmental and functional pressures. In so doing, they reinforce the idea that style is exclusively found in those selectively neutral, easily manipulatable, and therefore external aspects of the artifact, such as decoration or micromorphological features. Either explicitly or implicitly accepted, this concept of style leads archaeologists to focus on the most eye-catching facets of material culture, and in many instances, to subscribe to the "information exchange" theory proposed by Wobst (1977; for recent reviews of the literature, see Conkey 1990, Hegmon 1992, this volume; Wiessner 1990). To summarize, the "information exchange" theory postulates that style is a cost-effective strategy of communication, aiming at asserting cultural identity and at negotiating intersocial relationships. In other words, designs, morphological features, or any other visible attributes, constitute consciously emblemic messages of ethnicity, and from the archaeologist's point of view, convenient means for approaching social boundaries.

Needless to say that this rather narrow definition of style, as well as its subsequent applications, have received severe criticism (i.e., David et al. 1988; Dietler and Herbich 1989; MacEachern 1994; McIntosh 1989; Sackett 1982, 1990; Sterner 1989). Yet few archaeologists seem inclined to go a step further, to acknowledge that style might be present whenever there is a choice between equally viable options (Sackett 1990). If this is the case, then "style" could reside in every stage of the manufacturing process and thus in every technical feature of a manufactured object (Lechtman 1977; Lechtman and Steinberg 1979; see also Childs 1991; Dietler and Herbich 1989).

INTEGRATING TECHNICAL STYLE INTO POTTERY STUDIES

As challenging as it is, the concept of "technological (or technical) style" is not merely an intellectual construct. It is founded upon the ethnographically verified assumption that similar aims can always be reached in different ways, but that the choices that artisans make essentially proceed from the social contexts in which they learn and practice their craft. Moreover, this form of stylistic expression *could* prove to be quite stable through time and space, for it often relies on unconscious and automated behaviors. Unlike ornamental and formal styles that can serve different social purposes and be easily manipulated, imitated, or rejected, studying technical style offers an opportunity to explore the deepest and more enduring facets of social identity.

Before integrating this concept into the field of pottery studies, however, several questions need to be considered. First, we must ascertain that every stage in the manufacturing process is the locus of a stylistic expression. No one will deny that a potter can shape or decorate a vessel in a variety of ways, regardless of its intended function or the raw materials at hand. But can the same be said of critical operations such as clay processing or firing? Even if there is still room for technical choices at those levels, the existence of some environmental, manufacturing, and functional constraints could reduce the number of options that a potter has, and hence the stylistic salience of those stages of the *chaîne opératoire*. This is indeed a critical issue and I believe that the best way to avoid falling back in the trap of the Ceramic Ecology-Functionalist approaches is to begin by evaluating the *actual* performances and limitations that characterize each transformation process. This goal can be achieved with analytical and theoretical tools borrowed from other fields, as long as we rationalize their use and adapt them to our own materials.

Second, we must determine the factors that affect the stylistic dimension of technical behaviors. What factors influence the process of decision-making? Is the social context in which potters learn the craft more (or less) determinant than the context in which they practice it? Do all the stages of the manufacturing process display the same stability through time and space?

Third, we must explore the link between ethnicity and material culture patterning. Do technical behaviors consistently relate to particular facets of social identity? Does the scale of technical variations parallel that of other kinds of style? Are all the stages of the manufacturing process comparable in this regard?

In an effort to answer these questions, I undertook a wide and systematic ethnographic survey in southern Cameroon between 1990 and 1992. During these three years, I had the opportunity to work with a hundred potters belonging to 21 linguistic groups (Fig. 4.1) and living in contrasting environmental and cultural contexts (Gosselain 1993, 1995). Although my intention was to cover the entire territory occupied by each group, the spatial distribution of communities and individuals under study was not even. This is explained by the complete disappearance of pottery production in several areas, the inaccessibility of certain regions, and field season deadlines. The study area expands considerably, however, when considering the learning site of each potter observed (see Fig. 4.1).

ETHNOGRAPHIC BACKGROUND

The 21 groups under study belong to 7 linguistic entities: Narrow Bantu, Ring Grassfields, Eastern Grassfields, Mambiloïd, Tikar, Oubanguian, and Adamawa (Dieu and Renaud 1983). The first five groups are part of the Wide Bantu linguistic grouping and display varying degrees of affinities; the other two are part of the Adamawa-Ubangi grouping and must be clearly distinguished from the former.

Population size and density vary dramatically across the study area, from 260,000 individuals (Bamum) and 110 inhabitants per square kilometer (Bamileke Fe'fe') to 1, 100 individuals (Djanti) and 0.80 inhabitants per square kilometer (Gbaya). In most instances, groups consist of small, politically independent, and mainly exogamous chiefdoms that are either homogeneously distributed within the linguistic boundaries (e.g., Yamba, Eton) or concentrated in certain areas (e.g., Vute, Kepere, Gbaya). Other groups occupy a single autonomous chiefdom (Nsei), a series of small enclaves displaying a tendency to endogamy (Bafeuk), or are organized as a kingdom, with a pronounced centralization of political and economical power (Bamum and Tikar).

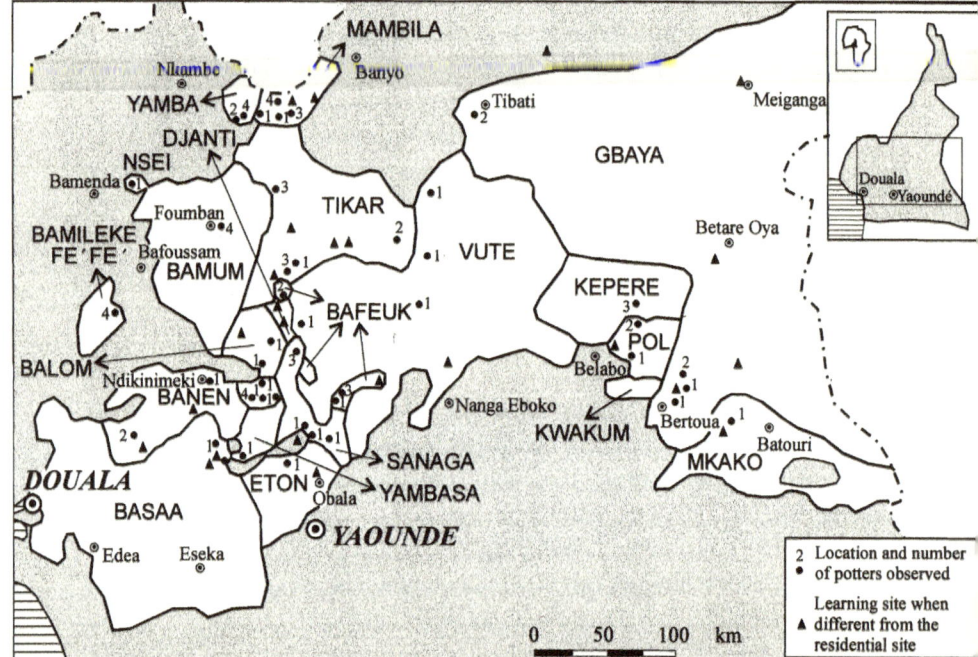

Figure 4.1. Location of potters and linguistic groups surveyed in southern Cameroon between 1990 and 1992. Linguistic boundaries drawn after Dieu and Renaud (1983).

Subsistence is mainly based on small-scale agriculture (corn, cassava, peanut, taro, sweet potato, sorghum), palm tree exploitation, livestock (chickens, goats, pigs), and, to a lesser extent, fishing and hunting. Some groups also rely on cash crops, such as cocoa and coffee (e.g., Mambila, Bamileke Fe'fe', Bamum, Basaa, Eton), and on the manufacture of specialized crafts (pottery, basketry, weaving). These products are either bartered for foodstuffs, sold in local markets, or distributed through large-scale exchange networks (Yamba, Nsei, Bamileke Fe'fe', Bamum).

Four contexts of pottery production were observed:

(1) In most instances, pottery-making is a secondary economic activity that is subordinate to the primary concerns of the artisan, such as farming and domestic tasks. Although open to anyone without any restriction of gender, age, or social status, pottery-making is generally practiced by a few women in a limited number of villages. Production is sporadic and usually restricted to the dry season (which, as the potters put it, is both a period of off-peak agricultural activity and a period of better accessibility to clay deposits). Yet some women manage to work at other times, when they receive orders for pots, when they need money, or when they consider the dry season as a potentially dangerous period for practicing the craft.[3]

Vessels are either used by the potter, sold, bartered, or given as gifts in her or his immediate community, or sold in nearby markets. Income is minimal and usually allotted to personal purchases.

(2) Among Bamum and Bamileke Fe'fe', pottery-making is a year-round activity that nearly all adult females practice in a single locality. Here again, the craft is subordinate to primary economic concerns, but women engage in pottery-making during every moment of leisure and sometimes rely on their daughters or other relatives to complete agricultural and domestic tasks. Finished products are either sold in local markets or to middlemen, who then resell these vessels in the craft markets intended for tourists in such major towns as Foumban, Bafoussam, Douala, and Yaounde. Income from pottery production remains minimal, however, and no potter can make a living out of it.

(3) Among Nsei of Bamessing, pottery-making is a year-round activity that nearly all female and male members practice in certain districts. Generally, women subordinate the craft to their primary economic concerns and specialize in the making of daily-use pottery. Men also manufacture those vessels, but primarily specialize in the production of ceremonial vessels, clay pipes, and clay figurines. According to Ueli Knecht (pers. comm. 1994), more than half of them practice the craft on a full-time basis. Finished products were traditionally bartered for palm oil, clothes, or luxury goods throughout the whole western and northwestern Cameroon. Nowadays, they are either sold locally or in the main craft markets of the country.

(4) Among Yamba, pottery-making is a year-round activity that only male inhabitants from two neighboring villages practice.[4] According to the potters interviewed, all the men of these villages used to make pottery on a full-time basis, bartering their vessels for food and utilitarian goods produced in other Yamba communities. Few of them owned a field and such fields were cultivated by the female members of their family. Nowadays, the preference for plastic, aluminum, and glass containers has forced the potters to abandon the craft or to rely on it as a secondary source of income.

RESEARCH METHODS

In the field, my first aim was to carefully observe and compare the techniques potters used at the different levels of the *chaîne opératoire*, from resource procurement to postfiring treatments and distribution of the products. I also tried to determine the reasons underlying potters' technical choices, first by questioning them about the origin of their knowledge, the way they selected raw materials, or the way they controlled the issue of the different technical operations, and second, by evaluat-

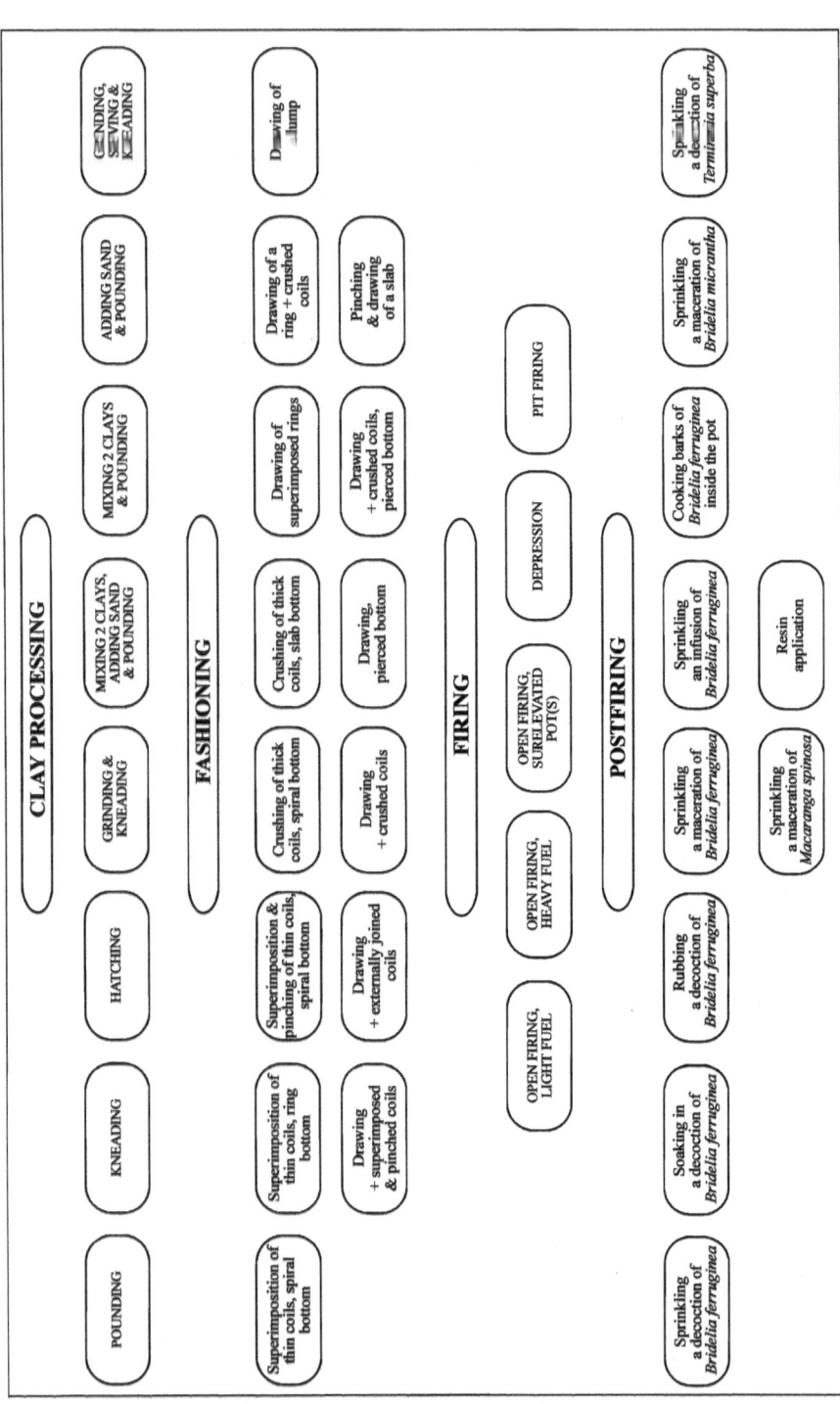

Figure 4.2. Techniques and materials used by south Cameroonian potters for completing the main stages of the manufacturing process.

ing the performances that characterized each transformation process. To this end, a series of measurements and experiments was made, some of which had never been attempted or systematized in traditional contexts of production (see details and results in Dialo et al. 1995; Gosselain 1992a, 1994, 1995): clay plasticity/workability before and after processing (with the help of a pocket shear device); exchanges of raw materials between potters using different techniques; drying modalities and duration; firing parameters;[5] contexts and methods of use of the vessels. These investigations were reinforced by a subsequent series of laboratory analyses and experiments that characterized raw material properties and potentials: granulometric composition of clays (some of them were also submitted for chemical and mineralogical analyses), paste behavior during drying and firing (using raw clays as well as processed ones), nature of fuel materials and stem barks used for postfiring treatments, chemical composition of the latter and assessment of their effective role by simulation.

TECHNICAL DIVERSITY AND CULTURAL SIGNIFICANCE

When considering the results of field observations alone, one is first struck by the variability of technical behavior in respect to the size of the study area. As shown on Figure 4.2, eight techniques are used for processing raw materials, fourteen for fashioning the vessels, five for firing them, and ten for treating their surfaces after firing. Depending on the stage of the *chaîne opératoire* and on the level of analysis, differences are either pronounced (e.g., processing of clay by grinding, sieving and kneading vs. simple pounding or sand-tempering and pounding; pit firing vs. open firing) or more tenuous. In the fashioning process, for example, the fourteen techniques can be grouped in four main categories: (1) pinching, (2) drawing of a lump, (3) coiling, and (4) drawing of one or several ring(s) of clay. A differentiation is made according to the way these categories are combined or the way clay is added and deformed when roughing out either the whole pot or only a part of it (see some examples in Figure 4.3). Other differences involve the use of contrasting raw materials in the scope of otherwise similar procedures: these include "light fuel" (palm fronds, grass, corn stalks, barks) vs. "heavy fuel" (logs, limb sections) for open firing, as well as plant species exploited for preparing organic coatings.

Field and laboratory investigations reveal that all these options allow one to achieve the same goals, from both a technical and functional point of view. These technical choices are, in other words, functionally equivalent. Using any of the fourteen fashioning techniques, for instance, potters can produce vessels of similar size and form (although a few choose to change techniques according to vessel dimensions). Concurrently, similar firing parameters are recorded in structures

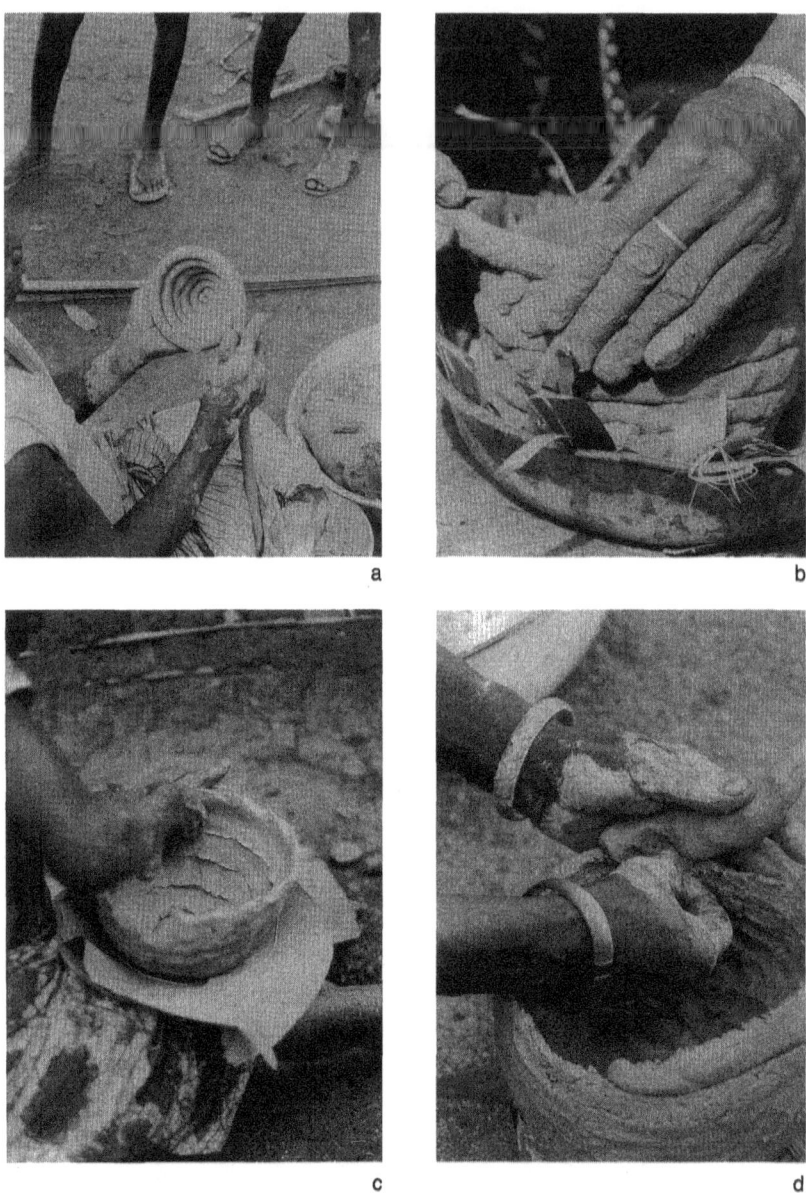

Figure 4.3. Several variations in manufacturing technique by addition of coils: (a) superimposition of thin coils (Mambila); (b) superimposition and simultaneous pinching of thin coils (Eton); (c) internal crushing of thick coils (Bafia); (d) internal crushing of externally joined coils (Gbaya). (O. Gosselain, photographer.)

that differ completely in form, size, type of fuel, and vessel's positioning. Postfiring treatments also illustrate this equivalence of choices, since the different tree species[6] whose bark is used for preparing organic coatings all contain the same chemical constituents (procyanidins), a group of condensed tannins whose pyrolysis allows a waterproofing of the vessels (Dialo et al. 1995).

A related, but nonetheless fundamental observation is that little interdependence exists between different stages in the manufacturing process. A choice made at one level does not automatically condition the choices made at the next levels, which means that all combinations can be envisioned. Potters who grind and sieve their clays, for example, can subsequently fashion the vessels by coiling, pinching, or drawing of a lump, fire them in the open or in a pit with either "light" or "heavy" fuel, or treat their surfaces by smoking or by sprinkling a decoction, a maceration or an infusion of fresh stem barks. Although the number of combinations encountered in the field remains somehow limited, the technical equivalence of options clearly ensures their potential interchangeability.

Finally, field and laboratory investigations show that no environmental, technical, or functional constraints determine the adoption of a particular behavior. Regardless of the materials at their disposal, the method of use and scope of products they manufacture, potters are free to choose among a very wide range of options.[7] This fact is especially apparent when relationships between raw material granulometric composition and techniques used for processing them are examined. As shown in Figure 4.4, most of the selected clays tend to concentrate in the range characterized by a large proportion of sand and gravel (40 to 80 percent), an intermediate proportion of silt (15 to 45 percent), and a low proportion of clay (3 to 30 percent).[8]

Despite the overall variation, different potters sometimes exploit materials of similar texture. They process them in a variety of ways (simple pounding; sand addition and pounding; grinding, sieving, and kneading), however, even though they occasionally use the same fashioning and firing techniques and consistently produce vessels that will be used for the same purposes, in the same contexts. With only one exception (but see discussion in Gosselain 1994), granulometric data also show that none of the available—and successfully exploited—clays is sufficiently coarse or sufficiently fine to render particular refining or tempering operations compulsory. These results, as well as the raw materials exchanges undertaken in the field, suggest (1) that we are faced with a number of technical and functional aims that permit substantial flexibility in the selection and subsequent processing of clays, and (2) that the notion of clay's appropriateness is not subject to consensus, but instead to individual appreciation.

The lack of any environmental pressure is further exemplified at the level of postfiring treatments: for instance, south Cameroonian potters who use an organic

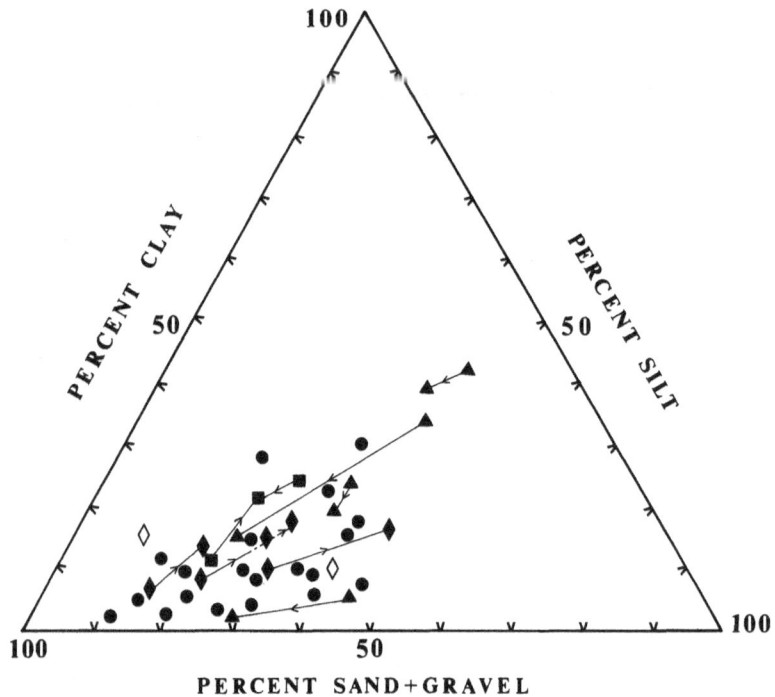

Figure 4.4. Granulometric distribution of clays exploited by south Cameroonian potters with respect to processing techniques. *Circle:* pounding, kneading, hatching, or grinding and kneading; *square:* mixing two clays and pounding; *triangle:* adding sand and pounding; *lozenge:* grinding, sieving and kneading. For the last three techniques, arrows indicate that the same material(s) is used before and after processing.

coating have the possibility of exploiting the bark, fruit, or root of at least nine plant species (Gosselain 1995:302–304). However, individuals select one or two of them at the very most and seem to completely ignore the existence or potentials of the remaining species.

Taken as a whole, field and laboratory results show that technical behaviors cannot be explained in purely materialistic terms and, therefore, are not as predictable as previously thought. This does not mean that procedures are randomly selected, however, or that complete interchangeability exists. Notably, a series of economic and symbolic pressures appears to influence the process of decision-making.

In a context where pottery production is just a part-time activity with weak economic returns, potters tend to subordinate their work to their primary economic concerns and to limit as much as possible their investment in time and energy. With respect to clay source selection, for example, one sees that (1) they never engage in prospecting tours, preferring to rely on accidental discovery;[9] (2) they only exploit those clay deposits that are situated in locations already frequented for other purposes (such as residential locations, fishing sites, fields, hunting paths, and marketplaces); and (3) they change sources as soon as they relocate their main activity areas or residential sites (Gosselain 1994). Similarly, these potters are accustomed to adopt as tools, devices, or facilities, objects and locations that they already employ in the realm of other production activities. Such tools may include (but are not limited to) hoes, grinding stones, wooden pestles and mortars, corn-drying places, and cooking hearths if pottery-making is in the hands of women. When pottery-making is in the hand of men, tools include banana and cocoa dibbles, jumper bars, pocket knives, and building materials.

While less conspicuous, symbolic concerns are probably as determinant as economic ones in pottery manufacture and use. In southern Cameroon, like in other regions of the African continent (Barley 1994; David et al. 1988; Gosselain 1995), vessels are often equated with persons while the manufacturing process is metaphorically associated with other kinds of transformation processes, either natural (gestation, sexual maturation, menstruation, germination) or cultural (care of new-born, initiation, wedding, funeral). The components of such a symbolic system are particularly explicit when questioning the potters about the prohibitions that they must respect in order to ensure the issue of the *chaîne opératoire* or to protect them, the members of their community, or the natural cycle from accidents metaphorically similar to those affecting the vessels.

Moreover, one sees that similar techniques or recipes may be applied to different realms, as long as they respond to the same symbolic logic. Among Bafia, for instance, the techniques and ingredients potters use for waterproofing the vessels are similar in every respect to those used by traditional healers for curing diseases characterized by discharges (diarrhea, open wounds, pustules) or for healing circumcision wounds (Gosselain 1992b, 1995). Other potters closely associate clay processing to the preparation of cassava porridge, a widespread daily meal. For this reason, they break clay lumps into small pieces (like cassava tubers), spread them on the ground until they are completely dry, grind them in a mortar or on a grinding stone, sieve the resulting powder, and hand knead the finer fraction in a basin, after having moistened it.

At a more general level, technical choices simply appear as the result of a learning process: potters select and transform the materials as they have been taught to do, being neither keen to modify their habits nor interested in other ways of doing it. Technical behaviors can thus be assimilated to traditions, or styles, and it would not be surprising to see them associated to some facets of social identity. Such relationships do exist in southern Cameroon; however, we will see that they are much more complex than we would at first expect.

TECHNICAL CHOICES AND SOCIAL BOUNDARIES IN POTTERY-MAKING TRADITIONS

Some stages of the manufacturing process, such as clay processing, firing, and postfiring, turn out to be poor cultural markers. Here, the distribution of technical variants appears random, crossing or overlapping geocultural boundaries, and corresponding either to individuals, localities, or microterritories. The only stage strongly related to social identity is that of the fashioning—or rather roughing out[10]—process. As can be seen when extrapolating field data on the ethnolinguistic map (Fig. 4.5), the distribution of fashioning techniques generally coincides with linguistic boundaries. There are, of course, several exceptions here: some different populations share similar technical traditions; and some potters use techniques that differ completely from those used by other members of their group. Moreover, the extent of some traditions within certain territories still needs to be assessed.

The situation becomes even more interesting, however, since a series of technical groupings and (apparent) affiliations also correspond to linguistic groupings and affiliations. Among Bantu-speaking people, for example, some members of the A40 and A60 groupings[11] (Banen, Yambassa, Sanaga) must be drawn together (Bastin et al. 1983) and clearly distinguished from their A40 (Basaa), A50 (Bafia, Balom, Djanti), and A70 (Eton) neighbors (Janssens 1993). I was not able to collect data among Yambassa, but both Banen and Sanaga potters essentially rough out their vessels by drawing of a lump, while potters of the remaining groups use several variants of the coiling method (with another important—and linguistically paralleled—distinction between the internal crushing of thick coils [A50 grouping] and the superimposition of thin coils [A40 and A70 groupings]). In other words, populations that are linguistically affiliated and share a common history tend to fashion their vessels in much the same way, or tend to use similar techniques that differ significantly from those of their nearest neighbors.

These observations have a clear archaeological implication since they show that the analysis of pottery fashioning processes may lead to a rather precise recognition of cultural boundaries. Furthermore, of all the stages of the *chaîne opératoire*, fashioning is amongst the easiest to reconstruct from archaeological vessels, although detailed studies remain curiously few (Courty and Roux 1995; Glanzman and Fleming 1985; McGovern 1986; Pierret 1995; van As 1984; Vandiver et al. 1991; Woods 1985). We need to be sure, however, that the situation observed in southern Cameroon is not coincidental. In other words, we need to untangle the mechanisms concerned in order to understand and interpret phenomena of technical homogeneity and of technical heterogeneity.

This question leads us to the problem of the spatial and temporal distribution of knowledge, or rather to the problem of learning processes and networks.

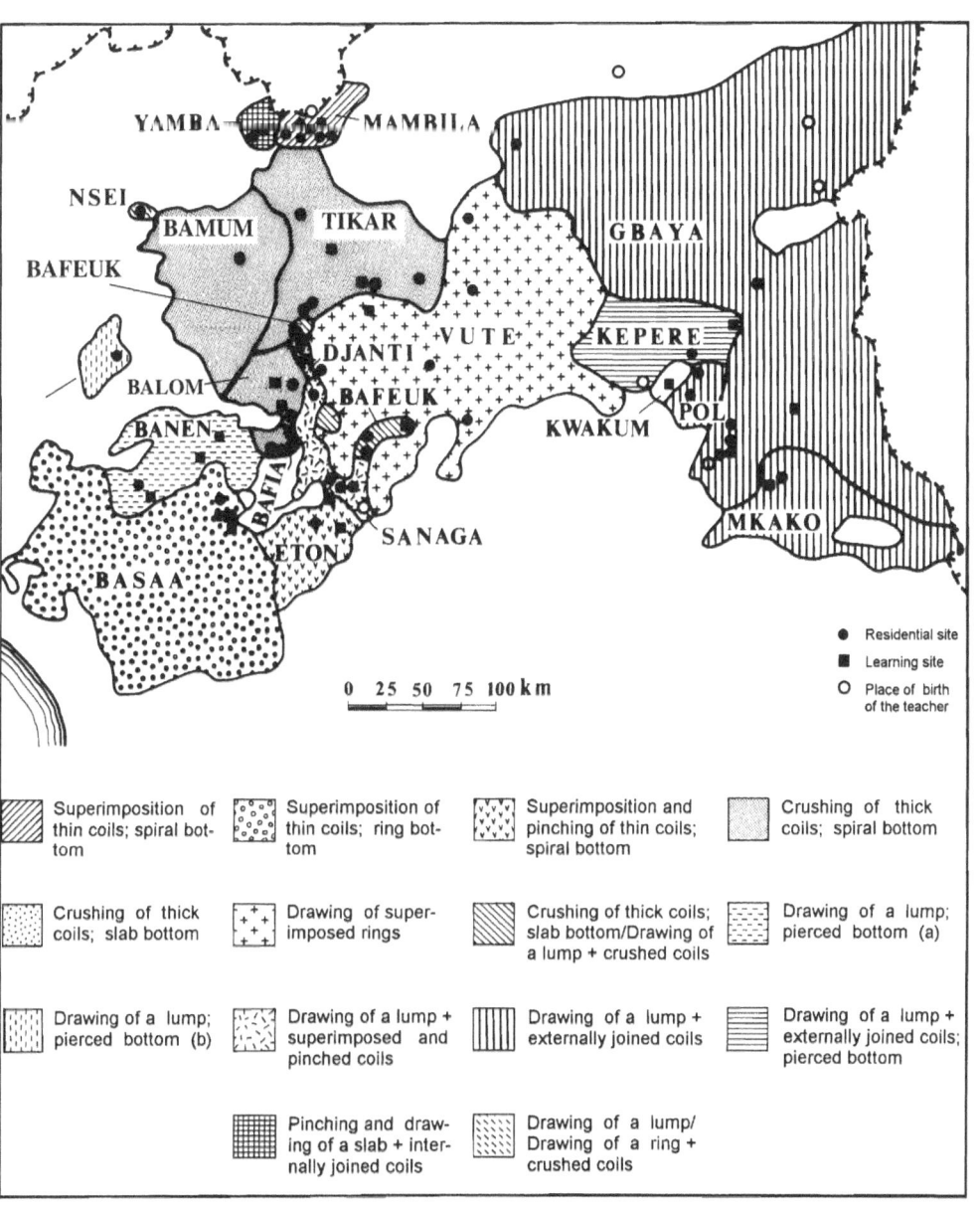

Figure 4.5. Extrapolated distribution of fashioning techniques with respect to linguistic boundaries.

DIFFUSION OF TECHNICAL KNOWLEDGE

The information gathered in the field shows that pottery-learning frameworks are not very formalized in southern Cameroon: anyone has the right to learn and to practice the craft, as far as she or he is willing to do so and she or he finds someone who is willing to provide instruction in the technique. The choice of a teacher is largely a matter of convenience and proximity, and the age of the apprentice as well as the time spent on learning are again not consistent. More practically, most will start the learning process as a child or adolescent, choosing a teacher within the nuclear family (mother, father, sister), the extended family (grandmother, aunt, uncle) or, less frequently, the collateral family (co-wife, mother-in-law, co-wife of the mother) (see Fig. 4.6). Only five potters (6.5 percent) in the entire sample have acquired their knowledge from a nonrelative. In these cases, the teacher was a neighbor, a friend, or a person with whom a few relations have been established. Whatever the teacher's identity may be, however, her or his linguistic affiliation almost always coincided with that of the apprentice. Among the 77 potters interviewed, only 6 of them claimed to have learned the craft from a person who belonged to a linguistic group other than their own. These people were all women originating from neighboring groups and had changed their residence after marrying.

Two phases must be established in the transfer of pottery manufacturing knowledge. During the first phase, the apprentice assists her or his future teacher with some stage of the *chaîne opératoire*: procuring, carrying and processing raw materials, building the firing structure, and treating the vessels after firing. This participation allows her or him to learn the materials, the different transformation processes, and the taboos to be respected when practicing the craft. One must remember, nonetheless, that pottery-making is just another household task that a young person is expected to fulfill and that both boys and girls are liable to assist their relatives, even if that activity has pronounced gender connotations for adults.

The second phase requires a more formal type of learning as well as an authentic motivation from both the teacher and the apprentice. The problem is how to learn to fashion the vessels, a process which generally starts as play but soon presents a series of difficulties: vessel walls get out of shape, pressures exerted are either too weak or too strong, parts of the profile disjoint, the final shape of the vessels remains irregular, and so on. In order to help the apprentice overcome her or his failures, the teacher must go beyond the part of a mere model, working beside her or him, correcting gestures, rectifying errors, and even taking her or his hands into her or his own, until the apprentice is able to work alone (Fig. 4.7). Potters claim that this phase of close interaction can take from two or three months up to a year. At the end of this period of time, all gestures and postures related to the

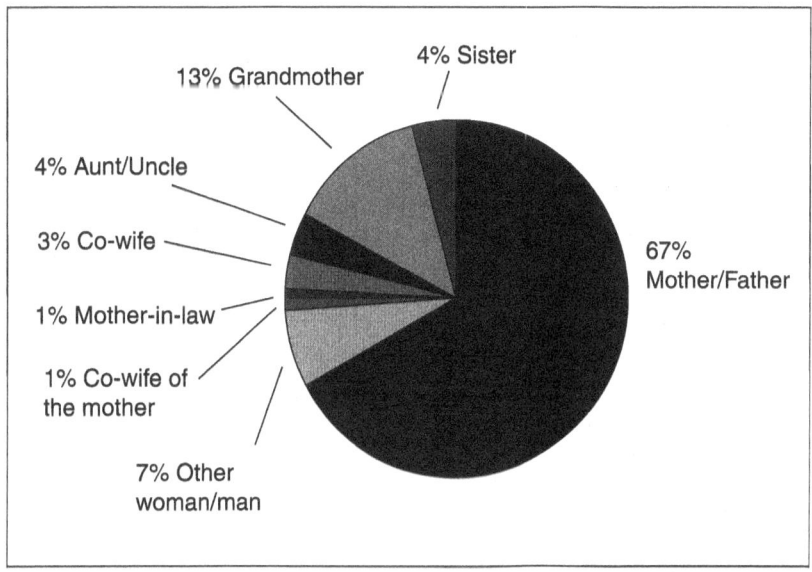

Figure 4.6. Identity of the person from whom potters interviewed (N = 77) learned the craft.

fashioning process will have become integrated as motor habits (see Arnold 1981, 1985 for the importance of this concept).

The foregoing summary provides a rough outline of the vertical transmission of knowledge from generation to generation. We must next consider its transmission through space, which is clearly related to the mobility of teachers and apprentices. As shown on Figure 4.8, over 80 percent of the individuals observed learned the craft in their home village, but only half of them still lived in their home village at the time of my survey. Among the remaining potters, most accomplished their learning after marrying and leaving their home village or, as a child, after their parents changed residence. Only three potters had to move to learn how to pot, and eventually settled down again in their home village. In that case, the learning process took place during temporary sojourns at a relative's home or in another community.

On the whole, 45 percent of the potters interviewed could have thus transmitted their knowledge through space, simply because they worked in a village different from that where they had learned their skill. But what is the importance of this spatial distribution? When looking at the distance between the learning place and their present residence (Fig. 4.9), it appears that these movements involve only short distances: more than half of the movements are within a 25 km range, and three-quarters occurred within a 50 km range. These figures do not differ sig-

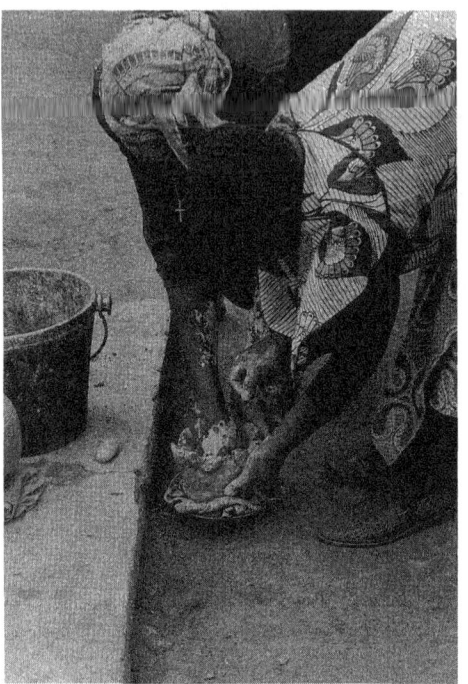

a

Figure 4.7. Typical stages in learning to fashion a vessel: (a) a Bafeuk potter (wearing a black dress in the photograph) shows her daughter how to hollow and enlarge a lump of clay; (b) she lets her daughter try by herself, but carefully watches the whole process; (c) unsatisfied, mother shows daughter again the appropriate gestures and rectifies simultaneously the shape of the rough out. (O. Gosselain, photographer.)

c

Social and Technical Identity in a Clay Crystal Ball   97

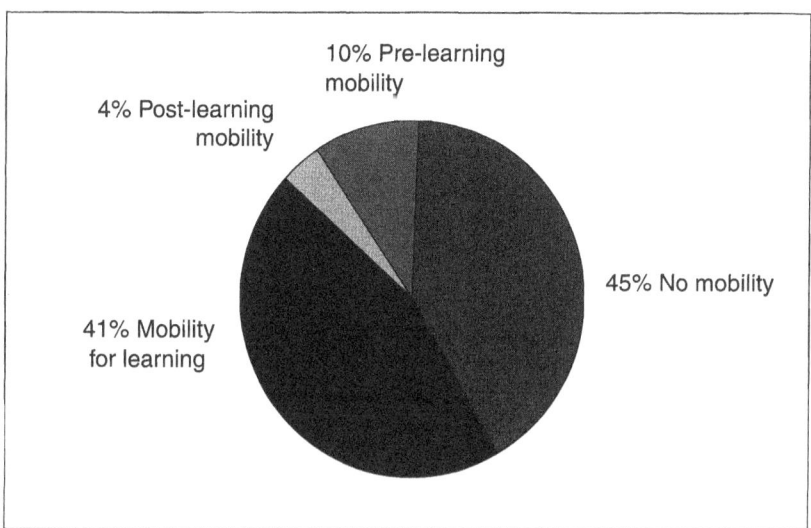

Figure 4.8. Mobility with respect to the learning site among the potters interviewed (N = 77).

nificantly from those of the previous generation (Fig. 4.10): some three-quarters of the teachers had moved within a radius of 50 km before transmitting their knowledge to their apprentice(s).[12]

When plotting all these movements on the linguistic map (Fig. 4.11), we see that most of them do not cross cultural boundaries, even though the territories concerned are small. In other words, technical knowledge and by extension traditions or styles appear to circulate mainly in an intra-ethnic manner.

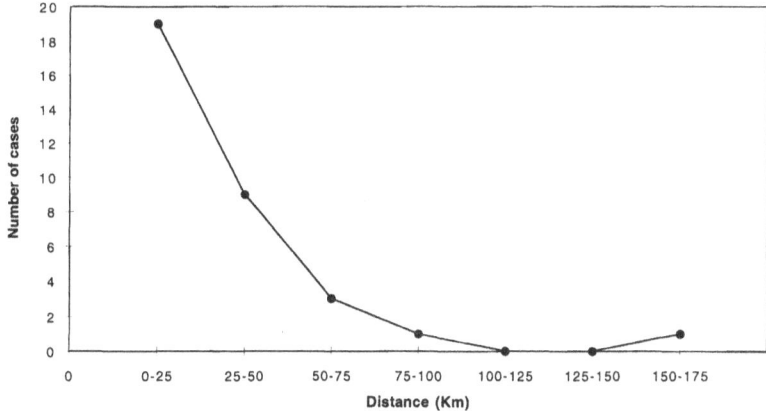

Figure 4.9. Distance between the residential site of the potters interviewed and their learning site (where different).

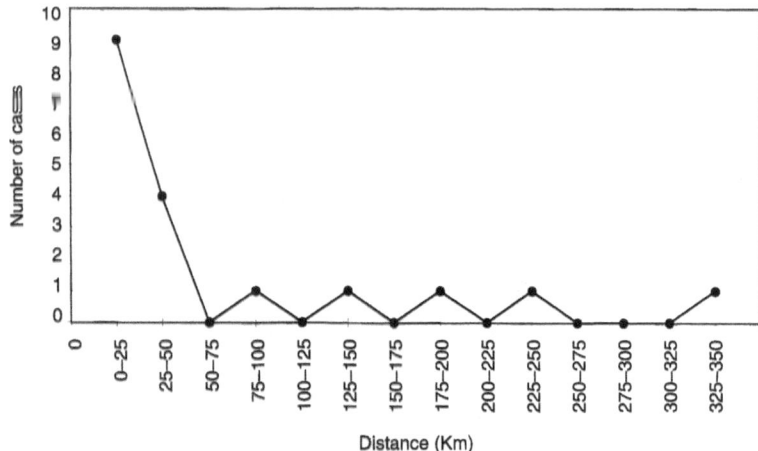

Figure 4.10. Distance between the place of birth of the teacher and her or his residential site at the time of my observations (where different).

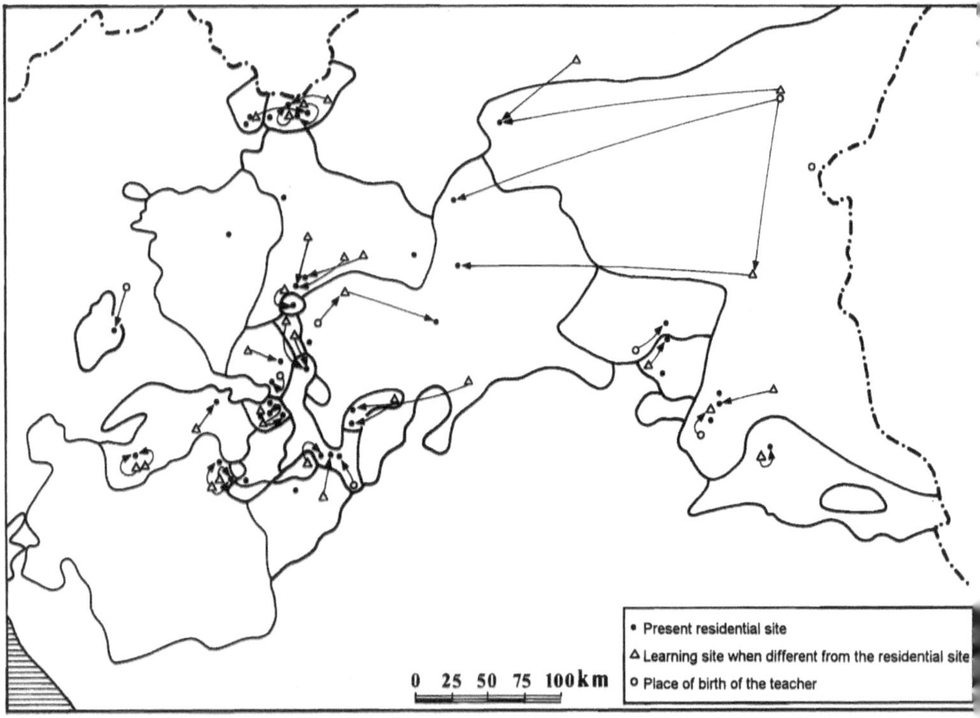

Figure 4.11. Movements related to learning within the study area.

- Present residential site
△ Learning site when different from the residential site
○ Place of birth of the teacher

Only Gbaya potters frequently cross their territorial boundaries, traveling greater distances than all other artisans. One should note, however, that Gbaya live in one of the more sparsely populated areas of southern Cameroon (distance between villages can exceed 100 km) and that they have the reputation of being a very mobile population (Burnham et al. 1986; Copet-Rougier 1987). Also, traveling distances are probably greater today than they were in the past, in part because road systems and means of transport have considerably improved, and in part because major changes (such as rural depopulation and the shift to plantation arboriculture) have affected the socioeconomic system.

DISCUSSION AND CONCLUSION

A careful examination of procedures used at the different levels of pottery *chaînes opératoires* shows that technical behaviors constitute full cultural productions rather than mere adaptations to environmental and functional pressures. Unlike Behavioral Archaeologists, Neo-Darwinists, or other trendy theoreticians, south Cameroonian potters are not obsessed by the ecological and functional fitness (or "adaptedness") of their behavior: their technical and functional aims are, in practice, more flexible than we have previously believed. While this flexibility ensures a high adaptability to contrasting environmental and social contexts (and hence the viability of technical traditions through time and space), it also allows potters to meet the primary (and mainly nontechnical) requirements of the activity: limitation of the time and energy devoted to pottery-making, subordination of certain operations to other production activities, use of tools and devices already employed for other purposes, capability of postponing or expediting some stages of the manufacturing process, and compatibility with symbolic conceptions associated to other technical (or cultural) processes.

Being thus completely embedded in social and economical systems, production steps in the making of pottery may provide a clue for approaching cultural identity, whereas the mechanical properties (or "performance characteristics") of the products only appear as a side effect. For instance, two potters who belong to the same technical tradition and who thus use similar tools and procedures, may produce vessels with very distinct physical properties, simply because they live nearby—and therefore exploit—different clay sources. Concurrently, if techniques such as slipping, burnishing, or applying an organic coating can either be easily adopted or remain unchanged for a long time, it is not so much because of the performance characteristics they allow as it is the visual aspect they confer to vessels and because of the symbolic meaning put into them by both producers and users. As summarized by Lemonnier (1993a:3), the logic and coherence of . . . technological knowledge—or whatever we call the information and mental operations that underlie individual action and behavior—are not related solely to the physical phenomena that are set in motion by a given technique. Social representations of technology are also a mixture of ideas concerning realms other than matter and energy. In short, the mental processes that underlie and direct our actions on the material world are embedded in a broader, symbolic system.

Most studies devoted to the examination of procedures and raw materials used by traditional potters remain unfortunately superficial, preventing any coherent comparison with other regions of Africa, or other continents. A series of examples collected in the ethnographic (and ethnoarchaeological) literature reveal, however, that the situation observed in southern Cameroon is not exceptional and that technical diversity must be explained in terms other than purely materialistic ones. For instance, one frequently observes the use of contrasting raw materials and processing techniques among potters who nevertheless live in the same environment and share similar functional aims (e.g., P. Arnold 1991; David and Hennig 1972; Gallay et al. 1994; Miller 1985; Reina and Hill 1978; Spriggs and Miller 1979). If, as stated by Braun (1983:112) and other scholars (Arnold 1985; Ericson et al. 1972; Rye 1976; Schiffer and Skibo 1987), cross-cultural regularities exist in the selection of mechanically—or ecologically—dependent attributes and behaviors, then patterns of technical distribution should not correspond to individuals, village communities, or other kinds of social groupings. However, such a correspondence is rather common.

Even if a comparison of ethnographic data reveals possible cases of adaptation to external pressures, a more extensive survey of the literature allows one to question the universality of the adaptations identified. In analyzing data collected in India by Saraswati and Behura (1966), for example, Arnold observes that the use of kiln and oven firing is restricted to "the areas which have the largest number of wet months in a yearly cycle. . . . Open firing, on the other hand, is widespread, but it constitutes the exclusive firing technique in [areas] where there are only four to five months with more than 50 mm of rainfall and thus seven to eight months with less than 50 mm" (Arnold 1985:215). Convinced that wet weather can seriously jeopardize the firing process if conducted in the open, he concludes that the adoption of kilns and ovens is born of necessity, for such structures permit "the potter to fire during conditions which would make firing ordinarily difficult or impossible" (Arnold 1985:213). If we turn to Africa, however, we see that the use of kilns and ovens is restricted to the driest areas of the continent (North Africa and Sahel), while open and pit firing are randomly distributed, but constitute exclusive techniques in the tropical area (even when pottery-making is a year-round activity) (Gosselain 1995:241–249). In other words, weather and climate could have an adverse effect on the craft, but the "regulatory feedback mechanisms" (Arnold 1985:17) they induce are not as universal as Arnold asserts.

Ethnographic studies also show how symbolic conceptions can influence potters' behavior. Among the Gurensi of Ghana, for instance, vessels used by women in their daily activities are always broken during the final rites of their funeral. According to Smith (1989:61), "the sherds preserve a link between the woman and her family on the one hand, and the Earth on the other. Many sherds are ground down to produce the grog used in making clay for pot. In this capacity, they are renewed, becoming part of another cycle of life." During recent fieldwork in northern Cameroon (Gosselain et al. 1996), Koma-Ndera informants consistently explained how the maceration of *Acacia nilotica* pods, which potters sprinkle on the pots after firing, was a powerful medicine used to "cool down" and "strengthen" objects and people that had passed through a process of change. For this reason, the mixture is also applied to a boy's penis after circumcision or to a child's first teeth. Similarly, the application of an organic coating sometimes aims at "aging" the vessel (in giving it an artificial patina) because it is said that old pots are better than fresh ones, in much the same way that an old man is better than an adolescent (Virot 1994). How far we are from the flatly materialistic explanations offered by Schiffer regarding similar treatments (Schiffer 1990; Schiffer et al. 1994).

Having illustrated the deep cultural rooting of technical behaviors and hence, their stylistic nature, we must next assess the meaning of technical choices in terms of social identity and boundaries. As previously shown, the relationships between social and technical identity are rather complex in southern Cameroon: while the distribution of fashioning techniques generally coincides with that of linguistic boundaries, most stages of the manufacturing process appear to be randomly distributed, corresponding either to individuals, localities, or microterritories. (And one should note that this situation also prevails in other regions of the African continent [Gosselain 1995].)

Regarding the fashioning process, the striking correlation between technical diversity and linguistic affiliation seems to be best accounted for by the context in which potters learn their skill. We have seen, indeed, that (1) learning mostly takes place within the nuclear or extended family, (2) the linguistic identity of the teacher generally coincides with that of her or his apprentice(s), and (3) post-learning movements imply only short distances (most of them fall within a 50 km range) and rarely cross linguistic boundaries. All these factors clearly ensure a coherency between the spatial range of technical traditions and that of linguistic groupings. In addition, since learning networks do occasionally cross geocultural boundaries, the presence of exogenous traditions within otherwise homogeneous groupings becomes explicable.

But one crucial question remains: why, of all the stages of the *chaîne opératoire*, is it precisely the shaping stage whose variations best match the geographic extent of learning networks and thus, the linguistic boundaries? Apparently, different types of technical knowledge that are acquired during the learning process do not necessarily evolve in the same way. During my fieldwork, for instance, I had the opportunity to work with two sisters. Each had learned the craft from the same person (their mother), at the same time, but upon their marriage, each moved to a different village. Although they shared identical fashioning techniques, they used different procedures when processing raw materials and firing the pots. These variations were shared by the other potters of their new village.

In fact, as far as processing, firing, and postfiring techniques are concerned, such an opposition between local homogeneity and regional heterogeneity is quite frequent in the study area, even though the potters involved belong to distinct learning networks. Similar examples have also be recorded in other parts of Africa or in other continents (e.g., DeBoer 1986; Dietler and Herbich 1989; Herbich 1987; Wahlman 1972). These patterns suggest that post-learning interactions can lead the potter to modify certain aspects of her or his behavior and thus, to abandon a part of her or his initial technical identity. Reasons underlying the changes are not all of the same nature: they may stem from a desire to conform to the norms of a new community or to improve a practice that is judged inappropriate, or even be a response to new social, economic, or symbolic pressures. The changes that do occur, however, only concern those stages of the manufacturing process that do not rely on motor habits, that are performed in the open, and that may sometimes involve a collaboration between potters or assistance from other social actors. If certainly true for the decoration process, it also applies to clay processing, firing, and postfiring. All these manufacturing stages constitute channels through which new ideas and attitudes can continue to be adopted, even after potters have completed the learning process.

On the contrary, techniques used for fashioning the vessels are more resistant to change, probably because they do essentially rely on motor habits (see Arnold 1981, 1985:235–237, 1989; Foster 1965; Nicklin 1971). Moreover, they are associated with one of the most personal and least symbolically invested operations of the *chaîne opératoire*. These reasons may explain their frequent correlation with larger and more pervasive forms of social ties, such as linguistic affiliation.

It follows that the different stages of the manufacturing process do not all have the same stylistic significance. Being more sensitive to post-learning interactions, some production steps could reflect a deliberate identification with or distinction from particular social groups (two processes that can be experienced at both a communal and individual level, as DeBoer [1990] convincingly demonstrated). Without being necessarily concomitant, their evolution could parallel that of designs or vessels' morphological features. Other stages in the manufacturing process prove more resistant to change and could, in a passive or unconscious manner, reflect profound social relationships. As for ornamental structures (or "grammars"; Gosselain and van Berg 1992), their study may allow recognizing communities of different origin and, should the occasion arise, reconstructing certain migratory movements.

It must be stressed, however, that the notion of "deep social relationships" does not equate perfectly with linguistic affiliation. Although frequently associated with language (e.g., Arnold 1981, 1985; David and Hennig 1972; Delneuf 1991; Frank 1993; Gallay et al. 1994; May and Tuckson 1982; Pinçon and Ngoïe-Ngalla 1990; Reina and Hill 1978), variations in the way that vessels are fashioned may also relate to larger or smaller spatial units than linguistic boundaries. In southern Cameroon, for instance, one sees that learning networks actually spread over areas that include several villages but remain smaller—and sometimes much smaller—than ethnolinguistic territories (Fig. 4.11). Although more extensive survey may reveal that these areas tend to be larger, this pattern of restricted distribution is probably typical of those societies where pottery technical knowledge is transmitted in an informal way along kinship/friendship/neighbor networks and through small-scale individual movements, as previously described.

The picture is completely different when social prescriptions govern the process of learning. Among the patrilineal and patrilocal Luo of Kenya, for example, new wives are submitted to a severe postmarital resocialization process under the supervision of their mothers-in-law. Those who marry into a homestead of potters will learn (or re-learn) to produce vessels according to the local norms. One observes, therefore, a series of potting microstyles whose distribution closely matches that of extended-family homesteads or clusters of homesteads (Dietler and Herbich 1989; Herbich 1987). In other instances, the right to learn pottery is restricted to people who live in, marry into, or originate from certain villages or homesteads (Bedaux 1986; Nicklin 1979; O'Hear 1986; Vernon-Jackson 1960). Learning networks can also follow a strict gender, clannish, or workshop pattern (Brown 1989; Delneuf 1991; Mohr-Chavez 1992; Priddy 1971); all these circumstances lead generally to a small-scale distribution of technical styles.

On the contrary, in those societies where pottery-making is restricted to female members of a caste (or *classe artisanale*; Gallay et al. 1994:29), potting traditions are distributed over large territories, for potters have to observe severe marriage prohibitions (e.g., Gosselain et al. 1996; Haaland 1978; MacEachern 1990, this volume). As for Sub-Saharan Africa, this situation could account for the widespread and cross-linguistic distribution of fashioning techniques such as pounding in a concave mold or molding on a convex mold (De Crits 1994; Gosselain 1995; Huysecom 1992).

Clearly, these different patterns of "socio-technical" distribution can be further complicated through population mixing (Delneuf 1991; Frank 1993; Gallay 1994; MacEachern 1994, this volume) and by spatial (and social) shifts in the contexts of production and consumption of vessels (Dietler and Herbich 1994a, this volume). We must thus beware of simplistic associations between social identity and material culture patterning.

Sackett (1977, 1982, 1990) has continuously maintained that stylistic expression is to be found in every aspect of an artifact and largely results from an enculturation process. I totally agree with him but, following David, MacEachern, and others (David 1992b; David et al. 1988, 1991; Dietler and Herbich 1989; MacEachern 1994, this volume), I would add that style is a polythetic phenomenon: although concomitant, its different components do not reflect the same cultural aspects and may even arise from very distinct processes. A careful examination of archaeological data will certainly allow one to recognize contrasting patterns of stylistic variation and, should large areas and long time periods be considered, to provide evidence of social interaction networks and population movements. Given the unpredictable nature of cultural behaviors, however, I doubt the possibility of relating these variants to particular facets of social identity or of specifying the exact context in which these interaction processes took place.

In an academic milieu where building unidimensional models and theories is still perceived as a scientific achievement *par excellence,* this somewhat mitigated conclusion will probably appear as a failed attempt. I remain convinced, however, that a reconstruction of ancient technical behaviors is of great potential in archaeology, so long as we acknowledge the inherent limits of the inferences we can make, and above all, as we abandon all our preconceived ideas about technology.

ACKNOWLEDGMENTS

My fieldwork in southern Cameroon, supervised by Prof. P. de Maret, was supported by a research grant from the University of Brussels. Research and analyses were conducted in collaboration with the following institutions and laboratories: University of Yaounde I, ISH and MESIRES (Cameroon); Section of Archaeology of the Musée Royal de l'Afrique Centrale; Laboratoire de Pharmacognosie et de Bromatologie and Laboratoire de Chimie Industrielle et Analytique (University of Brussels). I wrote the present paper while being a member of the Ceramic and Society Project, a project developed at the University of Brussels and funded by a grant Actions de Recherche Concertées provided by the Direction Générale de la Recherche Scientifique, Communauté Française de Belgique. This paper directly or indirectly benefited from the insightful comments of S. T. Childs, E. Cornelissen, N. David, J.-M. Decroly, P. de Maret, B. Dialo, M. Dietler, H. Doutrelepont, M. Graullch, I. Heibich, T. Ingold, B. Janssens, P. Lavachery, A. Livingstone Smith, S. MacEachern, S. K. McIntosh, J. Moeyerson, C. Perlès, J. Sackett, M. Stark, and M. Vanhaelen. I thank them, as well as the late J.-J. Ayiboto, who assisted me during the three field seasons in Cameroon. Finally, I express my gratitude to south Cameroonian potters and villagers whose welcome, patience, and interest constituted invaluable help. Figures 4.1, 4.4, 4.5, 4.10 were drawn by Y. Paquay. Figure 4.2 was generated by D. Van Aubel.

NOTES

1. Skibo (1992) has attempted a careful examination of pottery functions among the Kalinga of Philippines but, surprisingly, he did not record the temperatures to which vessels were submitted during daily cooking sessions.
2. Here, indeed, concepts and theories borrowed from ceramic engineering or natural sciences are laid down as laws that ensure an a priori explanation of technical diversity and change.
3. According to some Sanaga and Eton potters, noises generated by bush fires (quite frequent at that time of the year) can cause vessels to explode during firing.
4. In this context, women are not allowed to learn the craft or to take part in any operation of the manufacturing process.
5. Data were recorded with the help of ten thermocouples and included heating rate, temperature range, and length of exposure to temperatures.
6. *Bridelia ferruginea, Bridelia micrantha, Terminalia superba, Macaranga spinosa.*
7. But note that this notion of choice is purely theoretical, for only a few potters know of alternative ways to complete the different manufacturing operations and most of them express definite—and often functional—opinions about the procedures, tools, and materials they use.
8. The granulometric analysis of clay samples involved dry sieving the fraction above 63 μm, wet sieving the fraction between 63 and 32 μm, and decanting the fraction below 32 μm. In accordance with current classification systems, I consider gravel to be above 2,000 μm, sand to be between 2,000 and 63 μm, silt to be between 63 and 2 μm, and clay to be less than 2 μm. On Figure 4.4, these four granulometric fractions are broken down into three categories to render comparisons easier.
9. For example, while plowing fields, removing earth on riverbanks for the construction of fish barrages, observing soils brought up by crab-burrowing activities, or observing embankment erosion by a living area and along roads or tracks.
10. As shown by Roux (1994; Courty and Roux 1995:20), the fashioning process most often includes two distinct operations: (1) roughing out, (2) preforming. During the first operation, clay lumps are either grossly deformed or joined together, until obtaining a hollow volume

that does not present the final geometrical characteristics of the vessel. This operation is usually referred to as the "shaping process." During preforming, the potter uses either her/his own hands or a series of tools in order to shape the wall of the rough-out and to give the vessel its final geometrical characteristics. Although usually referred to as the "finishing process," this operation is often followed by a true finishing of the surfaces.

11. After Guthrie's classification (1967–1970).
12. But we are dealing here with distances separating the residential site from the place of birth, for no potter knew the exact place where her or his teacher had learned the craft.

5

Scale, Style, and Cultural Variation:
Technological Traditions in the Northern Mandara Mountains

SCOTT MACEACHERN

From time to time we shall refer to archaeological groupings like the Acheulean culture or the Magdalenian culture, which consist of the material remains of human culture preserved at a specific time and place at several sites; these finds are the concrete expressions of the common social traditions that bind a culture.

(Fagan 1995:22)

Archaeologists often assume that an association exists between artifact stylistic variation and some particular levels of human social organization, the latter most frequently defined as "peoples," "societies," or "ethnic groups." This has often manifested itself in a straightforward equivalence drawn between human collectives and artifacts that acted as *fossiles directeurs* for those collectives—"hand-axe cultures," *"gens de l'arc,"* the "Corded Ware/Battle Axe group," and so on. Such straightforward identifications are not as popular among archaeologists today as in the past, although they do exist. Still ubiquitous, however, is the association of an archaeological assemblage with a descriptive term that partakes of the characteristics of an ethnonym—Acheulean, Magdalenian, Kintampo, Dorset, and so on. In this case, a generative link is more or less explicitly assumed between some human group in the past and a particular material ensemble; "Magdalenian" people were responsible for the production of Magdalenian artifacts. The implication being made here is that the spatial and temporal extent of the artifact suite so denominated corresponded to the spatial and temporal extent of a particular human society.

Archaeologists know, of course, that this need not be the case. There is no a priori reason for such identification of artifacts with particular groups, either today or in the past. In the modern world, examples of artifacts that cross social-cultural borders abound, from chopsticks to Kalashnikov assault rifles to that perennial exemplar, the Coca-Cola bottle. Equally, no one would seriously claim that the Acheulean corresponds to a single culture in any usage of the term applicable to modern humans, covering as it did a large part of the Old World and persisting for hundreds of millennia. We admonish each other that treating technological traditions as if they were unitary, bounded societies is a mistake but, nevertheless, the habit persists. We learn these habits in our earliest undergraduate training (the usage is not confined to *People of the Earth*), and pass them on to succeeding generations of students in our turn. We name "archaeological cultures," "traditions," and "industries" with the professed intention of denoting only artifact distributions and associations, but our consciousness and these terminologies are continuously invaded by the presence of the humans who produced, used, and discarded those artifacts. We impose the names of our "archaeological cultures" upon ancient populations as stand-ins for their own ethnonyms, which we will never know.

This conceptual shift, in which archaeologists are drawn almost imperceptibly from analysis of material culture to conjecture about human groups, has serious implications for archaeological research. In this chapter I am concerned not primarily with the identification of patterning in material assemblages, but rather with the meanings that archaeologists attach to that patterning. A number of researchers, and particularly Lemonnier (1986, 1992) and Sackett (1985a, 1990), have over the last fifteen years wrought important changes in the ways that we think about technical and stylistic variation in material systems. However, rather less change has occurred in the ways in which we conceive of human analogues to those material systems, in what archaeologists expect to see at the end of their researches. My discussion will rely primarily upon African examples, but the questions identified may, I think, also be applicable in some other parts of the world.

ETHNOGRAPHY, ETHNICITY, AND STYLISTIC VARIATION

> This history has produced a world of separate peoples, each with their culture and each organized in a society which can legitimately be isolated for description as an island to itself.
>
> (Barth 1969b:11)

It is quite common within modern archaeological and ethnoarchaeological research to find that artifactual stylistic variation (which we may define for present purposes as the choices made among a set of equivalent functional or nonfunctional attributes of artifacts and technological processes) is expected to shed light on questions of ethnicity, the definition of ethnic boundaries, and at least some elements of the corporate behavior of ethnic units (Childe 1929; Clark 1994; Conkey 1990; Cordell and Yannie 1991; David et al. 1988; Hodder 1982; Lemonnier 1986, 1992; Otte and Keeley 1990; Renfrew 1974; Sackett 1985a, 1990; Wobst 1977). That artifact stylistic variation can inform upon social relations is self-evidently true in a general sense. The existence of an artifact assemblage co-varying through space or time implies that some sort of relationships between producers and consumers of the materials in question existed, but in and of itself this says little either about the nature of those relationships or about the characteristics of any human collective within which such relations were embedded. We may ask, then, why archaeologists so often assume that they see ethnic groups, rather than cultural patterning at larger and smaller scales, when examining distributions of stylistically similar artifacts. We may further ask whether it is possible to detect the conditions under which ethnic boundaries and material culture will coincide, and what possible alternative social levels of artifact production, use, and distribution might look like.

Within the North American research tradition, the dominant academic model of archaeological research locates such research as a subdiscipline of anthropology, with its primary goal the extension of the anthropological quest into the past. This has implications for the ways in which we conceive of our research. North American archaeologists at least have embraced the notion that the objective of archaeological research, the goal toward which professional archaeologists should strive, is an ethnography based upon materials, with essentially the same purposes and preoccupations as anthropological research centered in the ethnographic present (MacEachern 1992a). Our hero is the ethnographer, who spends long periods of time in the field, learns the local language, and gets to know the local population, and so finds out how societies really work. Archaeology is for anthropologists who can't handle contact with living people, and Willey and Phillips's (1958:2) dictum that "archaeology is anthropology or it is nothing" is glossed to "Archaeology is *ethnography* or it is nothing."

This leaves archaeologists with a dilemma. It is a truism that ethnographers are not usually particularly interested in artifacts and technologies, the stock in trade of archaeologists. Their concerns are more frequently with nonmaterial aspects of culture, and these are more difficult to approach through archaeological research. A frequent theme in archaeological reports and methodological formulations involves the impoverishment of archaeological data sets when these are compared to the richness and complexity of observable human behavior. This impoverishment stems from several facts: some cultural elements are not expressed materially; some cultural meaning associated with artifacts is lost when those artifacts enter the archaeological record; and we confront the vagaries of taphonomy, sampling, and interpretation within archaeological contexts. The Black Box of ethnographically defined cultural behavior may emit material for the archaeologist to recover, but its inner workings remain obscure. Social-cultural anthropologists of course also question the validity, representativeness, and general epistemological relevance of their data, but most do not appear to share in the archaeologists' assumption that their database is irretrievably flawed or impoverished—probably because they are not faced with an unattainable anthropological exemplar of what their work should be.

The question is, then, how can archaeologists pursue ethnographic goals if their data set is in some essential way inadequate for such a pursuit? What portions of ethnographies would be amenable to archaeological analysis? We can, of course, examine technologies and economic adaptations, but these are rather too reminiscent of the tiresomely material, and are not in any case the centerpieces of many ethnographic inquiries. Through the examination of variation in ancient material culture, we may, however, be able to contribute to an important area of ethnographic research. Systematic investigations of ethnicity, ethnic identification and the constitution of ethnic groups (or "tribes," to use an archaic, but still widely encountered term) have been key elements of social and cultural anthropology since the professionalization of the discipline in the nineteenth century, and they have become even more important at the end of the twentieth century, as questions of individual and group identity reassert themselves in different parts of the world.

Archaeologists, Ethnic Groups, and Tribes

Ethnographers have historically studied people as members of ethnic groups (Lewis 1991). They may investigate subgroups within such units, but the ethnic group itself acts as a common boundary of research, and ethnonyms remain convenient markers on the spine titles of reports, localizing such research within geographical and academic space. This is not surprising. Social, linguistic, and administrative constraints have often greatly increased the difficulties of effective research across ethnic boundaries; indeed, the linguistic competence and extended periods of fieldwork expected of investigators (Barnard and Good 1984) virtually mandate restricted research areas. The mapped distribution of ethnicities serves as a vital mnemonic in the effort to impose order on the areas that ethnographers work in. Like Adam, they have traditionally exerted dominion over the world by partitioning and naming it.

The groups so studied by ethnographers are often expressed in material culture, either because of the adoption of emblemic artifacts that explicitly signal group identity or because the regularities of production and use that derive from participation in a shared social system in many cases generate more-or-less coordinate artifact suites. It would seem quite reasonable to suppose that such material expression of social relations was also characteristic of past societies, at least during much of the period during which modern human beings have occupied the Earth. The long-standing ethnographic concentration on research at local scales leaves us with few broadly applicable models of social organization and production apart from the ethnic group/"tribe," at least before the appearance of state-level societies. More important, it means that there has been remarkably little systematic, cross-cultural investigation of cultural similarity and differentiation since the *Kulturkreis* school became unpopular. This is exacerbated by the ethnographic disinterest in material cultures noted above. Is it any wonder that, as would-be ethnographers, archaeologists often find it difficult to avoid the assumption that named material cultures are generally equivalent to named ethnic units as the latter term is traditionally used?

However, recent historical and anthropological research has raised serious doubts about the cultural, spatial, and temporal coherence of many of the ethnic groups that archaeologists use as conceptual models for ancient human groups (Ambler 1988; Comaroff 1987; de Vos and Romanucci-Ross 1975; Ranger 1983; Samarin 1984; Southall 1970; Vail 1991; Vansina 1990:18–21; in different contexts, see Miller and Boxberger 1994; Topic 1994; Vieira Powers 1995). In Africa and elsewhere, many of these ethnic units as we presently conceive them were the creations of colonial administrators and later intellectuals, including ethnographers. The motivations for this invention and imposition of identities were varied. They included, among others, colonial convenience and the establishment of easily governable entities that could be controlled and taxed; divide-and-conquer strategies, which sought to minimize resistance to foreign rule; and the creation of vehicles toward power among both local and foreign elites in various regions. To these ends, communities were split or forcibly amalgamated; "tribes" were reified or created out of whole cloth; and fluid and responsive social units were reconceived as static, bounded, homogeneous monoliths. Even languages, so intimately connected to group identities and social relations in the minds of ethnographers and archaeologists, were molded, distorted, and codified in order to support these manufactured identifications (Harries 1991:85–89; Hofmeyr 1987:95–123; Ranger 1991:125–137). Ethnography as the study of dominated peoples is well reflected in the concepts of ethnicity that anthropologists have historically used.

This does not imply that ethnographers and archaeologists can merely search for "authentic" precolonial ethnic identifications in Africa and elsewhere, to use as indigenous substitutes for the external identifications imposed by colonialists or manufactured in the crucible of the modern world. There is an active debate within anthropology concerning the status of ethnicity as a valid structuring mechanism for premodern societies. It has been argued that the construction of ethnic units is not, in fact, an act intrinsic to human culture, but rather ". . . a consciously crafted ideological creation . . ." (Vail 1991; see also Comaroff 1987; Vincent 1974), with origins usually to be found in relations of domination of some communities by more powerful ones. (This question is most often phrased in terms of European colonial rule and present-day political-social relations, but there is no reason that such processes should not have occurred at other areas and in other times when such relations of domination existed.) The debate between primordialists and circumstantialists will no doubt continue, but archaeologists are in the meantime faced with a problem: how might we conceive of material culture variation without a framework of discrete ethnic groups, given that our ethnographic analogies are organized within such a framework? Through what processes would patterned regularity in material culture be maintained under such circumstances?

Clark (1994) criticizes what he defines as an Old World-West European mode of archaeological research, which he perceives as historicist, strictly empiricist, and dominated to varying degrees by simplistic equivalences between arbitrary typological and social units. He contrasts this with a New World-American paradigm that is anthropological and realist, one that allows for more fluid and heterogeneous relations between artifactual and human groupings. His criticisms are well taken, but he may have underestimated the degree to which archaeological systematics in the New World have been distorted by historicist and empiricist biases as well. If European archaeology has been dominated by typological-chronological systems based upon mid-nineteenth-century conceptions of evolution and of artifact variation (Chazan 1995), both Old World and New World paradigms have been dominated by ethnographic reconstructions and theories of human interaction that are based upon eighteenth- and nineteenth-century European conceptions of nationalism (Hobsbawm 1990) and colonial exploitation.

Archaeologists are thus being forced to recognize that the ethnicities that they seek in the past are problematic entities even in the ethnographic present. However, this difficulty is often resolved with a reference to Frederik Barth (1969b) and the fluidity of ethnic boundaries, and then quietly let drop without further consideration. This avoids a basic question: What do we archaeologists actually mean when we talk about ethnic groups in the past? Do we search for tribes—prehistoric equivalents of *The Nuer* or *The Kwakiutl* or *Yanomamo: The Fierce People*, and equally suitable for the bookshelf? Do we look for groups of primary self-identification, which Barth thought was a particularly useful component in a definition of ethnicity in 1969? Are we trying to detect the cultural discontinuities, ambivalences, and inversions laid out in *Cosmologies in the Making*, another essay by Barth (1987) but one apparently less often quoted by archaeologists? Are we seeking the relationships of domination within which ethnicity is by definition enmeshed, according to John Comaroff (1987)? The latter three works differ in fundamental ways, but they hold in common a sense of ethnic identification that is multivariate, dynamic, and often ambivalent, one that resides in psychological states and changing social relations rather than in immutable historic givens. The psychological realm is perhaps the most difficult for archaeologists to approach in most circumstances, and so we fall back upon more tractable models of individual and group identification and action. Archaeologists still look, most often, for tribes.

Geographical Scale and Artifact Production

Is there any indication that stylistic variation in the archaeological (and ethnographic) record tends to occur on a "tribal" or ethnic level rather than on social or geographical scales greater or smaller than that? One might expect such equivalences to be manifested in archaeological groupings that have roughly the same spatial extent as do ethnographically defined ethnic groups in the same area, or at least groups with similar economies in similar environments. Such congruencies do occur (Cordell and Yannie 1991:99), but do not appear to be the norm; archaeological distributions of artifacts frequently occur over much larger territories than do the ethnic units that ethnographers study in the present. In Africa, archaeological constructs like Kintampo (Stahl 1994), Urewe (van Grunderbeeck 1983), or Iberomaurusian (Camps 1974) are spread over regions often larger than modern states, within which scores of ethnic groups are now found. Despite well over a century of quite intense research into European Upper Palaeolithic cultural variation, the same situation exists there (Clark 1994:333; see Otte 1990; Otte and Keeley 1990). Entities in Neolithic Europe, in the Americas, and in Asia are similar. We assume that, by the end of the Pleistocene, human mental organization and social interaction in different areas of the world were in general comparable to those in the ethnographic present. In that case, we should expect that archaeological distributions reflecting ethnic differentiation would be more restricted in extent, or at least more internally heterogeneous, than they actually are.

A common response to this difficulty (Lemonnier 1986:160, 182; Sackett 1990:40–41) has been to posit that real ethnic differentiation is reflected in the artifact samples recovered from sites, but that archaeological investigations have been insufficiently refined to detect that differentiation. It is assumed that, as more sophisticated means of investigating artifact and site variation are developed, we will be able to distinguish now-invisible differences within archaeological "cultures." Such "cultures" will then be broken down into the constituent assemblages that were produced by real, well-defined human groups. At this point we will have generated a window into the past, through which we can examine the working of human societies. It is no doubt true that in many cases patterned variation exists, so far undetected, within recovered archaeological samples, but it does not seem really useful to ascribe one of the most striking features of the global archaeological record simply to inadequate analysis. Such a degree of inadequacy speaks of more fundamental problems.

A number of issues exist here. In the first place, there exist serious doubts about the pervasiveness of ethnic units in the past, as noted above. In the second place, we must ask to what extent stylistic variability through space and time must reflect ethnic boundaries where such boundaries did exist. This question is directly related to characteristics of the artifact-producer groups in a region. If we assume that stylistic stability, in technological processes and in the material results of those processes, exists because producers form an interacting social unit that provides a learning environment and deviation-reduction mechanisms, then it may seem permissible to assume that such groups are most often ethnic groups or subunits of such groups (communities at various scales, for example). After all, enculturation usually takes place within ethnic units, as does a great deal of social interaction, and the motor habits that one learns as one grows up are extremely resistant to change (Young and Bonnichsen 1984).

However, this interpretation ignores the extent to which ethnic groups are encased within a matrix of similar units and the degree of interaction and human movement that can take place across ethnic boundaries where the latter occur. Exogamy, migration, and craft specialization provide mechanisms through which steady or intermittent population flow can occur; others exist. A number of studies in different disciplines have indicated ways in which artifacts may be manipulated across ethnic boundaries, expressing identities and affiliations along other social axes (David 1992a; Hodder 1982; Larick 1986; Wiessner 1984), but there have been relatively few ethnographic studies of the dynamics of such production and exchange over time. Similarly, there is a tendency to conflate ethnographic studies—in which the arrangement and cultural significance of often somewhat idealized artifact suites are emphasized—and our expectations of archaeological research results; the presence of an ethnically specific ceramic typology, for example, says rather little about trade across group boundaries or about taphonomic effects.

In the third place, will ethnically significant technological and stylistic variability be identifiable archaeologically? Questions of practicality have often been dismissed as somewhat unworthy objections to hypothesis construction, as if theoretical and methodological ambition can somehow make up for limitations in the available data. Perhaps, as Lemonnier says (1986:182), referring to Leroi-Gourhan, "... ethnic groups produce objects whose morphology or mechanical properties differ to the degree that the observer is precise." Even if this were true, there is no reason to assume that such differentiation will always, or even very often, survive to inform archaeologists. To use two of Lemonnier's (1986:182, fn 9; 1992:70, 116, fn 6) examples, the technical systems used to strangle sleeping eagles or build flying-wing bombers will tell future archaeologists very little about Anga or American ethnicity: in the former case the materials and techniques used to kill the birds are rare and will not preserve, and in the latter only about twenty such artifacts have been built. African archaeology at least is often carried out in regions where most organic materials do not preserve, where clay architecture dissolves in decades and where the majority of the artifacts recovered are durable but in many cases rather uninformative potsherds or stone tools. These limitations are not universal, but they are much more common than are environments where preservation and research intensity would allow the fine-grained description of artifact production systems that Lemonnier and others have assembled. What significance should we accord to details of technological systems and to degrees of artifact differentiation that will almost never manifest themselves in archaeological assemblages at any practical level of analysis?

THE MANDARA MOUNTAINS: REGIONAL TRADITIONS AND LOCAL VARIATION

The Northern Mandara Region

Programmatic criticisms are easy to make, but are of little value without a consideration of concrete examples. I now proceed to discussion of a region and of a set of ceramic assemblages that from my point of view encapsulate the problems that I have already discussed. Data are derived from twenty-six months of archaeological and ethnoarchaeological research in Cameroon and Nigeria, as a member of the Mandara Archaeological Project and Projet Maya-Wandala, between 1984 and 1995 (MacEachern 1990, 1992a, 1992b, 1993, 1994, 1995, 1997, 1998). I also include some preliminary conclusions about other ways of conceiving of stylistic variation under the specific circumstances of this research.

The northern peripheries of the Mandara Mountains (Fig. 5.1) encompass a huge degree of linguistic and cultural diversity. Within an area of 1,500 sq km, we find assorted Chadic, Saharan, and West Atlantic languages spoken in plains and mountain communities, with social systems that vary from small-scale acephalous societies to regionally powerful Islamic states—most particularly the Wandala state, with its capital at Mora in the northeastern extremity of the massif. In and around the Mandara massif, "ethnicity," as defined by Western researchers, is based upon criteria of language use and colonial convenience. According to these

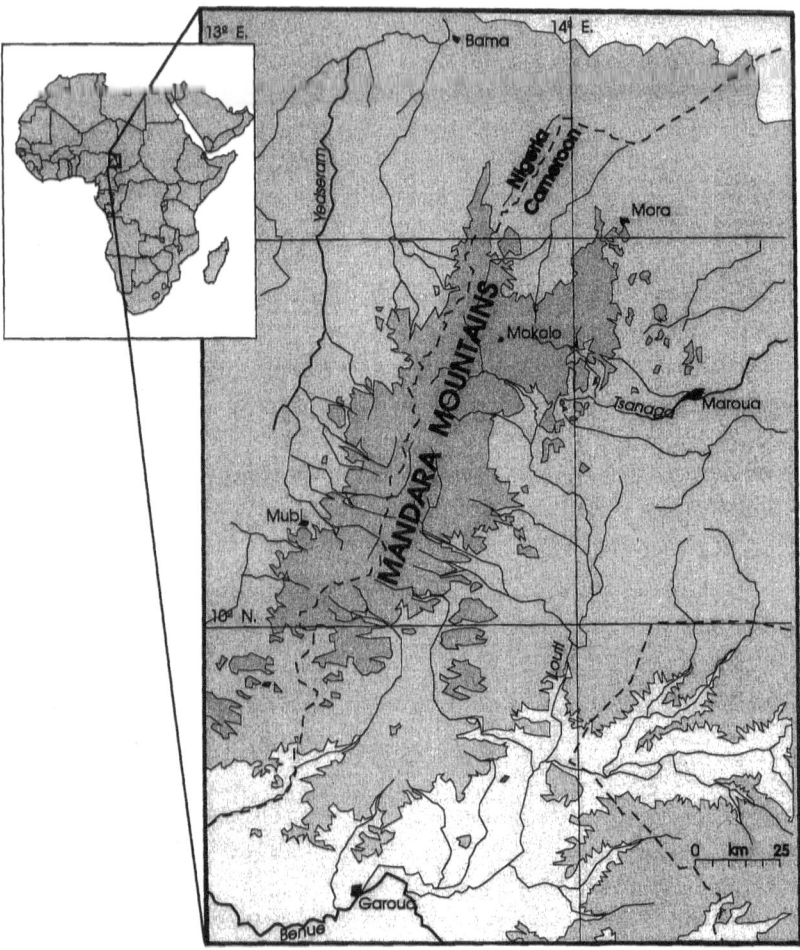

Figure 5.1. The Mandara Mountains.

externally imposed criteria, about 23 northern ethnic groups exist (Fig. 5.2). In fact, the group of primary self-identification and corporate organization among most northern Mandara communities, and particularly the montagnard ones, is the territorial patrilineage group, based upon the occupation of salient, defensible physical features and using conceptions of autochthony, kinship, and ritual cycle to express political relationships between different segments. According to these criteria of self-identification and organization, hundreds of Mandara ethnic units exist—but here the power of the ethnographer to control nomenclature breaks down, and investigators resort to a common research technique, using the im-

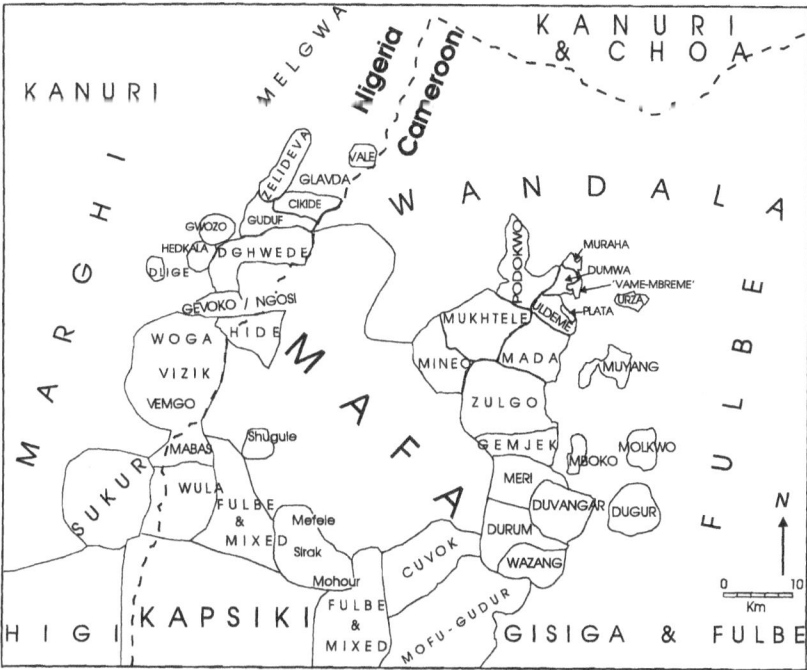

Figure 5.2. Ethnic groups in and around the northern Mandara Mountains.

posed ethnonyms to refer to regions outside of the communities that they themselves are conducting research in. (It should be noted that other forms of social and political organization exist in other areas of the mountains [David and Sterner 1993; Sterner and David 1993]. It should also be noted that nearly a century of colonial and national administration, exercised along "ethnic" boundaries, has made these groups more important than they earlier were. This has been the case throughout Africa, as people respond to the demands of changing political and economic systems.)

During our investigations to date, we have located about 200 archaeological sites, most of them dating to the Iron Age. As in many similar regions in West Africa, these sites yield pottery and comparatively little else. For example, about 99.5 percent of the approximately 200,000 artifacts recovered in 1992 and 1993 were potsherds, almost none of which were in a context that would allow reconstruction of vessels. The rest of the material recovered includes stone, metal, and organic artifacts, beads and some other items of small jewelry, rare clay statuettes, and other artifacts. In few of these latter artifact classes do we collect enough examples to make statistical comparison possible, and examination of recovered ceramics must be central to any analysis of technological variation.

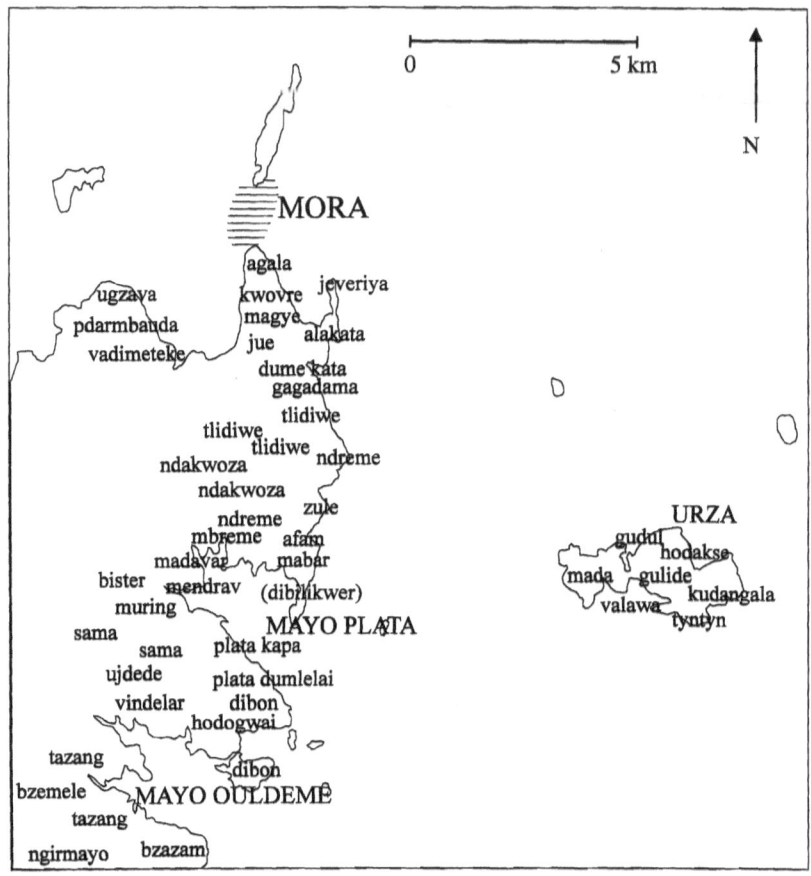

Figure 5.3. A partial listing of territorial lineage groups between Mora and Mayo Ouldemé.

Our ethnoarchaeological research has included analysis of present-day artifact variation and particularly the production, distribution, and discard of ceramic and iron artifacts. My own work with ceramics took place along the northeastern peripheries of the Mandara massif. Within this region, 10 acephalous montagnard "ethnic groups" divide themselves into about 70 territorial lineage groups and live in uneasy proximity to the capital of the Wandala state at Mora. Figure 5.3 gives a partial idea of this complexity, showing the lineages that were directly covered in my research. Pottery is produced by women of all ages, and within all of these communities; any woman with the interest and facility may produce and sell pottery, and between 30 percent and 50 percent of the women in most districts do so with varying degrees of commitment and enthusiasm. Pots are made for household use, are exchanged, sold and given as gifts within and between communities, and are sold at local markets.

Regional Ceramic Traditions

Within days of the first arrival of a group of Canadian archaeologists in Mora, in 1984, we noticed that two varieties of water-carrying pots—the most visible ceramic artifact in this Sahelian borderland, on display constantly as women fetch water from the scattered local wells—were used in different areas along the peripheries of the northeastern Mandara Mountains (MacEachern 1990:281–306; 1992a). These two varieties could be distinguished on the basis of: (1) general proportions, with statistically significant differences in various dimensional ratios on the vessels; (2) decoration (primarily different types of rouletting, incision, and appliqués), which was arranged in different distributions on different areas of the pots; and (3) differences in the type of handle used (Fig. 5.4a, b). These differences in handle type are quite striking and of disproportionate archaeological significance, since the solidity of handles means that they often survive as identifiable units. There are also some other differences in the techniques used to manufacture these vessels; for example, one variety has hand-molded bottoms while the other has coiled bottoms, although the rest of the pot body is built up through similar coiling techniques.

These two varieties of water-carrying pots, quite visible to an outsider, are each associated with different forms of a number of common vessel types found throughout the northern Mandara massif, and initially much less visible. Broadly, northeastern Mandara ceramics may be classified within four functional categories: (1) those used for carrying and storing water; (2) those used to store, cook, and serve food; (3) those used to produce, store, and serve beer; and (4) those used in rituals, for the maintenance of proper relations within the natural and human worlds. Each of these classes is subdivided into a limited number of functional pot types; thus, there exist water-carrying and water-storage pots, a pot type for cooking sorghum porridge and another for cooking sauces to go with that porridge and so on. Ninety-three percent of a sample of 1,015 montagnard vessels that I collected as part of my research (plus comparative samples gathered by other investigators) may be included in the first three of these broad functional categories, and the vast majority of the pots fall into one of only eight functional types.

Varieties of all of these vessel types exist within the same region, and they can to a greater or lesser degree also be separated on the basis of morphology and type and placement of decoration. This diversity parallels that of the two varieties of water-carrying pots. In archaeological terms, there exist two traditions of ceramic design and manufacture in this region and, in the time-honored habit of archae-

Figure 5.4. Northern Mandara water-carrying pots: (a) Maslava tradition; (b) Tokombéré tradition.

ologists, I named these the Maslava and Tokombéré traditions, after geographic-cultural regions where each is found (Fig. 5.5). To the south and west, other modern ceramic traditions exist (Barreteau and Sorin-Barreteau 1988; David et al. 1988; Gavua 1990; Sterner 1989), although I have collected a small amount of quantitative data on only one other, the Podokwo-Mukhtélé tradition found among montagnard communities west of Mora.

Pottery of these three northern ceramic traditions is used by people from at least 11 different montagnard "ethnic groups," living within more than 70 constituent territorial lineage groups. They are used as well by subjects of the Islamic Wandala state, who within the last century appear to have ceased indigenous ceramic production and relied upon purchase of pottery from montagnard producers at the local markets. As an optimistic graduate student, I searched for patterned variability in manufacture, form, decoration, and usage within these traditions that would allow me to delineate the ethnic boundaries that exist in the area today. In some cases, I did find variability at ethnic boundaries. Within the Maslava tradition, for example, some vessels made by women of the Urza ethnic group are distinguished by the particularly careful addition of a burnished red slip—a simple, but labor-intensive, technical effect. Some vessels made by Plata potters vary in proportion when compared to those of other Maslava groups, and a certain

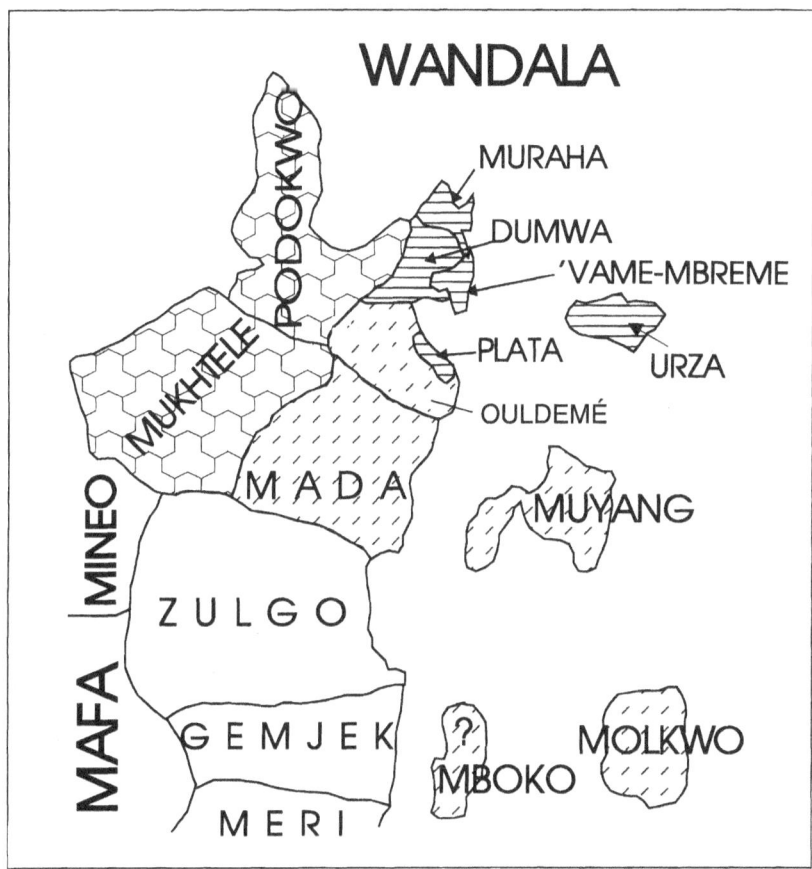

Figure 5.5. Ceramic traditions around Mora. Horizontal lines indicate the Maslava tradition, inclined angled lines indicate the Tokombéré tradition, and brickwork indicates the Podokwo-Mukhtélé tradition.

amount of similar variation in vessel proportions occurs within the Tokombéré tradition (MacEachern 1992b).

In general, however, differentiation along ethnic group boundaries is not important within these different ceramic traditions. Archaeologically, this lack of differentiation would be accentuated by taphonomic effects: burial would quickly degrade Urza potsherd surfaces to the point that care in slipping and burnishing would be impossible to detect; the fragmented nature of the sherds found on sites to date would make quantitative comparison of vessel proportions and decorative fields impossible; and sampling problems would make differentiation between ethnic groups even more difficult. Complicating the picture is the fact that variability in the ceramics held within households in all of the communities studied was quite high. Women experimented with new pottery forms and they also owned, and sometimes produced, pottery from traditions not native to the community that they lived in (see below). The local boundaries between these different traditions of artifact production are thus rather indistinct. In and around the northern Mandara Mountains, regional variation at the level of the ceramic tradition and local variation at the community level appears to be far more important than is variation at the intermediate level of the ethnic group, as the latter is traditionally defined by ethnographers. Archaeologically, it is likely that only regional variation would survive to be visible to researchers. This would, however, probably be impossible to confirm quantitatively without investigations at scales ranging from the neighborhood to the region and without the gathering of data on a truly enormous ceramic sample, far larger than the sample of vessels that I have examined so far—which illustrates some of the difficulties faced by researchers who want to study the social correlates of material distributions.

The technical systems involved in the production and use of Maslava and Tokombéré ceramics are generally similar; the quite complicated typologies of vessel function are arranged in the same ways, and differences in construction and use of ceramic artifacts between the two traditions were quite minor—the use of molded rather than coiled bottoms noted above, a tendency for Tokombéré water pots to incorporate molded head-pads while Maslava pots were used with a grass head-pad, and so on. Podokwo-Mukhtélé water-carrying pots are carried on the shoulder rather than on the head—a significant postural difference, given the daily visibility of these vessels—and this latter tradition appears to incorporate a somewhat different typology of vessel function, one that involves distinct kinds of cooking and beer-making pots and an expanded suite of ritual vessels when compared to Maslava and Tokombéré ceramics. In these characteristics, Podokwo-Mukhtélé ceramics are more similar to some of the more southern ceramic traditions (Gavua 1990; Sterner 1989) than they are to their northern neighbors. It is possible that we are at this point glimpsing either a regional hierarchy of artifact differentiation, one that also involves significant technological variation in ceramic production (Sterner, pers. comm. 1986), or simply a cross-cutting set of differentiations that implies no such hierarchy.

Characteristics of Ceramic Production

The primary unit of self-identification in this area is the territorial lineage group, and not the ethnic group as defined by European researchers. These territorial lineage groups are generally patrilineal and exogamous. There are few marriage restrictions functioning at the level of the ethnic group, although there are very rare cases where many of the territorial lineage groups within an ethnic group claim to be descended from a common ancestor, so that intermarriage is considered incest. The Plata are the only (quite small) ethnic unit on the northeastern peripheries of the mountain where this occurs. In general, marriage within and between ethnic groups involves considerations of logistics and convenience on the part of members of the constituent territorial lineage groups. Women tend to marry into adjacent lineage groups, whether these be of their own or another ethnic group. If their birth community is located near the center of an ethnic group territory, they will marry other members of that group; if it is on a boundary, they will very often marry members of other ethnic groups. Marriage and other types of social interaction across linguistic boundaries are mediated by the high levels of multilingualism to be found in this region (Barreteau et al. 1984:177–180; MacEachern 1990:235–262). Divorce is common; people may marry three to ten times during their lives (Richard 1977:284–285; van Beek 1987:121; Wade 1993), and women often have lived in as many different communities as they have had husbands.

Thus women, sole producers of pottery and also of nearly all archaeologically recoverable artifacts made in and around the mountains, move frequently between communities and across the boundaries of ethnic groups. Most of these women originally learned to pot from their mothers or from other female relatives—but most of those older women themselves will have married into the community at varying times in the past. Women who marry into a new community continue to make pots using techniques, morphologies, and decorations that they are familiar with for varying lengths of time, although they usually begin to imitate local pottery styles fairly quickly. (Maslava tradition ceramics [and particularly the well-made and well-decorated variants made in Urza communities] are more popular in modern markets than are Tokombéré or Podokwo-Mukhtélé ceramics. Women who know how to work in the former tradition and who marry into these latter groups sometimes teach their new relations how to make the more successful pottery, rather than—as would usually be the case—the other way around.)

Women often have problems in learning to make new types of ceramics, and may not become proficient in producing vessels with correct proportions or in making local types of roulettes for decoration. If they divorce and move away from a community within a few years, they may well never have had the opportunity to become proficient at making local pottery. Not surprisingly under these circumstances, there is a high degree of tolerance for production and use of pottery from different ceramic traditions within most communities. Under the circumstances, it is difficult to see how ethnically specific suites of motor habits could develop without intentional and intensive training, which does not exist.

To what degree, then, do the ceramic traditions persist? The stability of these traditions through time seems to be relatively high. Pottery produced 1,000 years ago on Iron Age sites just north of the Mandara Mountains is generally similar to present Maslava tradition ceramics from the same region (David and MacEachern 1988; MacEachern 1990, 1993; Marliac 1991; Wahome 1989), although there is as usual a great disparity in our knowledge about artifactual distributions in ancient and modern communities. Boundaries between the traditions are today to some extent co-located with physical barriers, and it is notable that stylistic differentiation most closely corresponds to "ethnic" differentiation among the Urza potters within the Maslava tradition. These people live in an isolated inselberg close to a number of Wandala centers, and historically their contact with their montagnard compatriots was rendered dangerous by slave-raiding across the intervening plains. The various groups producing ceramics of the same tradition tend also to speak languages belonging to the same Chadic language groups, although this correspondence is not complete (see below; MacEachern 1990:309–311).

Pots and History

In many cases, women marry into communities where they are expected to produce pottery not too different from the pottery that they learned to make when they were growing up. This has, however, little to do with ethnicity and much to do with considerations of proximity, terrain, and local culture history. The differentiation of ceramic traditions appears to be related to geography, the effects of which are reasonably easy to grasp, and to historical and social processes occurring over very long periods of time. These latter are rather more difficult, and it is necessary to at least try and be precise about what we mean by such events.

Researchers working in and around the Mandara Mountains have often conceived of regional culture history as the sum total of a set of population movements—an hydraulic model of ethnogenesis writ small, with cultural configurations determined by which wavelet lapped up where. This is certainly far too simplistic. The cogent critiques made by Clark (1994) and others indicate that archaeologists must be careful in invoking migration as an explanatory concept, given that many of the historical migration analogues used by archaeologists exhibit characteristics—relatively high population densities, relations to state-level societies—that might not have been present through much of prehistory. The extension of state rule through the southern Lake Chad Basin during the last thousand years, and the archaeological evidence for substantial plains populations in the first and second millennia A.D. (MacEachern 1993, 1994, 1998) indicate that this objection is not fatal in this case, but it must be kept in mind.

More important is our understanding of how population movements and ethnogenesis interact. In the northern Mandara at least, ethnic units cannot be understood as simple aggregates of territorial lineage groups claiming diverse origins. Population movements in this area should be conceived of as long-term tendencies of movement from particular regions to particular regions by many small groups of people, often individuals or families, and not as unitary, mass population shifts. Under these circumstances, the arbitration, assumption, and divestment of ethnic identities in the new territories of immigrants will probably have been a long-term process, one where such newcomers lacked power (often expressed in terms of relations to local spirits of the land) and where the standings of a variety of cultural elements were traded off against one another and renegotiated over time. Given the central role played by ceramics within traditional material cultural, especially within contexts of sociopolitical interaction involving ritual, sacrifices, and the sharing of beer, adoption of local pottery by small immigrant groups would have been a valuable way of signaling acceptance of local norms, just as it is for in-marrying women today. (Conversely, scattered references to the breaking of pots in raids upon neighboring communities might be considered the communal equivalent to the similar, individual destruction sometimes wrought by a wife as the culmination of a particularly bitter divorce [MacEachern, field notes 1986, 1992; 1991]—a ritual and social, as well as an economic, attack.) In this case at least, it is probably more accurate to conceive of the differences between Mandara ceramic traditions—whether these be in form, decoration, manufacture, or use—as accommodations and inheritances within powerful sociopolitical systems, rather than as arbitrary choices used to express similarity and difference.

The stability of Mandara ceramic traditions through multiple processes of ethnogenesis and ethnic differentiation may thus be to some extent understandable, almost as a case of historical inertia. At the same time, change does occur and material traditions do differentiate. Perhaps the most visible and potentially important difference between modern ceramics and material from Iron Age sites in this region is the relatively high frequency of tripod legs (and so presumably of tripod-legged vessels) in the archaeological samples. Tripod pots are rare in the area today, used only in a few ritual vessels, but are much more common within ceramic traditions of the southern Mandara Mountains, where they are used in the expanded set of ritual vessels found in those regions and also in meat-cooking pots used by males. Increased use of such tripod pots in the past may then indicate uses of ceramics not visible today and possibly some degree of differentiation from a more homogeneous Iron Age context.

There is, as noted, some correspondence between ceramic traditions and the linguistic affiliations of Chadic-speaking groups that practice these different traditions, and lexicostatistical analyses may place the differentiation of these languages in the early to mid first millennium A.D., a period of cultural differentiation and change within the regional Iron Age (Barreteau 1987; MacEachern 1995). However, languages are not primordial characteristics of peoples. Their use is as conditioned by social factors as are other cultural elements, and ceramic distributions in the region cannot be "explained" by reference to ancestral linguistic groups. We may also note the probable effects of the expansion of slave-raiding and the Wandala state over the last 500 years (which drastically changed the cultural map of the region and contributed greatly to immigration into the mountains [MacEachern 1997]), and of the expansion of a local market system over the last hundred years (which has been to some extent responsible for the disappearance of Wandala potting and the rise in popularity of Maslava-tradition ceramics made by Urza and other women) on Mandara ceramic traditions. Again, however, there is no evidence that the Maslava and Tokombéré traditions originated in these processes, but only that these processes, and no doubt many others, have modified those traditions in multifarious ways over time.

As yet, we do not have enough data to choose between a number of possible explanations for the origins of these regional ceramic traditions. These traditions have been acted upon by a multitude of natural and social factors over time, and have in the process been greatly changed. It may be somewhat unsatisfactory to say that distributions of this archaeologically important artifact type in and around the Mandara Mountains result from the interaction over centuries and probably millennia of a number of natural and social factors, some of which are historically contingent and all of which under-determine the final configuration in material culture, but that appears to be the case in this instance. Searching for origins may be less useful than trying to understand the processes of change and inertia that have affected these traditions through time.

Do these regional ceramic traditions correspond to any particular human groups, either today or in the past? Montagnard identifications with regional groupings do exist in this area today, although they are by no means universal; they are based upon traditions of common immigration from a particular territory of origin. However, none of these regional groups corresponds with the different ceramic traditions; there is simply no evidence that these regional distributions of material culture correspond to any feelings of self-identification among local people. In the southern Mandara massif and in other areas of Africa, equivalent distributions may be the signatures of regional interactions among populations of technical specialists (see below). Equally, northern Mandara distributions may be the traces of the interaction between environmental change and human adaptation in the early to mid-first millennium A.D. noted above, an interaction that manifested itself in the accumulation of larger, deeper archaeological sites, changes in local economies (MacEachern 1995), and perhaps expansion into the areas where the regional ceramic traditions now occur.

Other Axes (But Mostly Hoes) of Variation

Potsherds may be the most ubiquitous material trace found on archaeological sites around the Mandara Mountains, but other types of artifacts are also recovered. Iron tools are far rarer than ceramic artifacts, and we know a little less about modern-day stylistic variation in iron tools through the region than we know about pottery. Along the northeastern edge of the Mandara massif, hoes, the most common indigenous iron artifact and the basis of traditional agriculture, exist in named types that are identified with certain ethnic groupings (Robertson 1992). It appears, however, that this conceptual schema is not straightforward, nor is it necessarily reflected in production, purchase, or use. Different hoe types are bought and used according to their perceived utility in different soils. Local knowledge about soil quality in the territories of different communities provides the link between environment and utilization; people thus buy a *matal* (Mukhtélé) hoe not because they define themselves as Mukhtélé, but because their fields have characteristics (rockiness, for example) associated with the fields of the lineages that speak the language *matal*. In any case, we do not generally recover iron artifacts in sufficient numbers in this area to make quantitative comparison possible. Architectural elements may also furnish material indications of ethnic identities today—but they do not do so reliably, and many will not be recoverable archaeologically. There is now no area of material culture in which stylistic analysis would allow the consistent delineation of ethnic group boundaries as usually defined in this region. Archaeologically, I would be happy to be able to distinguish the regional ceramic traditions already described.

The picture grows more complicated. To the south in the massif, there exist regions within which we find castes of technical specialists, often known simply as "blacksmiths." In these societies, ceramics and iron tools, as well as facilities like forges and graves, are produced by men and women of this caste, usually between 2 and 5 percent of the local population. The ceramics that will make up 99.5 percent of the archaeological sample will thus be produced by perhaps 1 to 2.5 percent of the population—the casted female potters. No data exist on when this caste system originated in this area. Relatively small numbers of marriageable partners within many communities will mean that these specialists are more prone to cross ethnic boundaries than are nonspecialists, and will travel longer distances when they do so (MacEachern 1990). However, ethnic groups as traditionally defined in these areas, like the Mafa and the Marghi, are themselves often much larger than are the equivalent units on the northern peripheries of the massif. They are also even more diverse internally, and so material homogeneity may be correspondingly attenuated.

Distributions of stylistically related artifacts may in these cases disproportionately reflect (relatively) long-distance interactions by specialists rather than local production by the general population. This is a phenomenon that has occasionally been examined by researchers working in different parts of West Africa (Brooks 1993; McNaughton 1992; Neaher 1979; Perinbam 1980), but never in this area of Central Africa. We simply do not know over what distances within the mountains and from the plains the trading networks of local merchants extended; neither do we know how far other occupational specialists traveled in search of the resources and knowledge needed for their crafts (MacEachern 1997). Under these circumstances, I again think it unlikely that stylistic variation will be a useful retrodictor of "ethnic variation," tracking as it would a relatively small proportion of occupational specialists. On the other hand, the spheres of cultural and social interaction of these specialists might well correlate with archaeological distributions in material culture at a slightly larger scale—something similar to the "symbolic reservoirs" that have been posited for various areas of West Africa (McIntosh 1989; see also MacEachern 1994).

These considerations of distributions at various social and geographical scales yield a picture of material variation intriguingly similar to that described for the Sepik coast of New Guinea by Welsch and Terrell (this volume). In that area, artifact distributions are complex and multivariate, and it is not possible to differentiate neat culture areas (still less neat "ethnic groups") on the basis of these distributions. The same is true of the northern Mandara Mountains. As Welsch and Terrell say, the social fields and their material correlates found along the Sepik coast will not necessarily exist in other areas of the world, although archaeologists should arguably pay more attention to long-lasting ties of amity between individuals and communities, even over relatively long distances. Hereditary friendship-exchange relations do not appear to exist at the same degree of complexity or ubiquity around the Mandara massif, although they have not been particularly looked for by ethnographers either. In this area, female mobility, trade, military alliance, and occupational specialization seem to be much more significant, although a number of friendship relations between local men and visiting Wandala traders did occur and were in some cases reinforced by intermarriage. It appears, however, that in both regions we are beginning to see the traces of important regional systems of movement, interaction, and exchange, and to think about what these mean for archaeological interpretation.

CONCLUSIONS

Northern Mandara "ethnic groups," as traditionally defined by anthropologists, are based on linguistic and colonial administrative criteria, act only as an intermediate level in the determination of local identities and are in most cases eclipsed in importance by membership in the smaller territorial lineage groups or, at a larger scale, by the montagnard-plains dweller, animist-Muslim, specialist-nonspecialist dichotomies that govern regional interactions today. Under many circumstances, "ethnic group" membership is simply irrelevant, and this was still more often the case in the past. This irrelevance is also expressed in the decisions that men and women make about marriage and settlement and in the dynamics of artifact production and use. Ethnic frontiers will not be very visible archaeologically, because in a great number of circumstances the producers and consumers of archaeologically significant materials take little notice of those boundaries. In this area at least, archaeological research that attempts to delineate ancient ethnic boundaries and to study ancient ethnic groups, as these groups and boundaries would be defined by ethnographers, will probably be rather unrewarding.

Efforts to understand Mandara culture history at a regional level, where detectable material variation is actually expressed, will be considerably more useful. However, it is important to emphasize that the results of such investigations would not necessarily correspond to the results of traditional ethnographic research on Mandara ethnic groups. Studies of the human interactions that underlay these material traditions might well owe more to demography and geography than to ethnography, given the relative unimportance of any shared cultural identity among the relevant populations. The people responsible for these regional traditions do not speak the same language; they belong to no self-conscious corporate unit; they encompass a host of different social, cultural, and political relationships. The nucleus of norms and behaviors that lends coherence and stability to such traditions is general and ill-defined. Such a core might encompass variously shared beliefs about exogamy and divorce, autochthony, magical protection, the importance of beer in ritual, and relations with the ancestors, but studies of such large-scale cultural distributions are lacking in Africa and elsewhere.

In this case at least, archaeology and ethnography seem to be to a great extent divorced one from the other. However, I certainly would not argue that fine-grained ethnoarchaeological studies of material culture, durable or not, lack applicability to archaeology. Such research plays a vital role in the generation of data about the ways in which people build actions and artifacts into a human world, and under some circumstances it can offer very direct insights into archaeological problems. This should not, however, tempt us into the belief that we can always—or even very often—undertake ethnographic analyses on archaeological sites. The archaeological past and the ethnographic present differ in patterned ways and it seems to me more useful to direct limited resources toward areas of disciplinary strength—regional, long-term studies of social and cultural processes—rather than weakness. Research on ethnicity over the last thirty years locates it in a realm where social actors create, dispute, and shuffle identities; it is a multifaceted and situational phenomenon, linking individuals not only to social collectivities, but also to "... notions of 'life,' 'society' and 'the world,' as well" (Fishman 1989:39). It is not an anthropological synonym for "artifact variation."

This in no way implies that artifact stylistic and technological variation never occurs at the level of the ethnic group, either today or in the past. In many circumstances, such variation may well occur and then ethnicity may be fruitfully studied through archaeological research. The difficulty lies in determining when such co-variation may happen. We require detailed investigations into the social contexts within which artifact design, manufacture, and distribution take place, on local and regional levels, comparable to that undertaken by Robert Welsch and John Terrell. We require comparative studies on the effects of population densities, postmarital residence rules, market systems, and political complexity on material culture. We require detailed research on a micro-level of the contexts within which learning and production take place, of the kind undertaken by Olivier Gosselain and others. As these progress, we may be able to make predictions about relationships between ethnicity and artifact distributions. In the meantime, we must be attentive to the differing spatial and temporal scales at which social interaction and artifactual variation take place. Social systems, today and in the past, work at a number of levels, and it is a great mistake for archaeologists to emphasize one of these levels—the "ethnic"—at the expense of others, especially given that this may be one of the levels least suited to archaeological detection.

Archaeologists may not be especially well equipped to study past ethnic relationships. However, neither are ethnographers always well equipped to study human interaction on a regional level, especially over significant periods of time. The distributions of pottery and other artifacts in and around the Mandara Mountains, and the archaeological patterning resulting from those distributions, are not just poor reflections of the ethnic milieu of the region. They reflect different sets of relationships occurring constantly between different montagnard populations, relationships that involve people, techniques, artifacts, and innovations. Those relations are not epiphenomena secondarily derived from an irreducible bedrock of ethnicity, but rather an integral part of the diverse economic, social, and political interactions that people constantly engage in.

ACKNOWLEDGMENTS

An earlier version of this paper was presented at the 1995 Society for American Archaeology meetings in Minneapolis. I would like to thank Miriam Stark for her very useful comments on this paper and for her hard work in organizing contributions and editing the resulting book. I would also like to thank the two anonymous reviewers of this paper for their helpful comments. I would like to thank Genevieve LeMoine, Nicholas David, and James Helmer for their comments on earlier drafts of this paper; I have not followed all of their suggestions, but they were very much appreciated. The research behind this paper has been carried out through grants from the Social Sciences and Humanities Research Council of Canada (grants 410-83-0819, 410-85-1040, 410-88-0361, 410-92-1860 and 410-95-0379) and from the Department of Research Services of the University of Calgary.

6

The Cultural Origins of Technical Choice:
Unraveling Algonquian and Iroquoian Ceramic Traditions
in the Northeast

ELIZABETH S. CHILTON

In northeastern North America, the Late Woodland period (ca. A.D. 1000–1600) is defined and described largely on the basis of variation in ceramics. Archaeologists in the region have constructed fairly rigid stylistic typologies for ceramics in order to infer ethnicity and chronology (e.g., Engelbrecht 1978; MacNeish 1952; Rouse 1947; Smith 1947). While these typologies have been useful in some cases for the construction of culture histories, they have often become an end in themselves. Often these stylistic types are taken to be a direct reflection of group affiliation: for example, when a certain "type" of ceramic is found outside of its "homeland," it is interpreted as either stylistic copying, trade, or "female capture" (e.g., Brooks 1946; Byers and Rouse 1960; Engelbrecht 1972; Lavin 1988). While Northeast archaeologists have placed a premium on decoration for discerning such things as ethnicity (e.g., Engelbrecht 1978; Plog 1980a; *contra* Brumbach 1975; Goodby 1992), little attention has been paid to the wide variety of choices available to potters during ceramic production and use. Certainly the level at which we isolate aspects of material culture determines the patterns of behavior we are able to see (Lechtman 1977:12). Therefore, an overemphasis on decoration in Northeast ceramic studies has inhibited a deeper understanding of the technological and social contexts of ceramic manufacture and use.

The goal of the research presented here is to underscore and examine the complex relationships among technical choices, historical context, and society. These relationships are examined through an *attribute analysis of technical choices* for three Late Woodland archaeological sites in the Northeastern United States. Before I discuss the details of the archaeological context, I present a theoretical background for the notion of style in technology.

TOWARDS A THEORY OF CHOICE

As Dean Arnold rightfully points out "material culture is nowhere near as simple as it once seemed to be" (D. Arnold 1991:345). Thanks to the creative ethnographic work of archaeologists and other social scientists (e.g., Dietler and Herbich 1989; Hodder 1982, 1986; Lechtman 1977; Lemonnier 1989; Plog 1990; Sackett 1990), ceramic ecologists (D. Arnold 1985, 1993; P. Arnold 1991; Stark 1995c), and feminist archaeologists (Gero and Conkey 1991; Spector 1993), our understanding of the meaning of materials is much more complex, perhaps, than it was a decade ago. It is clear that we must move beyond a simplistic division between style and function if we are to gain any critical understanding of the social dimensions of material culture (Dietler and Herbich, this volume). If, for example, we accept Hill's (1985:374) definition of style—the "characteristic manner of expression, execution, construction or design"—we recognize style as permeating all aspects of variation in material culture. Indeed, style is a multilayered phenomenon, with different layers of style reflecting different cultural processes (Gosselain, this volume).

A focus on *how* things are made is somewhat of a departure from traditional anthropological archaeology. In fact, the study of techniques is often regarded as an area of inquiry outside anthropology (Mahias 1993:157). But the concept of "technological style" (Lechtman 1977) focuses on the relation between techniques and society—not on techniques in their own right (van der Leeuw 1993:240). According to Lemonnier (1992:3):

> In most cases, technological systems are summed up merely as static *constraints* without considering the social aspects of material culture. And in the few cases where the social aspects are explored, technological systems are reduced to statements about the shape of artifacts, or worse, their decoration . . . [emphasis in original].

Recent research shows that, in certain contexts, decorative style may be less indicative of social identities than are technological traditions (Childs 1991; Dietler and Herbich 1989; Gosselain 1992b; Lechtman 1977; Pfaffenberger 1992; Stark 1995c; Steinberg 1977:78; Sterner 1989). For example, Gosselain (1992b), in his ethnographic study of the Bafia and other Cameroonian groups, suggests that the vessel shaping *process* reflects ethnicity, more than does the end result. Likewise, Miller (1985) in his ethnographic study of pottery manufacture in central India, suggests that shaping *techniques*—not the shapes themselves—reflect social divisions of caste. Here the emphasis is on *choice*—rather than on the materials or tools—as critical in determining the final product (van der Leeuw 1993:241). The natural environment, rather than constraining choice, serves only as a backdrop or context for social relations (Dobres and Hoffman 1994:231; Lechtman 1977:14). Therefore, social agency is critical in "defining, determining, and articulating particular technologies and the operational sequences" (Dobres and Hoffman 1994:231).

The basic premise to theories of technological choice is that societies choose between a number of equally viable options; given a technical problem, choices transcend mere material efficacy or technical logic (Lemonnier 1989:156, 1993a:16; Mahias 1993:177). There are more subtle informational or symbolic aspects of technology that involve arbitrary choices about techniques and materials, and that are a part of a larger symbolic system (Lemonnier 1992:3). For example, when designing an airplane, an engineer is influenced by what she or he thinks an airplane "should be like" based on already existing designs, and her or his education and cultural experience (Lemonnier 1989:170).

A theory of technical choice does not simply replace an overemphasis on decoration with an overemphasis on technology or "function." The concept of technical choice is more comprehensive than other concepts of style (e.g., the social interaction [Deetz 1965; Engelbrecht 1978; Plog 1976] or information exchange theories [Plog 1980a; Wobst 1977]), because it requires as much attention to the sequence and context of manufacture and use as it does to what the finished product "looks like" or conveys. Thus, in this analysis "style" is viewed as the way an artifact is made, as much as, for example, the way it is decorated.

CERAMIC PRODUCTION IN NEW ENGLAND: OBJECTIVES

Since there is latitude for choice in virtually every technical aspect of human existence (Pfaffenberger 1992:499; see also Lechtman 1977:14 and Lemonnier 1986), in this study I emphasize the choices that are made by potters throughout the production sequence in order to move beyond a priori assumptions about the evolution of technology. I consider the entire sequence of decision-making involved in artifact production and interpret it in its specific sociocultural context. This approach is very much related to ceramic ecology, which emphasizes the interaction between ceramics and their natural and sociocultural context (e.g., D. Arnold 1985, 1993; Kramer 1985; Krause 1985; Longacre and Skibo 1994; Skibo 1992).

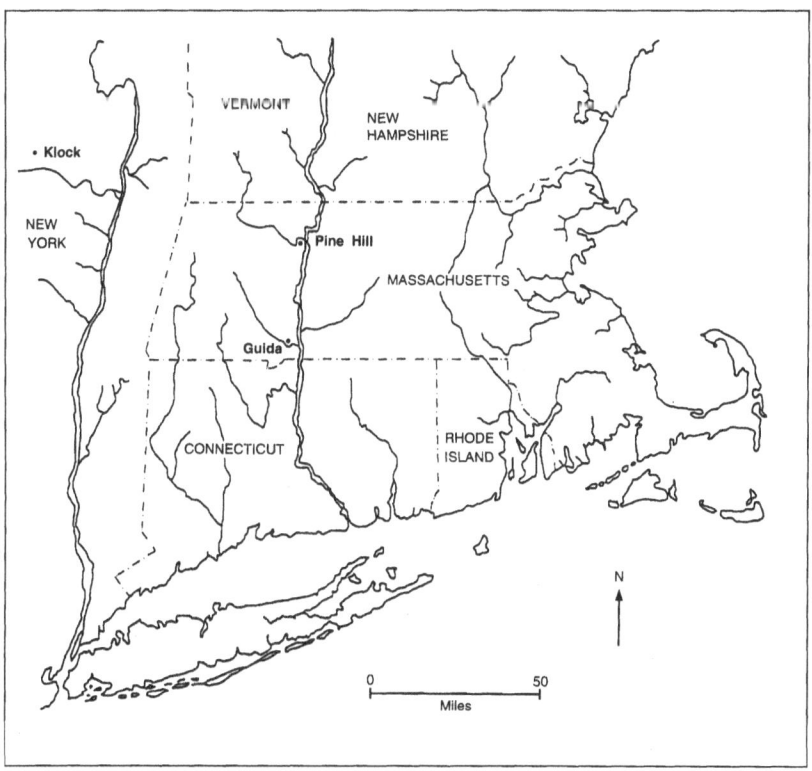

Figure 6.1. Southern New England and eastern New York, showing the location of key sites in this study.

The specific objectives of this research are to: (1) contribute to archaeological theory in the interpretation of material culture; (2) contribute to the refinement of archaeological method by exploiting alternatives to the typological approach—particularly by refining attribute analysis for archaeological ceramics; and (3) provide a more complete reading of New England prehistory—one that does not simply mimic the culture history of better-known groups, such as the Iroquois.

As a means to achieve these objectives, I examine ceramic variability at three Late Woodland sites in the Northeast (Fig. 6.1): two in the middle Connecticut River Valley in western Massachusetts (the Guida Farm site in Westfield and the Pine Hill site in Deerfield), and one in the Mohawk Valley in New York State (the Klock site in Ephratah, which lies approximately 80 km west of Albany). The two Massachusetts sites are thought to have been inhabited by Algonquian-speaking peoples of the Connecticut Valley (most likely the Pocumtucks at Pine Hill and the Woronocos at Guida Farm). The Klock site was most likely occupied by a

Mohawk community. All of the sites have components that date to the latter part of the Late Woodland period (A.D. 1300–1600). I chose these assemblages in order to evaluate the expectations that Algonquian and Iroquois ceramics differ with respect to: (1) the intended use of vessels; (2) the variables affecting decoration; (3) the scale of ceramic production; and (4) technical style.

THE DATA SET

The Guida Farm site is located just east of Westfield, Massachusetts. It was partially excavated by William Young in the early 1960s, and was also tested by numerous other professional and amateur archaeologists (see Byers and Rouse 1960). The two major excavations at the sites were conducted by Byers and Rouse (1960) in 1952 and William Young in 1958. Byers and Rouse excavated four trenches and three test pits (Byers and Rouse 1960:9). Apparently, none of the excavations uncovered evidence of settlement patterns or structures (see Byers and Rouse 1960:12). On the basis of the analysis of material culture, the major occupations of the site date to the Late Woodland and possibly early contact periods (ca. A.D. 1000–1700). It is a large site on a major river and likely represents a place on the landscape where people repeatedly returned. The site yielded large amounts of pottery from the Late Woodland period both on the surface and in association with features. It is unclear how much, if any, of the site still exists, due to farming, bulldozing and historic dumping (John Pretola, pers. comm. 1993).

I analyzed the ceramic assemblage from Guida that was excavated by William Young in 1964 (Young 1969). Young worked with the local Massachusetts Archaeological Society (MAS) for many years and excavated a portion of the site while working at the Springfield Museum (John Pretola, pers. comm. 1993). The collection contains about 1,000 sherds that were sufficiently complete to be used in this analysis.

Pine Hill was tested by the University of Massachusetts Archaeological Field School in the summers of 1989, 1991, 1993, and 1995; excavations were co-directed by the author and Arthur S. Keene in 1993 and 1995. No other documented collections are known to exist from the site, although there has been a great deal of looting. The site is located in an open deciduous forest on private land and is quite likely the best preserved Late Woodland site in western Massachusetts. The area is currently used as a wood lot, and the only plowing apparently took place in the early nineteenth century. Although the site analysis and interpretation is not yet complete, our current interpretation is that the site represents a seasonal encampment where small groups coalesced for hundreds, if not thousands, of years (Keene and Chilton 1995). However, the site's major occupation seems to have occurred during the Late Woodland period. Twenty possible storage or food processing features (11 of which contained Late Woodland ceramics) and 50 scattered postmolds have been identified since our excavations began in 1989. Two radiocarbon dates from pit feature lenses containing ceramics are: (1) cal A.D. 1230–1430; and (2) cal A.D. 1420–1520 (*p*=.67) or cal A.D. 1568–1627 (*p*=.33) (Chilton 1996). (Calibrated at 1 sigma with the program CALIB 3.0.3 [Stuiver and Pearson 1993].)

For the purpose of this study, I analyzed all of the ceramics collected from 1989 to 1993. Several hundred ceramic sherds were recovered from the Pine Hill site, 500 of which are sufficiently complete to be analyzed.

The Klock site is located in Ephratah, New York, 15 km north of the Mohawk River on Garoga Creek. The site was excavated under the direction of Robert Funk in the summers of 1969 and 1970, after preliminary testing by William Ritchie in 1950. Based on analysis of material culture by Kuhn and Funk (1994), the site is thought to date to the mid-sixteenth century. Two maize samples from a single pit feature at the site have been radiocarbon dated to cal A.D. 1483–1649 and cal A.D. 1326–1439 (calibrated using one sigma; Snow 1995). Klock is a stockaded village site with evidence of at least seven longhouses, and the site is thought to have been occupied for approximately ten years (Kuhn and Funk 1994). Over the two field seasons in the late 1960s, 157 features were encountered; 76 of these features were at least partially excavated.

The Klock assemblage contains more than 15,000 ceramic fragments. The Klock collection is much larger than those from the Connecticut Valley sites because: (1) on the basis of our knowledge of the large population size of the sixteenth-century Iroquois villages, the site was probably larger than contemporaneous Connecticut Valley sites; and (2) a much larger area was excavated. A random sample of enough sherds (n = 214) to comprise 100 vessel lots was chosen from the Klock assemblage for analysis (the means of establishing vessel lots is described below).

Before I discuss the details and results of the ceramic analysis of these three assemblages, I will provide some background for the Late Woodland period in the Northeast.

LATE WOODLAND CULTURAL DYNAMICS

Based on ethnohistoric and linguistic evidence, the Late Woodland communities of southern New England spoke dialects belonging to the Algonquian language family. The tribal groups in the Mohawk Valley, on the other hand, spoke dialects belonging to the Iroquoian language family. The Iroquois of the Late Woodland period resided predominantly in central and western New York and were surrounded by Algonquian-speaking groups. Current linguistic and archaeological evidence supports the theory that the Iroquois migrated into the region sometime between A.D. 900 and A.D. 1300 (Denny 1994; Parker 1916; Snow 1994; Swihart 1992; *contra* MacNeish 1952, 1976) and either intermingled with or wedged themselves between resident Algonquian groups in the Mohawk Valley. As I will discuss below, these linguistic differences between Algonquian and Iroquois groups were accompanied by a series of cultural distinctions.

In the greater Northeast, and especially for the Iroquois, the Late Woodland is perceived as a culturally dynamic period: agriculture became important for subsistence, communities became more sedentary, and population and the incidence of intergroup conflict increased (see Fenton 1978). However, the Late Woodland archaeology of the middle Connecticut Valley Algonquians is poorly known. In the middle Connecticut Valley, unlike areas to the south and west, no evidence exists for large, permanent, fortified settlements and intensive agriculture (see Thorbahn 1988).

Concerning the lack of evidence for large, Late Woodland villages in the region, archaeologists used to claim that they had not yet been found, or, as Ritchie claimed for the Hudson Valley, that they had been obliterated by the large-scale destruction of sites as a result of Euroamerican settlement and digging by amateurs (Ritchie 1958:7; see also Snow 1980:320). Certainly, the looting of sites has been, and continues to be, a serious problem (Jordan 1975). Also, due to the large, dynamic floodplain of the Connecticut Valley, some Late Woodland sites may have been buried or destroyed. Nevertheless, the seeming invisibility of Late Woodland villages in the eastern Algonquian area may reflect a high degree of mobility for the small groups resident in the valleys of the interior (Ritchie 1958:108).

Horticulture

While maize horticulture was present in the greater New England area by A.D. 1000, it was not practiced to the same degree across the region (George and Bendremer 1995:14; see also Cassedy et al. 1993 and Heckenberger et al. 1992). Certainly the timing and importance of maize horticulture in New England is controversial (see Ceci 1979, 1990; Demeritt 1991; Silver 1980). There is no evidence that maize was anything more than a dietary supplement in the New England interior, at least prior to European settlement (Dincauze 1991:30; McBride 1984:144; cf. George and Bendremer 1995; Snow 1980:333). Whether the Europeans were the direct cause or not, most New England archaeologists agree that *intensive* maize horticulture did not occur in New England until after the arrival of Europeans (see Ceci 1982; McBride and Dewar 1987; Thorbahn 1988).

Seventeenth-century accounts of the native New England diet belie claims of maize specialization. Wood (1977:86) in 1634, recorded for the Massachusetts Bay region: "In wintertime they have all manner of fowls of the water and of the land, and the beasts of the land and water, pond-fish, with catharres and other roots, Indian beans and clams. In the summer they have all manner of shellfish, with all sorts of berries." Josselyn (1988:93), reporting on his journey to the coast of Maine in 1674, echoes this diverse menu:

> Their Diet is Fish and Fowl, Bear, Wild-cat, Ratton and Deer; dryed Oysters, *Lobsters* rosted or dryed in the smoak, *Lampres* and dry'd *Moose-tongues*, which they esteem a dish for a *Sagamor*; hard egges . . . their *Indian* Corn and Kidney beans they boil . . . they feed likewise upon earth-nuts or ground-nuts, roots of water-Lillies, Ches-nuts, and divers sorts of Berries [original emphasis].

Based on his observations in 1643 of the Narragansett Bay region, Roger Williams stressed the importance of hunting and trapping of numerous animals and the collecting of acorns, chestnuts, walnuts, strawberries and cranberries (Williams 1963).

In contrast, the Iroquois of upstate New York were directly dependent on maize for subsistence. According to Parker (1968:9), maize was so important to the Iroquois "that they called it by a name meaning 'our life' or 'it sustains us.'" So important was maize to the Iroquois, that European invaders commonly burned Iroquois maize fields and maize stores as a warfare tactic (Parker 1968:17).

The type of horticulture practiced by the Iroquois was "shifting horticulture" (Niemczycki 1984:3; see also Morgan 1901). In order to prepare an area for horticulture, tracks of forest were cut and burned. The locations of the fields were shifted periodically to maintain the fertility of the soil. The type of maize cultivated by the Iroquois was Northern Flint or closely related varieties (Fenton 1978:325). This type of maize is unlike modern sweet corn, since it requires cooking for a long period of time over a hot fire. Overall, a great deal of time was devoted to the cultivation, harvest, storage and preparation of maize for consumption. Parker (1968) also indicates that there were many customs and rituals related to the cultivation and consumption of maize.

While maize was a dietary staple, other cultivated and wild plants were important for subsistence, such as beans, melons and squash, fungi and lichens, fruits, berries, nuts, roots, and bark foods (Parker 1968). Also, there is archaeological evidence that the Iroquois hunted deer, elk, bear, and turkey, and collected fresh water mussels (Funk 1976).

Settlement Patterns

Based on both ethnohistoric and archaeological accounts, we know that the traditional dwelling throughout New England was the wigwam (Fig. 6.2). The size of wigwams was apparently small; Williams (1963:121) describes a dwelling for two families as "a little round house of some fourteen or fifteen foot over." Likewise, Higgeson (1629:123) states: "Their houses are verie little and homely, being made

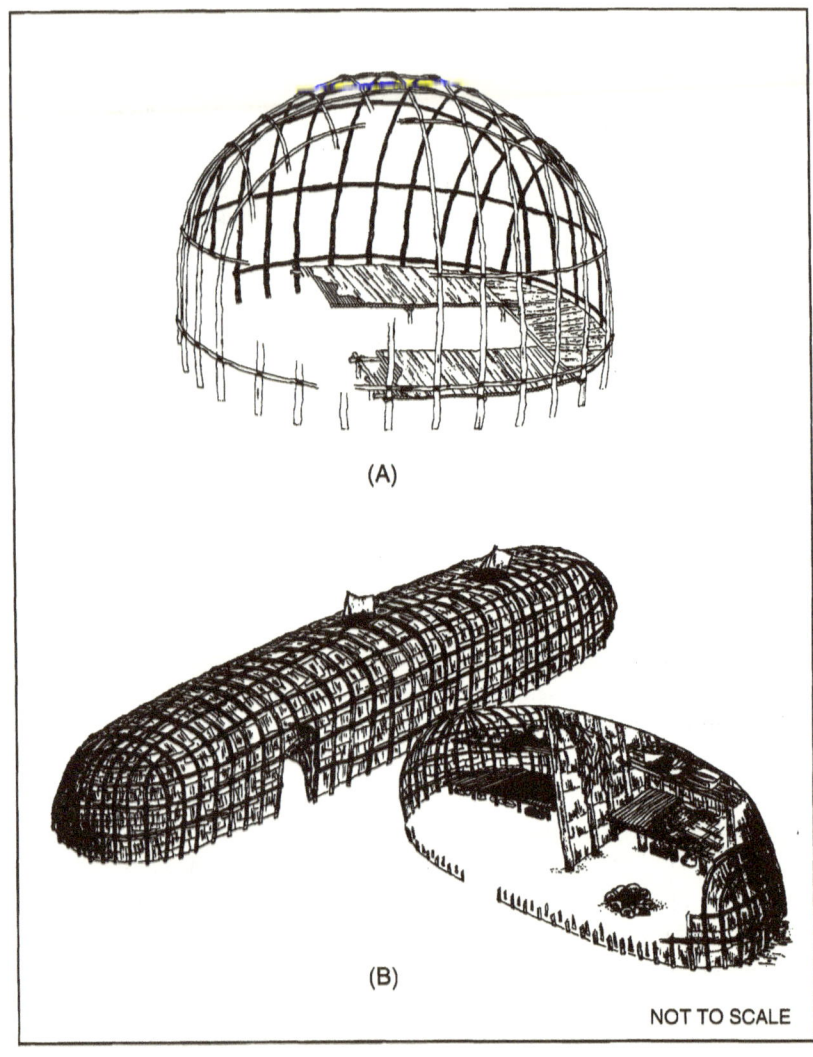

Figure 6.2. Reconstructions of indigenous architectural forms: (a) cut-away of a wigwam (after Sturtevant 1975, fig. 2c); (b) longhouse (after Kraft 1975:83).

with small poles pricked into the ground." Each house was likely shared by one or two related families (Morgan 1965:124; Williams 1963:61). For the Hudson Valley, Johan de Laet (1625–1640 [Jameson 1909:57]) says of the Algonquians living there that "some of them lead a wandering life in the open aire without settled habitation.... Others have fixed places of abode." Thus, it is clear that there was diversity in settlement practices even within a particular valley. For some groups, the size and shape of dwellings would change, depending on population density (e.g., small wigwams in the summer, multifamily longhouses in the winter; Cronon 1983:38).

Williams (1963:135) also comments on the Algonquians' seasonal movements and the flexibility of their habitations:

> In the middle of summer . . . they will flie and remove on a sudden from one part of their field to a fresh place . . . Sometimes they remove to a hunting house in the end of the year . . . but their great remove is from their Summer fields to warme and thicke woodie bottoms where they winter: They are quicke; in a halfe a day, yea, sometimes a few houres warning to be gone and the house up elsewhere.

Similarly, Josselyn (1988:91) notes the impermanence of New England communities: "Towns they have none, being always removing from one place to another for conveniency of food. . . . I have seen half a hundred of their Wigwams together in a piece of ground and they shew prettily, within a day or two, or a week they have all been dispersed."

In terms of archaeological evidence, while McBride (1984:322) claims that "(m)ost New England archaeologists report an increase in artifact . . . and site density as well as a trend toward fewer, larger sites after A.D. 1000," this is apparently true only in the lower Connecticut Valley (see George and Bendremer 1995). In fact, Ceci (1979) reports that there is *no* evidence of village-based settlement patterns on Long Island. McBride's claim of increasing site density and size is, likewise, *not* evident in the New England interior (see Thorbahn 1988).

In contrast to the small and impermanent settlements of the New England Indians, the Iroquois resided in villages of 30 to 150 ft, multiroomed longhouses (Fenton 1978:306; Fig. 6.2). These villages were inhabited for 25–50 years at a time (Tuck 1978:326). Morgan states that these longhouses accommodated up to twenty families (Morgan 1965:64). Each household was comprised of a group of kin related through the female line (Morgan 1965:64). Many of the Iroquois villages of the Late Woodland period were palisaded for defense, which is likely a reflection of the intercommunity conflict that arose as a result of sedentism and intensive horticulture (Hasenstab 1990:1; see also Morgan 1901:306).

Political Differences

The five and, later, six nations of the Iroquois shared a cultural base and "a well-developed tribal level sociopolitical organization which distinguished them from their Algonquian neighbors" (Niemczycki 1984:3). Population size apparently grew and settlements became more nucleated during the Late Woodland period, as the importance of agriculture and the availability of surplus food increased. As a result of increasing population growth and density, the size, power and rigidity of matrilocal-matrilineal groups increased as a means of controlling intergroup relations (Whallon 1968:242–243).

This sociopolitical organization differed greatly from the "loosely organized" Algonquians of New England (Fenton 1940:162). Here, no political unit was larger than the village, and there was no central authority to force political conformity (Thomas 1979:400). Since groups were fissioning and fusing seasonally, patterns of residence and the reckoning of kin needed to be more flexible. As E. Johnson (1993) proposes, mobility may have been a political strategy of resistance to authority—that is, the authority of certain native political leaders, and, later, the English. By maintaining flexibility and mobility in their settlement practices, the Algonquians of the interior could literally "vote with their feet," which may explain the infrequent occurrence of warfare prior to European contact.

Algonquian-Iroquois Interactions

Thus far I have presented differences between the Algonquians of the Connecticut Valley and the Iroquois of the Mohawk Valley as dichotomous. However, there is increasing evidence of interaction between the Mohawk and the Connecticut Valley groups, at least in the early historic periods (Haefeli and Sweeney 1993; Salisbury 1993). Trade networks associated with the immigration of Europeans may have served to augment or even reduce earlier ties between Algonquians and the Iroquois, but these ties were likely forged in the prehistoric period. There is also archaeological evidence for contact and trade prior to the arrival of Europeans. For example, New York chert tools and debitage are commonly found on archaeological sites in New England. While these "exotic" materials are more numerous in certain periods (such as the Late Archaic, about 6,000–3,000 years before present), they are present to some degree for most of the prehistoric period. Also, despite the differences between Algonquian and Iroquoian subsistence, settlement, and ceramics, the differences are mostly in *degree* and not in kind. For example, the two groups shared numerous cultigens: maize, beans, squash, and tobacco. Also, certain design motifs reoccur across the Northeast, such as the "ladder motif" (see MacNeish 1952:159) and geometrically designed, zoned collars. These similarities are the main reason why New England archaeologists have attempted to import New York typologies whole stock. While, in some cases the similarities may be the result of direct trade, stylistic similarities are often the result of social interaction and the occasional movement of people. There may, indeed, have been a symbiotic relationship between the Iroquois as settled horticulturalists and the Algonquian mobile farmers.

Ethnohistoric Accounts of Northeast Ceramics

There are few references to ceramic manufacture in the ethnohistoric literature for New England, and in the scant material available there is contradiction. For example, while Williams (1963:179) recorded that the "women make all their earthen vessels," according to Gookin (1792:151), men made pottery.

For the Iroquois there is more of a consensus that women produced pottery vessels (Morgan 1901:280; Whallon 1968:230). According to Sagard (1968:109), in his seventeenth-century account of his travels among the Huron, it was women who were firing pots "in their hearths" (*en leur foyer,* incorrectly translated as "ovens" in this English translation). There is little to no archaeological evidence for ceramic firing features in the northeast. Therefore, it is likely that most native peoples of the Northeast fired their pottery in multipurpose hearths. Thus, direct evidence for ceramic production and firing is absent.

According to Engelbrecht (1978:141), groups of related Iroquois women cooperated to make pottery. Since the Iroquois resided in semipermanent villages, the context and timing of ceramic manufacture would have been fairly consistent and predictable. On the other hand, for the Algonquians of the New England interior, small groups were likely fissioning and fusing throughout the year, and the smallest ceramic production unit was likely the nuclear family. Ceramics would have been produced in variable environments and in cooperation with various personnel. While there are no ethnohistoric references to the quantity of ceramic vessels produced, it is clear from the archaeological record that Iroquois ceramic production was conducted at a much larger scale than that of the Connecticut Valley Algonquians. These different contexts and scales of ceramic manufacture have implications for ceramic homogeneity and heterogeneity, which I discuss below.

Archaeological Ceramics

In general, native ceramics from the Late Woodland period are more elaborate than earlier ceramics in the Northeastern United States in both decoration and form (e.g., globular bodies, applied collars, castellated rims, and constricted necks [Goodby 1992:4]). There has been little research on native ceramic traditions in Massachusetts; thus, archaeologists in the state often rely on ceramic sequences developed for southern and coastal New England and New York (see Luedtke 1986:113; MacNeish 1952:98; Ritchie and MacNeish 1949; Lenig 1965). In southern New England, ceramic classifications are largely based on the work of Smith (1944,

TABLE 6.1
Late Woodland Ceramic Traditions in Southern New England and New York

	Windsor	East River	Shantok	Guida	Owasco	Mohawk
General description	Red-orange or tan color; roughened surface (brushed and stamped); later collared and incised; vessels are "thick" and "coarse"	Gray to brown color; smooth interiors, roughened exterior; shell impressed or incised decoration; mostly grit temper	Mostly shell-tempered, incised, collared and incised, commonly with appliqués or modeled nodes	Gray to black color; fine, micaceous temper; surface is "flaky" or "silky"	Cord-impressed decoration and surface treatment collars rare at first, later with castellations; grit temper	Incised smooth surface treatment, castellated collars, fine grit temper
Geographical placement	Coastal New England and eastern Long Island, Connecticut River to Westfield, Massachussetts	Southeastern New York and western Long Island	Southeastern Connecticut and eastern Long Island	Western and central Connecticut and Massachussetts	Most of New York	Eastern New York
Type names	Clearview, Sebonac, and Niantic types	Bowmans Brook and Clasons point types		Guida Incised, Guida Cord-marked, Plain, Stamped, Misc.	Numerous Castle Creek and Owasco types	Numerous incised and notched types, differentiated by motif
Time period	A.D. 1000–1700	A.D. 1100–1700	A.D. 1400–1700	A.D. 1500–1700	A.D. 1000–1400	A.D. 1400–1700

Sources: Byers and Rouse (1960), Lavin (1980, 1988), Lavin and Kra (1994), Lenig (1965), MacNeish (1952), McBride (1984), Ritchie (1980), Rouse (1947), Smith (1947), and Snow (1994).

Algonquian and Iroquoian Ceramic Traditions in the Northeast   145

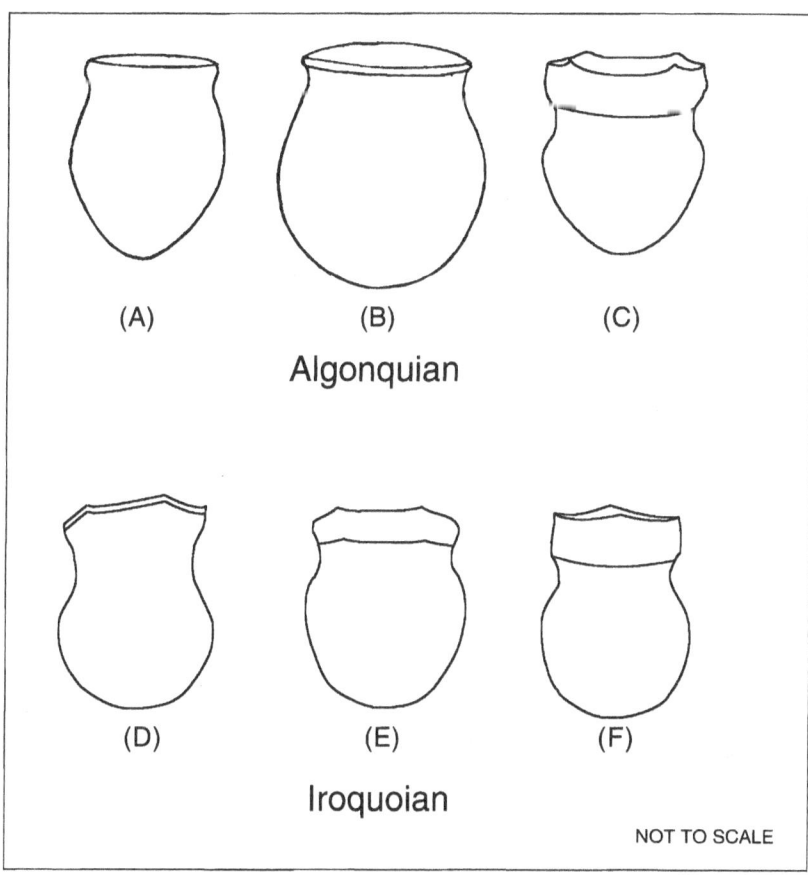

Figure 6.3. Vessel shapes of the Late Woodland period: (a) East River tradition (after Smith 1947, fig. 2); (b) Niantic vessel of the Windsor tradition (after Smith 1947, fig. 2); (c) Narragansett Bay area (after Fowler 1966, fig. 18); (d) Oakfield phase of western New York; (e) Oak Hill Corded type from eastern New York; (f) Late prehistoric Iroquoian vessel from central New York (d–f after Ritchie 1980, plates 104, 105, 11).

1947, 1950), Rouse (1945, 1947), and Lavin (1986, 1988) (see also Fowler 1945 and Pope 1953). Ritchie's extensive work in New York (e.g., Ritchie 1944) on the development of classification schemes was probably the impetus behind the development of similar schemes for coastal New York and Connecticut (McBride 1984:4; e.g., Smith 1947, 1950; Rouse 1947). Smith and Rouse defined three broad ceramic traditions (Windsor, East River, and Shantok) based on certain "diagnostic features" or attributes such as inclusion type, thickness, color, surface treatment, and decoration. Types have been defined within these broader traditions (Table 6.1 and Fig. 6.3). Lavin has continued to build on the typological work of Smith and

Rouse, often adding to or expanding the existing type names (see Lavin 1986 and Lavin and Miroff 1992). While attribute analysis and typological analysis have sometimes been presented as mutually exclusive forms of analysis (e.g., Petersen 1985), attribute analysis is often employed in southern New England as a *means* to assign sherds to previously known or new types. For example, Lavin (1986:3) views attribute analysis and typology as "complementary stages in the ordering of ... data." Others, however, have turned to attribute and technological analyses as means to move *beyond* the typologies that have constrained archaeological interpretation in the region (e.g., Chilton 1991; Dincauze 1975; Dincauze and Gramly 1973; Finlayson 1977; Goodby 1992; Kenyon 1979; Kristmanson and Deal 1993; Lizee 1994; Luedtke 1986; McGahan 1989; Pendergast 1973; Petersen 1985; Ramsden 1977; Stothers 1977).

CERAMIC ATTRIBUTE ANALYSIS

Attribute analysis involves the descriptive comparison of specific artifact features (Lavin 1986:3), such as temper or clay type, surface treatment, color, and decoration. Thus, two pots can share some attributes, but not others. In the method of analysis used here, *an attribute analysis of technical choice,* the goal is to look for variation *and* covariation across objects—not between groups of objects. An important component to this attribute analysis is that vessels, rather than individual sherds, are the units of analysis. Historically, the use of rim or sherd frequencies to describe ceramic assemblages has been a common practice in Northeast archaeology (Petersen 1985:10). However, researchers in the Northeast have increasingly employed vessels as units of analysis (Petersen 1985:10; see Dincauze 1975, 1976; Dincauze and Gramly 1973; Luedtke 1980; Petersen 1980; Wright 1980). The use of vessels as units of analysis is very important in the interpretation of human behavior because vessels were likely the most common units of meaning in prehistoric societies (Carr 1993; Skibo et al. 1989a).

In this analysis, I define an attribute as *one aspect or variable of a ceramic vessel,* such as surface treatment, color, inclusion type, or rim shape. Thus, each attribute has an infinite number of possible values. This definition of attribute differs from that used by Cowgill (1982), Rouse (1960, 1964), and Petersen and Sanger (1991), who define an attribute as the specific state of a variable—not the variable itself. Using my definition, an attribute represents a technical choice, such as vessel shape, inclusion size, or color. Therefore, the emphasis is placed on the *criteria* for selecting between various technical options—not the specific choices themselves.

Using this method, in order to establish a vessel lot—that is, a group of potsherds determined to be *minimally* from the same vessel—at least nine attributes are recorded for each sherd: modal thickness, inclusion material (e.g., temper),

TABLE 6.2
Sample Data from the Attribute Analysis of the Pine Hill Assemblage

VESSELLOT	CATNUM	# SHERDS	TESTUNIT	STRATUM	LEVEL	FEATURE	THICKNESS (mm.)	TEMPER1	TEMPER2	GRITCOMP	GRITCOMP2	GRITCOMP3	T1MIN	T1MAX	T2MIN	T2MAX	T1DENSITY (%)	T2DENSITY (%)	INTCOLOR	EXTCOLOR	SFTREAT	SFTREAT2	PORTION	RIMFORM	LIPFORM	ORIFICE	COLLARWDTH (mm.)	RESIDUES
6	93.1617	2	N498E524NE	201			8	QT	FE	NA	NA	NA	C	P	VF	C	25	2	10YR52	10YR53	WI	SM	B	NA	NA	0	0	N
6	93.1626	1	N498E524NE	202			6	QT	FE	NA	NA	NA	C	P	VF	C	25	2	10YR52	10YR52	SM		B	NA	NA	0	0	N
6	93.1762	1	N498E524NW	202			6	QT	FE	NA	NA	NA	C	P	VF	C	25	2	10YR52	10YR42	WI	SC	B	NA	NA	0	0	N
6	93.1765	1	N498E524NE	202			10	QT	FE	NA	NA	NA	C	P	VF	C	25	2	10YR52	10YR63	WI	SC	B	NA	NA	0	0	N
6	93.1765	1	N498E524NE	202			8	QT	FE	NA	NA	NA	C	P	VF	C	25	2	10YR52	10YR53	WI	SM	B	NA	NA	0	0	N
6	93.2017	1	N487E510SW	202			7	QT	FE	NA	NA	NA	C	P	VF	C	25	2	10YR52	10YR53	WI	SC	N	NA	NA	0	0	N
6	93.2114	1	N498E528SE	202			6	QT	FE	NA	NA	NA	C	P	VF	C	25	2	10YR52	10YR53	WI	SM	B	NA	NA	0	0	N
6	93.2222	1	N498E530	202			8	QT	FE	NA	NA	NA	C	P	VF	C	25	2	10YR53	10YR53	WI		B	NA	NA	0	0	N
7	91.939	1	N500E524	1	1		6	QT	FE	NA	NA	NA	C	P	VF	C	30	2	10YR52	10YR53	SM	NO	R	EV	FL	7	0	N
7	93.11	1	N498E528SE	201			7	QT	FE	NA	NA	NA	C	P	VF	C	30	2	10YR54	10YR53	SM		B	NA	NA	0	0	N
7	93.1193	1	N498E530SW	201			8	QT	FE	NA	NA	NA	C	P	VF	C	30	2	10YR53	10YR54	SM		N	NA	NA	0	0	N
7	93.1252	1	N498E530SW	202	2		7	QT	FE	NA	NA	NA	C	P	VF	C	30	2	10YR54	10YR53	SM		B	NA	NA	0	0	N
7	93.1451	1	N494E530SW	202			7	QT	FE	NA	NA	NA	C	P	VF	C	30	2	10YR54	10YR53	SM		B	NA	NA	0	0	N
7	93.1472	1	N498E522NW	220			8	QT	FE	NA	NA	NA	C	P	VF	C	30	2	10YR54	10YR53	SM		B	NA	NA	0	0	N
7	93.1626	1	N498E524NE	202			6	QT	FE	NA	NA	NA	C	P	VF	C	30	2	10YR54	10YR53	SM		N	NA	NA	0	0	Y
8	91.31	1	N500E512	1	1		4	GR	NA	FE	MI	QT	VF	C			20		0	10YR31	10YR53	SM		N	NA	NA	0	0	Y
8	91.314	1	N500E512	2	1		4	GR	NA	FE	MI	QT	VF	C			20		0	10YR21	10YR53	IN	SM	C	NA	NA	0	0	Y
8	91.321	1	N500E512	3	1		4	GR	NA	FE	MI	QT	VF	C			20		0	10YR41	10YR52	SM	WI	B	NA	NA	0	0	N
8	91.33	1	N500E512	2	3		4	GR	NA	FE	MI	QT	VF	C			20		0	10YR53	10YR53	IN	SM	R	ST	IN	7	0	Y
8	91.331	1	N500E512			22	3	GR	NA	FE	MI	QT	VF	C			20		0	10YR41	10YR53	SM		B	NA	NA	0	0	N
8	91.335	1	N500E512	2	3		3	GR	NA	FE	MI	QT	VF	C			20		0	10YR41	10YR53	SM		B	NA	NA	0	0	N
8	91.343	1	N500E512	2	3		4	GR	NA	FE	MI	QT	VF	C			20		0	10YR53	10YR53	SM	WI	N	NA	NA	0	0	N
8	91.347	1	N500E512	2	4		4	GR	NA	FE	MI	QT	VF	C			20		0	10YR31	10YR41	IN	SM	R	ST	IN	7	0	Y
8	91.551	1	N499.5E510	2	2		4	GR	NA	FE	MI	QT	VF	C			20		0	10YR53	10YR53	IN	SM	C	NA	NA	0	0	Y
8	91.674	1	N500E516	2	1		3	GR	NA	FE	MI	QT	VF	C			20		0	10YR53	10YR53	SM		B	NA	NA	0	0	N
8	91.68	2	N500E516	2	2		3	GR	NA	FE	MI	QT	VF	C			20		0	10YR53	10YR53	SM		B	NA	NA	0	0	N
8	91.683	1	N500E516	2	3		3	GR	NA	FE	MI	QT	VF	C			20		0	10YR53	10YR53	SM		B	NA	NA	0	0	N
8	91.771	1	N500E516	3			4	GR	NA	FE	MI	QT	VF	C			20		0	10YR53	10YR53	IN	SM	C	NA	NA	0	0	Y
8	93.1279	1	N496E528NE	202	2		2	GR	NA	FE	MI	QT	VF	C			20		0	10YR42	10YR53	SM		B	NA	NA	0	0	N
8	93.1361	1	N498E522SE	202			2	GR	NA	FE	MI	QT	VF	C			20		0	10YR42	10YR53	SM		B	NA	NA	0	0	N

Temper (TEMPER1, TEMPER2, GRITCOMP, GRITCOMP2, and GRITCOMP3):
FE = Feldspar   GR = Grit
MI = Muscovite   NA = Not available
QT = Quartz

Size (T1MIN, T1MAX, T2MIN, and T2MAX):
VF = Very fine (1/16–1/8 mm)
C = Coarse (1/2–1 mm)
P = Pebble (4–6 mm)

Surface Treatment (SFTREAT, SFTREAT2, and SFTREAT3):
NO = Notched   SC = Scraped
SM = Smooth   WI = Wiped
IN = Incised

Portion (PORTION):
B = Body   C = Collar
N = Neck   R = Rim

Rim Form (RIMFORM):
CL = Collared   EV = Everted
IN = Inverted   NA = Not available
ST = Straight

Lip Form (LIPFORM):
FL = Flat   IN = Inverted
NA = Not available

size, and density, exterior and interior color and surface treatment, and location of the sherd on the vessel (Table 6.2). Inclusions were identified using 10× magnification. Since it is extremely difficult to identify rock minerals in fired ceramic pastes, my inclusion designations were consistent, if not exact. Inclusion density was estimated using comparative charts (Terry and Chilingar 1955:229–234); be-

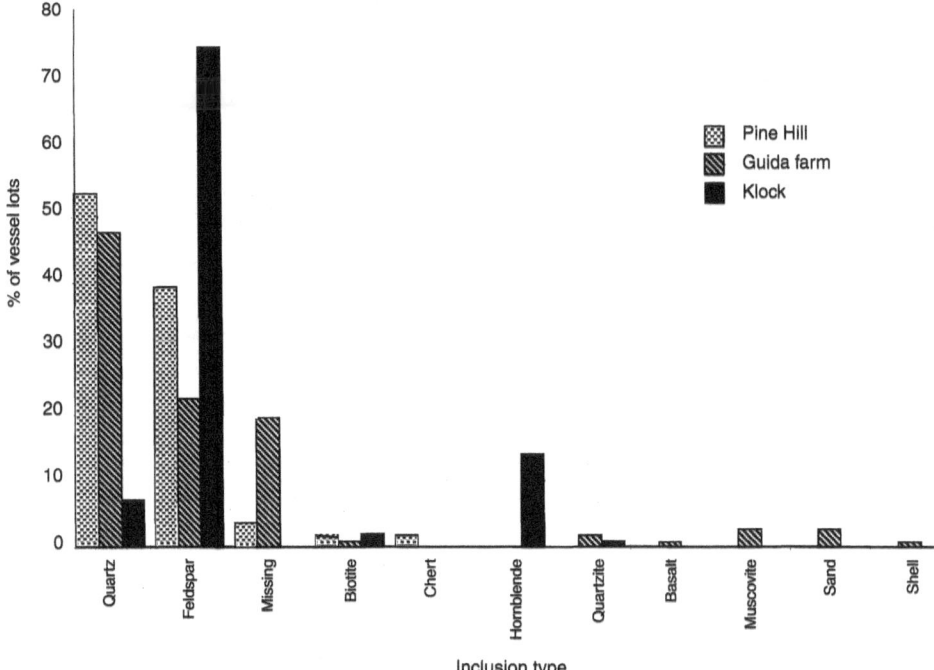

Figure 6.4. Primary inclusion type by site.

cause the amount of inclusion varies a great deal within ceramic pastes of handbuilt pots, estimates of inclusion density are sufficient. The final vessel lot determination is based on overall similarity in the attributes analyzed. As a result of this analysis, 56 vessel lots were identified from the Pine Hill assemblage, 108 from the Guida Farm assemblage, and 100 from the Klock site sample.

RESULTS

In this vessel lot analysis there are striking differences between the ceramics from the Algonquian sites and the ceramics from the Iroquoian Klock site. These differences fall into two broad categories: technical and decorative.

Technical Differences

In this section I will discuss only a few of the more important technical attributes. Other attributes analyzed included construction techniques, rim and lip form, collar size, and interior and exterior color (see Chilton 1996).

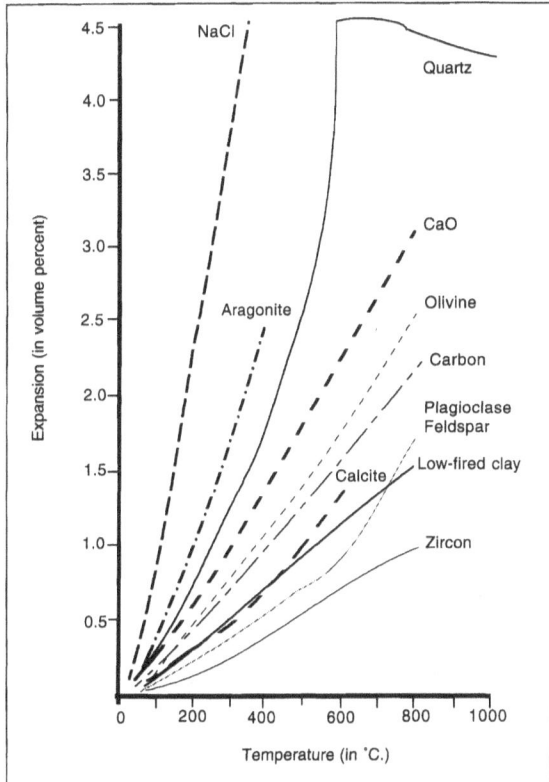

Figure 6.5. Thermal expansion curve (after Rice 1987, fig. 7.11; and Rye 1976, fig. 3).

PRIMARY INCLUSION TYPE   The primary inclusion type at Pine Hill and Guida was crushed quartz, followed by feldspar (Fig. 6.4). In contrast, the most common inclusion types at the Mohawk Valley Klock site were feldspar (mostly plagioclase) and hornblende, which are both present in the granitic and anorthositic rocks of the nearby Adirondack Mountains. It is important to note here that the optimal inclusion types for cooking vessels have thermal expansion coefficients similar to or less than that of clay (Rice 1987:229; Fig. 6.5), such as grog, calcite, crushed burned shell, feldspar, and hornblende. Quartz, on the other hand, is not an optimal inclusion type for cooking pots; it expands much more quickly than clay and can lead to crack initiation. Therefore, on the basis of inclusion materials used, the Connecticut Valley ceramics were not ideal cooking pots, on the whole. They may have been sporadically used for cooking, but they were not designed to cook maize over long, hot fires.

Not only do the assemblages from the Connecticut Valley have different kinds of inclusions, but they also show a higher diversity of inclusion materials used. I

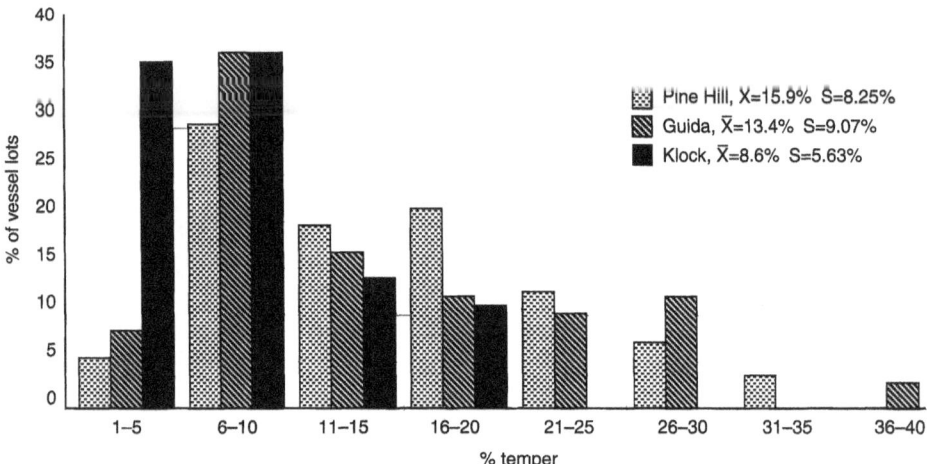

Figure 6.6. Temper density by site.

suggest that a higher diversity in inclusion type for the Connecticut Valley vessels is due to one or more of the following factors: (1) potters were living in a highly dispersed settlement pattern and were mobile throughout the year; therefore, they were using diverse sources of inclusions (see D. Arnold 1993:236); (2) potters did not select for specific kinds of inclusions (i.e., inclusion type was not considered to be an important attribute); (3) different inclusion types were used for vessels with different intended functions; and (4) few pots were made from the same batch of clay. Conversely, less diversity in inclusion type for Iroquois vessels may indicate that: (1) potters were relatively sedentary and, therefore, had access to a smaller range of inclusion types (or had consistent access to preferred materials); (2) potters selected for certain inclusion types because of their use properties (such as low thermal expansion); and (3) more pots were manufactured with the same batch of clay.

INCLUSION DENSITY   Inclusion density follows a similar, yet much more striking, pattern. The mean inclusion densities for vessel lots from Pine Hill and Guida are similar: 15.6 percent and 15.5 percent, respectively, with standard deviations of 7.7 percent and 8.9 percent (Fig. 6.6). The mean inclusion density for the Klock site is much lower at 8.6 percent—nearly half that of Pine Hill and Guida. The standard deviation for Klock is much lower at 5.1 percent, indicating less absolute variation from the mean.

A densely tempered paste is usually stronger. However, the more temper in a paste, the more potential problems as a result of thermal expansion—especially if the temper is quartz (see Braun 1983, 1987). Therefore, the Connecticut Valley ves-

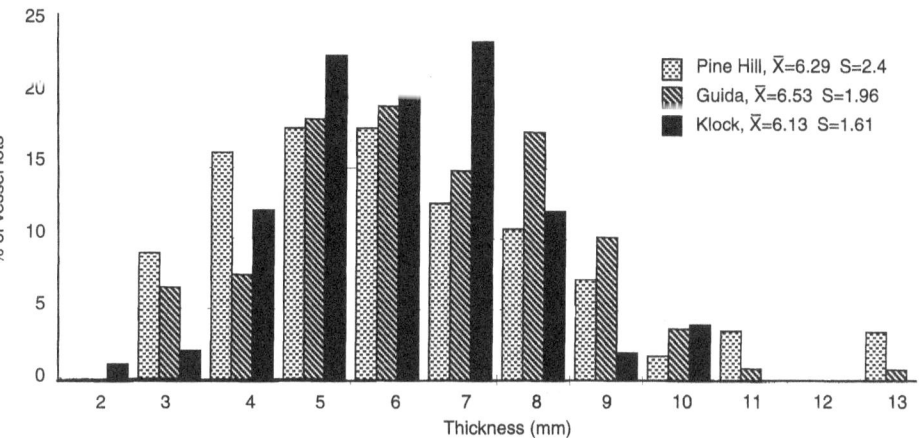

Figure 6.7. Vessel wall thickness.

sels would have been less resistant to thermal shock than the Klock site vessels, but more resistant to mechanic shock. Another advantage to dense inclusions, aside from increasing resistance to mechanical stress, is that it reduces drying time prior to firing; thus, ceramic could have been produced in a wider range of environments, even in colder seasons (D. Arnold 1985:97). One disadvantage to a densely tempered paste is that it may lose its plasticity and, therefore, its workability (Aronson et al. 1994).

WALL THICKNESS AND VESSEL SIZE   Means and standard deviations for vessel wall thickness are both slightly lower for the Klock site, as compared to the Connecticut Valley sites (the mean is 6.13 mm for Klock, and 6.29 mm and 6.53 mm for Pine Hill and Guida, respectively; Fig. 6.7), but the difference is not statistically significant. However, on the basis of body sherd curvature, the Klock vessels are, on average, 70 percent larger than those of Pine Hill and Guida (29 cm vs. 17 cm mean diameter; Fig. 6.8). Because larger vessels are expected to have relatively thicker walls in order to support the additional weight, the vessel wall thickness of the Klock assemblage *is* significantly thinner in proportion to vessel size. Morgan (1901:9) indicates that the Iroquois pots were "usually of sufficient capacity to contain from 2 to 6 quarts." Larger vessels may also reflect "communal dining" (Snow 1994:13) by the relatively large residential kin groups that existed among the Iroquois in the Late Woodland period (Tuck 1971; Whallon 1968).

Vessel wall thickness directly affects resistance to thermal shock: vessels with thinner walls are less apt to crack when used for cooking. Therefore, since the Klock pots had significantly thinner walls (relative to overall size), potters were ap-

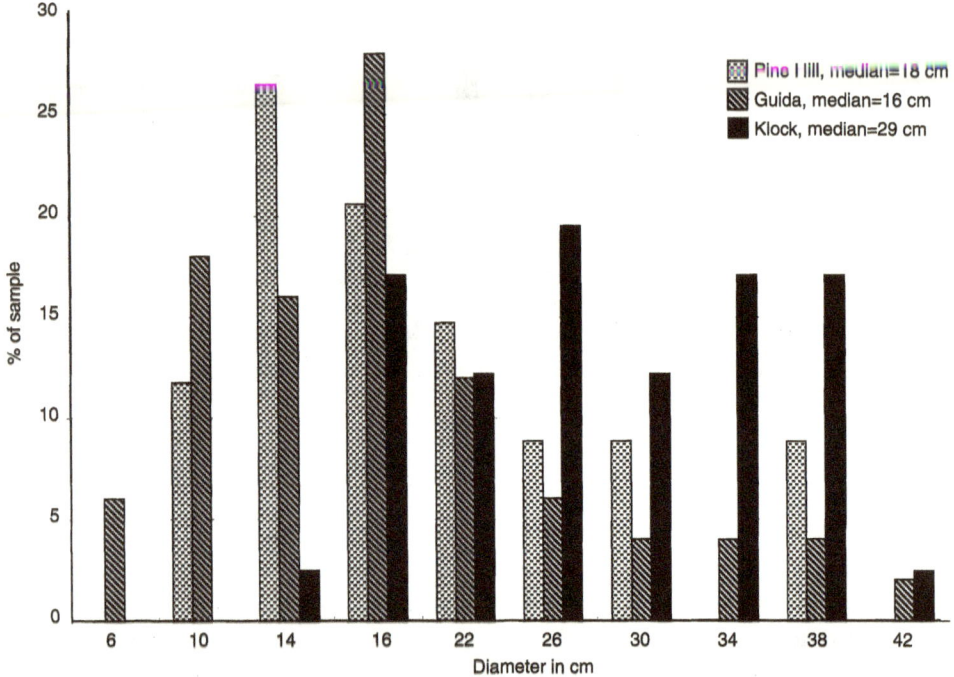

Figure 6.8. Body sherd curvature.

parently constructing pots with walls thin enough to withstand the thermal stresses of cooking maize over long, hot fires. Since susceptibility to thermal shock increases with vessel size (see Kingery 1955 and Searle and Grimshaw 1959), other attributes were used to compensate, such as wall thickness and inclusion kind, size, and amount.

The Connecticut Valley vessels, on the other hand, had thicker vessel walls relative to overall size. Pots with thick walls are less fragile (more resistant to mechanical shock) but are more likely to crack when exposed to heat. Therefore, I suggest that the Connecticut Valley potters were producing vessels that were intended to withstand mechanical stresses, which would be important for nonsedentary groups (cf. D. Arnold 1985:110). The overall small size of the vessels would have made them easier to transport; many of these small vessels could easily have served as portable containers for water or food.

Decorative Differences

Surface treatment consists of impressions, or evidence of scraping or smoothing on the exterior and interior surfaces of the fired clay. Examples of surface treat-

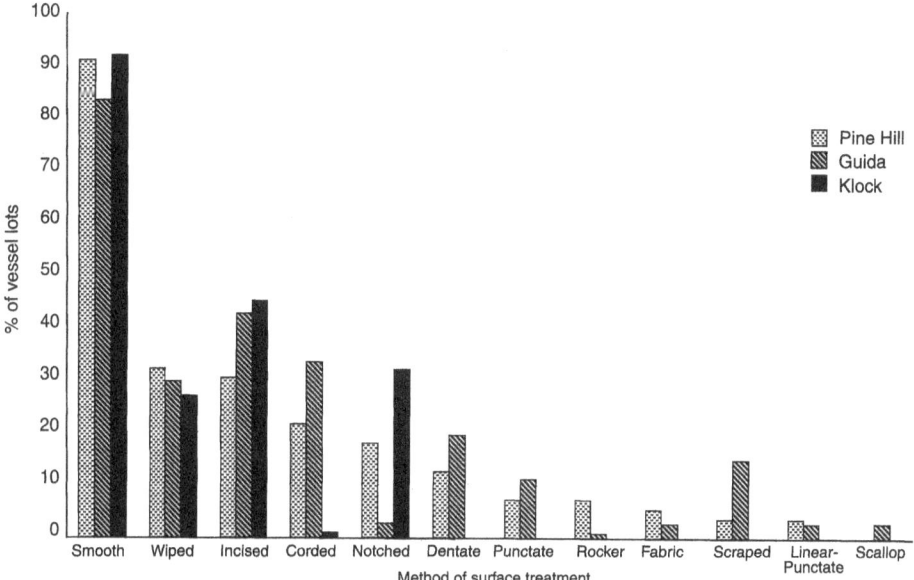

Figure 6.9. Surface treatment by site.

ment include smoothing, wiping, incising, and cord-marking (Table 6.2). Surface treatment can be an artifact of manufacturing technique, or purposeful "finishing" for either technical or decorative purposes.

There are striking differences in surface treatment between the sites analyzed. Surface treatments from the Klock site are exclusively smoothed, wiped, incised, and notched (Figs. 6.9, 6.10); body sherds are mostly smooth and wiped, and collared rims are exclusively incised and notched. The Pine Hill and Guida assemblages show much more diversity in terms of surface treatment. Surface treatments for these sites include: cord-marked, dentate-stamped, punctated, rocker-stamped, fabric-impressed, scraped, linear punctated, scallop shell-impressed, as well as smoothed, wiped, incised, and notched (Figs. 6.9, 6.11, and 6.12).

On the basis of conventional assumptions that dentate-stamping and fabric-impressing date to the Middle Woodland period (A.D. 0–1000) and therefore predate incising, the assemblages from Pine Hill and Guida might appear to have Middle Woodland components. However, there is evidence that dentate stamping may also be a Late Woodland and even a Contact period trait, at least in New England. For example, at Pine Hill one sherd exhibits both incising and dentate stamping (vessel lot 33), and two vessel lots from Guida have dentate-stamped collars (vessel lots 7 and 29); collars are thought to be an exclusively Late Woodland trait. It is clear from these examples that in the Connecticut Valley unilinear evolutionary changes in surface treatment cannot be assumed.

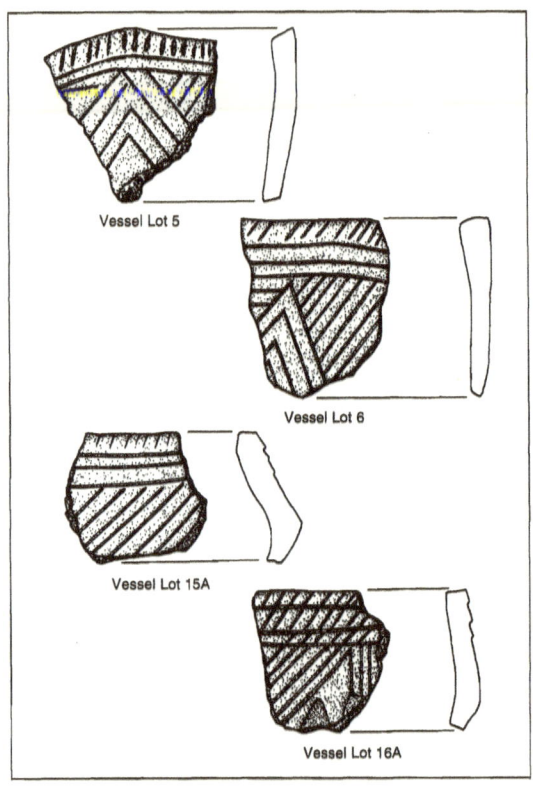

Figure 6.10. Vessel lots from the Klock site. (Illustration by Maureen Manning-Bernatsky.)

Surface treatments can profoundly affect vessel performance (see Schiffer et al. 1994). For example, a roughened surface, such as those produced by cord-marking, fabric-impressing, or scraping, can increase thermal shock resistance and reduce thermal spalling (Schiffer et al. 1994). A rough surface also provides a more secure grip (Rice 1987:232), and may increase heat absorption and the evaporation of liquids (Charlton 1969; Herron 1986). Therefore, surface treatments are not chosen automatically; a potter makes choices and compromises along numerous axes according to various personal, social, and technological criteria. It is possible, therefore, that the Connecticut Valley pots were more often given roughened surfaces to compensate for the fact that other attributes, such as vessel wall thickness, may have reduced the ability of vessels to transfer heat. Therefore, for the occasions when pots *were* used for cooking, a corrugated surface would have been advantageous. As previously mentioned, a roughened surface would also make the vessel less slippery, and, therefore, easier to transport.

How else can we account for the greater diversity in surface treatment at the Connecticut Valley sites? Since Algonquian groups were quite mobile, and had rel-

Algonquian and Iroquoian Ceramic Traditions in the Northeast 155

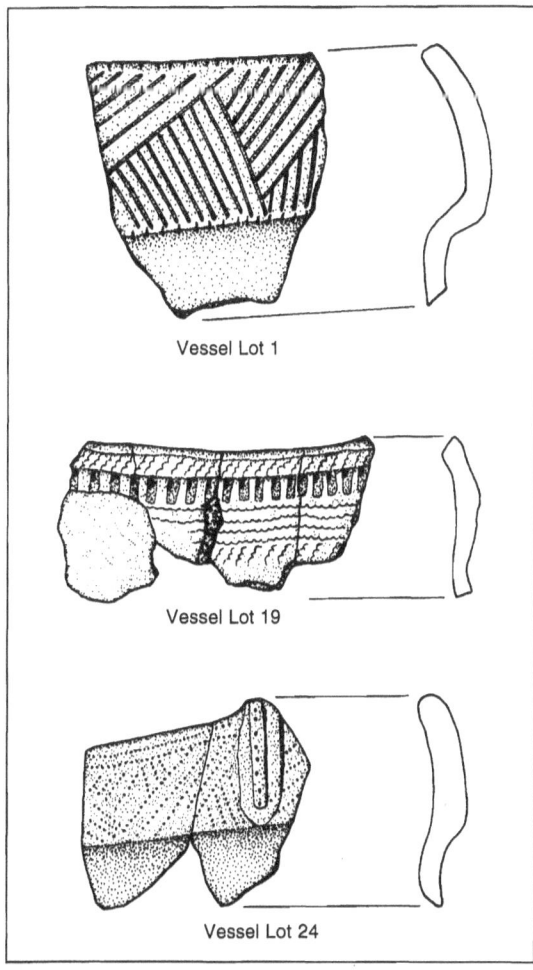

Figure 6.11. Vessel lots from the Guida Farm site. (Illustration by Maureen Manning-Dernatsky.)

atively fluid social boundaries, the social and environmental contexts of ceramic manufacture and use would have been quite variable. It should come as no surprise, then, that the Late Woodland ceramics of the Connecticut Valley show great variability in surface treatments, as well as other technical choices. For the Iroquois, who were living in more permanent, structured communities, pots would have been made in similar social and ecological contexts each time, under more predictable circumstances; there was both stability and continuity in ceramic traditions. Accordingly, in the present study the Iroquois ceramics show much more internal homogeneity in terms of decorative and technological attributes. The ceramics from the Klock site have a limited range of decorative types and motifs. All of the rimsherds from the site fit neatly into the established stylistic typologies for

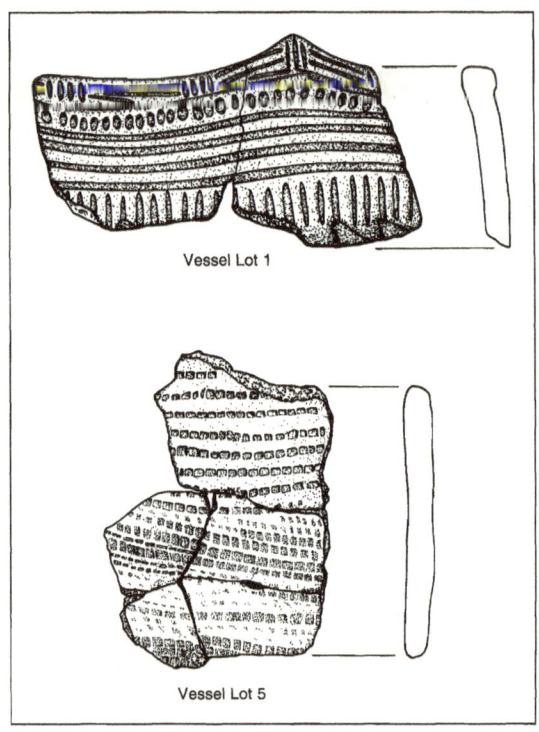

Figure 6.12. Vessel lots from the Pine Hill site. (Illustration by Maureen Manning-Bernatsky)

Iroquois ceramics: 84 percent of the collared rim sherds from the site have been typed by Kuhn and Funk (1994) as "Garoga Incised," and 7 percent have been typed as Wagoner Incised. The remaining 9 percent were typed as Chance Incised, Deowongo Incised, Martin Horizontal, Cromwell Incised, and Thurston Horizontal (Kuhn and Funk 1994:9).

There is certainly merit in applying the information exchange model in this case, in terms of technological style *and* decorative motifs. First, as Wobst (1977: 323) points out, "only simple invariate and recurrent messages will normally be transmitted stylistically." Mohawk ceramic designs *are* simple, fairly invariant, and recurrent. They are also restricted in time and space, which is why they "work" so well as chronological and geographical markers (Kuhn and Bamann 1987:41). With their fairly rigid social organization, Iroquois pottery iconography may have been used to signal group identity.

In contrast, very few vessels from Pine Hill and Guida fit neatly into the established typologies. While many of the vessel lots from Guida resemble traditions defined for southern New England, particularly the Windsor and East River traditions, there is so much overlap in the New England traditions as they are defined (Table 6.3), that much of the collection cannot be assigned to known types. The

Algonquian pattern (or lack thereof) may indicate individual expression, fluid social boundaries, and a flexible social organization.

Production Scale and Diversity

Algonquian and Iroquoian ceramics apparently differ not only in their intended uses, social milieu and production contexts, but also in production scale. More specifically, Algonquian ceramic production is most aptly described as "household production" (van der Leeuw 1984:722). Household production is carried out exclusively by members of the household for household consumption (D. Arnold 1985:225). Household potters need to devote little time to ceramic manufacture (P. Arnold 1991:92). For New England Algonquians, variability may have been maintained within household production because of: (1) the infrequency of the activity; (2) the low number of producers involved; and (3) the lack of controls over access to resources and information (Rice 1991:273).

Iroquois ceramic production, on the other hand, is closer to what is termed "household industry" (van der Leeuw 1984:722; *contra* Allen 1992). Household industry is characterized by part-time production for group use. In household industry more people are involved in the production sequence (D. Arnold 1985); for the Iroquois, production would have been carried out by groups of related women (see Engelbrecht 1978:141). The major differences between household production and household industry are that in household industry: (1) production is conducted more frequently; (2) there in an increase in the amount of production, and, therefore; (3) there is an additional investment in labor (P. Arnold 1991:92). Production becomes more regularized, and, archaeologically, increased output leads to greater ceramic densities (P. Arnold 1991:93). Household industry is often correlated with population increase, when there may be may an increase in demand for ceramics (D. Arnold 1985:226). The archaeological evidence for the Iroquois supports a population increase (or, at least, clustering), increased pottery production (see Tuck 1978), and a certain amount of pottery specialization during the Late Woodland period. This "household industry" does not refer to production for market. Iroquois society is characterized by a high degree of communal living. Therefore, items such as pots were produced and shared within and between lineages and clans. It is likely that a certain amount of standardization was achieved simply through repetition and routinization of the production of large amounts of pottery (see Rice 1991:268).

Certainly, as Allen (1992:154) points out, there is "no evidence to support the presence of a higher level of organization for ceramic production [among the Iroquois] than the household level." Indeed, I suggest that production was happening at the household level, but it was happening on a larger scale, and with more

TABLE 6.3
Common Ceramic Types for Southern New England and New York, A.D. 1000–1700

Traditions	Windsor	East River	Guida	Mohawk
Common Ceramic A.D. 1100–1400	Sebonac (A.D. 1100–1400): simple conical vessels, shell temper, interior and exterior surface brushing, stamping and dragging decorations (Smith 1947).	Bowmans Brook (A.D. 1100–1200): smooth interior, cord-marked exterior; rims decorated with horizontal rows of cord-wrapped stick impressions; incising not common (Smith 1947).		
A.D. 1400–1700	Niantic (A.D. 1400–1700): thin walled, shell-tempered, globular pots, with constricted necks, collars and complex stamped scallop shell designs; sometimes smooth interior surface (Lavin and Kra 1994; McBride 1984:127; Rouse 1947). Niantic Incised: smooth, constricted necks, collared rims, incised decoration (Keener 1965). Lavin (1988:15) suggests that these are "strongly reminiscent of Hudson and Mohawk Valley types." Other names: Niantic Stamped, Niantic Linear Dentate, Niantic Punctate, etc.	Clasons Point (A.D. 1200–1700): collared rims, globular bodies, stamping decreases as incising increases; plain surface treatment common; grit temper frequent, but shell tempering increases throughout; at the end of the period vessels "approximate forms of the eastern Iroquois" (Smith 1947).	Guida Incised (A.D. 1500–1700): gray to black; fine, micaceous temper; surface "flake" or "silky." Designs of finely incised lines, very close together on narrow collars; Byers and Rouse (1960:17) suggest that the motifs are "Iroquoian." Other Guida types: Guida Cord Marked, Guida Fabric Marked, Guida Plain, Guida Stamped, and Guida "Miscellaneous" (Byers and Rouse 1960).	Chance Phase (A.D. 1400–1500): incising replaces cord-impressing for collar and neck designs; fine grit temper, semiglobular shape; check-stamped and cord-marked replaced by smooth surfaces. Gavoga Phase (A.D. 1500–1700): large, globular vessels, incised linear patterns on castellated collars, notches at base of collar (MacNeish 1952; Ritchie 1980).

stability and continuity, than in New England communities. In this sense, I view ceramic specialization and production not as a series of stepped categories, but as a continuum of choice.

CONCLUSIONS: THE ROLE OF CHOICE

In this analysis I have tried to underscore both social and technological-mechanical factors that are likely to have influenced the paste composition and vessel morphology of the ceramic assemblages discussed. If Algonquian and Iroquoian people were interacting and sharing information, then the Connecticut Valley Algonquians had access to the knowledge and technology necessary to: (1) become sedentary farmers; and (2) make large, thin-walled, globular, smooth-bodied pots. However, they did not do either of these things. I argue that they were capable of implementing theses changes, had they chosen to do so. Instead of assuming that for some strictly ecological or evolutionary reason they had not reached the same evolutionary stage as the Iroquois, I suggest that we view them as active agents of their own social change. As such, they made choices concerning subsistence, settlement and social structure. In terms of house form (Prezzano 1992), settlement patterns, subsistence, politics, and ceramics, they made choices within the "spectrum of . . . equally viable options" (Sackett 1990:33). As potters, they made choices in ceramic production that both reflected and affected these decisions (see Miller 1985:205 and Wobst 1978:307).

As demonstrated in this study, variation in ceramics occurs at two levels. First, variation exists in what Sackett calls *instrumental form* (Sackett 1990:33) which, in this case, includes choice of temper, temper density, vessel wall thickness, and vessel size. This type of style is "built in, rather than added on, to the pots" (Sackett 1990:33). The second kind of variation, *adjunct form*, refers to variation that is added on to the pot, such as decoration or surface treatment (Sackett 1990:33). It is the analysis of both aspects of variation at the same time—the "functional" and the "aesthetic"—that provides the view of material culture as part of the wider system of meaning (Lemonnier 1992:99).

I would like to end with a word of caution: Algonquian and Iroquoian ceramic traditions are only one component of a much broader technical system. As discussed previously, there is archaeological and ethnohistoric evidence for interaction between Algonquian communities in New England and the Iroquois; therefore, the two groups had access to similar kinds of technological information.

It is also important to acknowledge that we cannot assume the same "meaning" of technical choices in different societies. For example, some societies may be more prone to "technological routine" than others (Lemonnier 1992:22). It is, therefore, possible that a certain *Algonquian conservatism* played a role in the choices made—or *not* made, in the case of the routinization of ceramic production, sedentism, and social hierarchy. In this paper I have focused on "logical" reasons for choices in ceramic production. Nevertheless, I would like to underscore the fact that there is a certain amount of arbitrariness involved in both technical and nontechnical or adjunct choices (see Lemonnier 1993a:16). There may not be "logical" or obvious social or technical advantages to specific choices, at least not from within a strictly ecological or evolutionary framework. Therefore, archaeologists must also turn to historical explanations of meaning in material culture in order to understand the complexity of technical choice.

ACKNOWLEDGMENTS

I am tremendously grateful to Miriam Stark, who organized the symposium from which this volume stemmed and whose comments and suggestions had a profound effect on this paper and my views on material culture in general. I would also like to thank the members of my dissertation committee, Dena F. Dincauze, Arthur S. Keene, H. Martin Wobst, and Kevin Sweeney, for their thoughtful comments on my writing and research. This research was aided by a Grant-in-Aid of Research from Sigma Xi, and a Graduate Research Award from the Department of Anthropology, University of Massachusetts, Amherst. Pottery sherds were drawn by Maureen Manning-Bernatsky. Finally, I wish to thank David K. Schafer for his midnight editing, computer wizardry, and undying support. Any flaws in this work are solely my responsibility.

7

Technological Patterning and Social Boundaries: Ceramic Variability in Southern New England, A.D. 1000–1675

ROBERT G. GOODBY

Social boundaries are abstractions and ideological constructs, recognized differently and for different reasons by people on the basis of their perceived identity, interests, and social context. The importance of this point in the search for social boundaries in material culture has been underappreciated. The technological and decorative styles of seventeenth-century Native American ceramics from southeastern New England reflect the extent to which social boundaries and social identity were contested and transformed in response to English colonialism and growing internal tensions within and between native societies, and illustrate the complexity of the relationship between social boundaries and technological patterning.

Archaeologists have long used the study of style to identify social or cultural boundaries and discrete social or ethnic groups in the past. Recently, the concept of style itself has become the topic of debate, with the recognition that style is consciously manipulated by people and not a passive reflection of social identity, that some style served as a medium of communication (Wobst 1977), and that "style" resides in many levels of material culture, not just in form and decoration. Recognizing the problematic relationship between "expressive" style and social groups, archaeologists have recently begun to search for less consciously employed styles, styles of technical choice (e.g., Chilton 1996) or isochrestic variation (Sackett 1990) that are held to reflect unconscious norms held by members of particular societies. Since technological style or patterns of isochrestic variation are (presumably) not manipulated for individual ends but reflect the learned and ingrained behaviors behind basic technological behavior, it is hoped they will be more appropriate for detecting social boundaries in the archaeological record.

Ironically, in their attempt to study these relationships in ethnographic contexts so that prehistoric archaeological contexts could be better interpreted, ethnoarchaeologists have developed explanatory models that emphasize how highly contextualized the relationship between style and social boundaries really is (e.g., Hodder 1979a; 1982; Lechtman 1977; Lemonnier 1986; Wiessner 1983), and how much of what shapes this relationship is in the realm of the nonmaterial and archaeologically undetectable. Convincing models of this relationship have been developed for particular ethnographic contexts, but their applicability to prehistoric contexts, ethnohistoric contexts, or other ethnographic contexts is always in question. This paper assumes that the relationship between style and social boundaries is problematic and embedded in the dynamics of particular historical and social contexts. Technological patterning is seen as one of a number of types of variability that may inform us about social boundaries, but only if such patterning is studied in relation to other lines of evidence. Technological patterning alone may tell us little about how social boundaries functioned, how they were perceived, and how they shaped individual behavior; nor can we assume such patterning will even reveal the existence of social boundaries in all cases.

The very concept of social boundaries has come under increasing scrutiny, especially when it is applied to nonstratified societies. Bounded social or "ethnic" groups recorded by Western ethnographers were often brought about by one or another form of colonial intervention (see MacEachern, this volume), and the "tribe" itself as a form of social organization is now seen as a common indigenous response to colonialism (Fried 1975; Wolf 1982). While historically or ethnographically known social boundaries may have lacked precolonial antecedents, or been highly fluid, this does not diminish the significance of the question at hand. Archaeologists need to be able to discover the material culture correlates of social boundaries when such exist, and to make convincing cases against their existence for times and places where they were not developed. There seems, however, to be no single methodological or theoretical approach to this question that will be suitable for all contexts.

As anthropologists have come to appreciate the effect of colonialism on the formation of social boundaries among many indigenous peoples, it would seem necessary to situate an interpretation of style in seventeenth-century ceramics in its own historical context, drawing on archaeological and historical data to assess the ways material culture reflects the development and transformation of social boundaries. Interpretation of ceramic variability and its relationship to local communities, social boundaries, tribal entities, and the like are problematic even in well-documented ethnographic contexts (MacEachern, this volume). Ethnoarchaeologists have demonstrated the complexity of these relationships through the study of traditional ceramic-producing communities in many parts of the world (e.g., Longacre 1981, 1991b), and have identified aspects of ceramic variability whose spatial distribution corresponds to boundaries of various kinds of "groups"—but what these attributes are, and what sort of "group" they correspond to, vary from one context to the next. In a situation where ethnoarchaeological research is not possible (which includes, of course, the entire "prehistoric" and early historic period), contextual information, while equally important in shaping material culture variability, is considerably more elusive. In such circumstances, context must be built by using a broad spectrum of archaeological and ethnohistoric data. An effort at contextual interpretation is presented here, but one that has as its goal a more accurate approximation of the nature of social boundaries, and the meaning of archaeological data, than has been offered by previous studies in this geographic area.

SOCIAL BOUNDARIES IN SOUTHEASTERN NEW ENGLAND

In the seventeenth century, European observers consistently described three main tribes or "nations" in southeastern New England: the Pequot-Mohegan,[1] Narragansett, and Wampanoag (Gookin 1806:147-148; Morton 1969; Williams 1973). The territories of these tribes were clearly demarcated in European accounts, with the boundaries of the Pequot extending east to the modern border of Connecticut and Rhode Island, and the boundary between the Narragansett and Wampanoag located in the vicinity of the Seekonk River (Fig. 7.1).

The Pequot-Mohegan, Narragansett, and Wampanoag shared a similar culture, spoke closely related dialects of Algonquian (Goddard 1978; Williams 1973:174), followed a common economic strategy in which horticulture supplemented broad-based hunting, fishing, and gathering, held similar religious beliefs, and shared a common political system with leaders (known as sachems) whose powers were sharply limited and contingent on achieved status (Marten 1970; Salwen 1978; Simmons 1978; Starna 1990). Patterns of intermarriage reinforced these similarities in culture; vestiges of this are seen in the seventeenth-century intermarriage of Pequot and Narragansett, even as relations between the two groups had begun to markedly deteriorate (Goodby 1994:108–111).

As in other colonial contexts, English colonists preferred to deal with centralized political entities at the same time they sought to encourage divisions among these entities. By crediting individual sachems with the authority to direct and speak for all "their" people, the English could enter into negotiations with whole "nations," and pit one nation against another as they sought to prevent pan-tribal

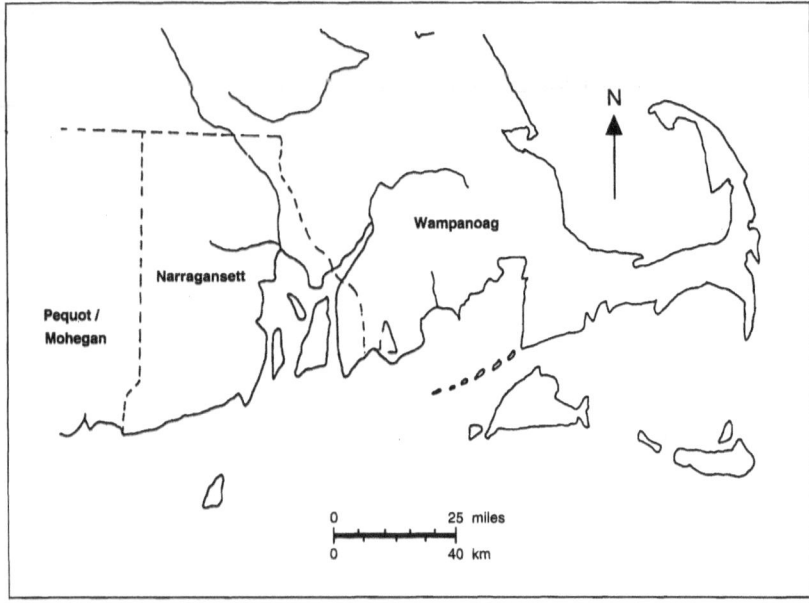

Figure 7.1. Tribal territories in southeastern New England.

alliances from forming. In practice, the tribe was first and foremost a political alliance among the sachems of various local communities. Relations among the sachems were often problematic and unequal, and the exact composition and overall cohesion of tribes varied considerably. The tribe as such was a continually contested entity, debated internally (Robinson 1990) and even in war rarely functioned as a unified whole (LaFantasie 1988:126).

The historical record suggests the process of boundary formation among the three tribes began by the earliest years of the contact period (Salisbury 1990:84). The Wampanoag sachem Massasoit claimed to have fought against the Narragansett with his father prior to the establishment of Plymouth Colony in 1620 (LaFantasie 1988:621), and historic sources emphasize a long-standing enmity between the two groups (Gookin 1806:148; Heath 1963:58). The existence of a distinct, albeit shifting boundary between the Narragansett and Wampanoag is seen in accounts of the Narragansetts extending their frontier to the east following the devastation of the Wampanoag by an epidemic in the years 1616–1617 (LaFantasie 1988:178, 451). As a result, Massasoit was reportedly obliged to cede land and "subject" his people to the Narragansett sachems (LaFantasie 1988:620–621). The early formation of an alliance with Plymouth Colony was an attempt by Massasoit to escape the domination of the Narragansett sachems, and marked the beginnings of the English "divide and conquer" policy, to be pursued with considerable success until the onset of King Philip's War in 1675.

Despite this long-standing conflict, the Narragansett were reported to have affinal ties to the Wampanoag, and are known to have sheltered Wampanoag refugees during King Philip's War, in which many Narragansett joined the Wampanoag and others in a generalized uprising against the English colonists (Chapin 1931:78; LaFantasie 1988:694). There was also regular interaction between peoples of the two tribes (Wood 1968:71), although the visits were sometimes marked by intimidation of the Wampanoag by visiting Narragansetts (Morton 1969:44–45). The boundary between the Narragansett and Wampanoag was the subject of contention throughout the seventeenth century, as rival sachems defended conflicting land claims before the increasingly powerful English colonial authorities (LaFantasie 1988:610).

Relations between the Pequot-Mohegan and the Narragansett followed a similar, if more violent outline. The conflict between the Narragansett and their western neighbors continued seemingly without interruption from the 1630s to the 1670s. This tension was encouraged at various times by the Massachusetts Bay Colony, the United Colonies, and Rhode Island's Roger Williams (LaFantasie 1988:79). Tensions peaked in the 1630s with the Pequot War, in which a large force of Narragansetts joined with the English and the Mohegan to attack the Pequot fort at Mystic, Connecticut, resulting in the destruction of the Pequot as a regional power. Almost immediately after the war, conflict erupted between the Narragansett sachems and the Mohegan sachem Uncas (LaFantasie 1988:126; Salisbury 1990:87) and continued until King Philip's War. Narragansett-Mohegan tension became particularly pronounced after the murder of the Narragansett sachem Miantonomi by the Mohegan in 1643, an act encouraged by the English authorities at Plymouth (Bradford 1968:365), which angered the Narragansett for many years thereafter (LaFantasie 1988:116).

Despite this extended period of conflict and tension, it is also clear that strong social ties connected the Narragansett and the Pequot-Mohegan. Following the Pequot War, at least some Pequot refugees sought, and found, refuge with the Narragansett. Genealogical information suggests the families of the Eastern Niantic (an important constituent group of the Narragansett) and Pequot-Mohegan sachems had intermarried on a regular basis (LaFantasie 1988:613), and the Mohegan sachem Uncas claimed descent from Narragansett sachems (Simmons and Aubin 1975). Roger Williams mentions being visited in Rhode Island by a Mohegan man named Wequamugs, whose father was a Narragansett and mother was a Mohegan, who was consequently "freely entertained by both" (LaFantasie 1988:164). The traditionally close relationship between these two peoples is also suggested by the existence of Pequot "defectors" who aided the Narragansett in the Pequot War (LaFantasie 1988:77, 95).

In sum, the relationship between the Narragansett and the Pequot-Mohegan was characterized by ambivalence and conflicting loyalties. While strongly similar in culture and language, and connected through ties of kinship, events of the seventeenth century drove a wedge between political leaders, and English intervention helped to maintain, and at times exacerbate, political divisions. The ambivalence of this relationship is suggested by the Narragansett reaction to the English slaughter of hundreds of Pequot in the Mystic attack. The Narragansett were appalled by the slaughter of women and children (LaFantasie 1988:73, 85), an understandable reaction particularly if some of them were considered kin. Immediately after witnessing the slaughter, the Narragansett deserted their English allies and returned home, urged the English to treat their Pequot captives kindly, and proceeded to take in Pequot refugees (LaFantasie 1988:86). The Narragansett sachem Miantonomi particularly sought custody of the daughter of a Pequot sachem he claimed was a close friend of his (LaFantasie 1988:113). The wars between the Pequot-Mohegan and Narragansett were in many respects civil wars, with all the conflicting loyalties such wars produce. While a distinct boundary emerged between the two, it did not negate the significance of cross-boundary ties.

In sum, the events of the seventeenth century, driven by a colonial policy of fomenting political divisions, resulted in the emergence of sharply contested social and political boundaries among native peoples in southeastern New England. These boundaries were transformed both by English colonial policy and the increasing ambitions and rivalries of male sachems into dangerous frontiers, the crossing of which often led to violence, intimidation, and murder. These boundaries were consistently recognized by Europeans and Native Americans alike, and were the ongoing subject of negotiation, threats, invasion, and war between 1630 and 1675 (Bradford 1968; Gookin 1806; Heath 1963; Morton 1969; Wood 1968). It remains to be seen, however, how these boundaries were reflected in the material culture of native people.

TECHNOLOGICAL STYLES IN SOUTHEASTERN NEW ENGLAND CERAMICS

Until recently, the study of prehistoric ceramics in southern New England has been dominated by a culture-historical framework developed in the 1940s (Lavin 1980, 1986, 1987; McBride 1984; Rouse 1947; Smith 1950). This framework has focused more attention on classification and chronology than on spatial variability and the search for social boundaries. A central culture-historical construct, the

Windsor Tradition, was a ceramic tradition extending over a broad but poorly defined area encompassing much of southern and central New England, dating from 3,000 B.P. to the time of European contact and generally corresponding to the distribution of historically known Algonquian-speaking peoples in southern New England (Lavin 1987; Rouse 1947; Smith 1950). Efforts to describe the general attributes of Windsor ceramics reveal their highly variable nature:

> The pottery is either grit or shell tempered. They vary from tan to brown in color. . . . both conical and round bases are characteristic. . . . the bodies are either straight-sided or provided with shoulders. The shoulders . . . either end at the rim or are surmounted by necks. . . . some vessels apparently had a collar . . . while others did not. . . . Designs are present only on a portion of the sherds, many vessels apparently having lacked them. (Rouse 1947:11–12)

The Windsor Tradition was held, in various formulations, to be distinct from other ceramic traditions in the Northeast, although the possibility that this reflected significant cultural or social boundaries has not been addressed until recently (Lavin et al. 1993).

During the Late Woodland period (ca. A.D. 1000–1500), Windsor Tradition assemblages are highly variable, and include the "types" Windsor Brushed, Shantok Cove Incised, Sebonac Stamped, and shell-stamped Niantic phase vessels (Lavin 1980, 1986, 1987, 1988; McBride 1984; Rouse 1947; Salwen and Ottesen 1972). In a recent analysis of Late Woodland vessel lots from both Narragansett and Wampanoag territory numerous vessels consistent with Windsor "types" were found, and a comparable degree of intra-assemblage diversity was noted (Goodby 1994:101–102). Technological attributes exhibited greater variability than their seventeenth-century counterparts, with more diversity in temper type, percentage of temper relative to paste, temper sorting, surface treatment, and prevalence of fractures along coil lines (Table 7.1). There is considerable variability in the "quality" of Late Woodland vessels, with some vessels manufactured with coarse, unsorted temper and exhibiting a marked tendency to fracture along coil junctures. This variation in quality may reflect the "expedient" nature of some ceramic vessels, significant variation in the skill or experience of individual potters, or both. Decoration of these vessels was relatively simple, and typically limited to bands of impressions created with the edge of a marine shell, combed or brushed lines created by dragging a toothed implement across the surface of the vessel, or simple incising (Table 7.2, Fig. 7.2). Effigies were not placed on Late Woodland vessels, and the relatively few collared vessels of this period lacked prominent castellations. Neither Windsor Tradition ceramics in general, nor the distribution of particular

TABLE 7.1
Technological Attributes of Narragansett and Wampanoag Vessels

Attribute	Attribute Categories	Narragansett (N=24) (percent)	Wampanoag (N=19) (percent)
Temper type	Shell	92	94
	Grit	8	6
	Other	0	0
Temper size	Fine	63	63
	Fine-Medium	25	19
	Medium	4	6
	Medium-Coarse	4	0
	Coarse	4	12
Temper density	≤10%	21	25
	10–20%	67	67
	≥20%	12	8
Temper sorting	Well sorted	67	83
	Poorly sorted	29	17
	Unsorted	4	0
Surface treatment: collar	Smoothed	95	100
	Burnished	5	0
Surface treatment: neck	Smoothed	91	100
	Textile impressed/ smoothed	9	0
Surface treatment: exterior body	Smoothed	80	92
	Textile impressed/ smoothed	13	8
	Burnished	7	0
Surface treatment: interior	Smoothed	92	94
	Channeled/smoothed	4	0
	Channeled	4	6
Average rim thickness (mm)		6.67	6.21
Average body thickness (mm)		4.99	5.57

TABLE 7.2
Decorative Attributes of Narragansett and Wampanoag Vessels

Attribute	Attribute Categories	Narragansett (N=24) (percent)	Wampanoag (N=19) (percent)
Rim form	Collared	68	0
	Straight	21	100
	Everted	11	0
Lip form	Square	50	47
	Round	45	53
	Pointed	5	0
Lip decoration	Undecorated	47	53
	Incised	37	42
	Shell stamped	11	0
	Notched	5	5
Collar decoration[1]	Incised	80	79
	Dentate stamped	5	0
	Punctation	35	16
	Notching	5	21
	Cord impressed	10	0
	Undecorated	5	0
Effigies (no. of examples)	"Corn-ear"	4	6
	"Phallus"	1	5
	Human face	0	1

[1] Totals exceed 100% as more than one decorative technique is frequently employed on a single vessel.

types within this tradition correspond to historically known tribal boundaries (Chilton 1994, 1996; Goodby 1992, 1994). Instead, they are distributed across the territories of many distinct tribal groups known from the seventeenth century.

The traditional typological approach, epitomized by the Windsor Tradition framework, has obscured the real nature of ceramic variation in this area, in which a high degree of decorative and technological variability is present in ceramic assemblages. The boundaries of this variability (for example, limits on temper size, material used for temper, possible forms of surface treatment, coiling as a method of vessel construction) reflect a single, coherent tradition or technological style of ceramic manufacture common to the native peoples of this area. This pattern is consistent with a broader cultural unity and shared history, facilitated by centuries

Figure 7.2. Vessel from the Titicut site, Bridgewater, Massachusetts. (Courtesy of the R. S. Peabody Museum of Archaeology, Phillips Andover Academy.)

of intermarriage, trade, and other forms of reciprocal relationships. Recent studies of ceramic variability in New England suggest that sharp spatial disjunctions are less characteristic than clinal variation (Kenyon 1983; Luedtke 1986). The lack of sharp boundaries in Late Woodland ceramics reflects the historically close links and lack of well-defined social boundaries among the peoples of southeastern New England. It is in this historical context that the appearance of distinct social boundaries in the seventeenth century and their affect on the choices made by native potters must be addressed.

CERAMICS AND SOCIAL BOUNDARIES: THE SEVENTEENTH CENTURY

Ceramic vessels were not produced in great numbers by native women in the seventeenth century. This is reflected in the small numbers of these vessels recovered from seventeenth-century village and cemetery sites. Portions of 107 vessels were recovered from Fort Shantok, the main Mohegan village, only 35 of which were complete enough to determine vessel morphology and overall decorative motif (Williams 1972:135). Only 13 vessels recovered from the contemporaneous Fort Corchaug site on Long Island were complete enough to make these determinations (Williams 1972). Ceramic vessels also appear in low frequency in seventeenth-century cemetery burial sites, being present in no more than 10 percent of graves from the Wampanoag cemeteries at Titicut and Burr's Hill and the Narragansett cemeteries RI-1000 and West Ferry (Goodby 1994; Mrozowski 1980; Robinson et al. 1985; Simmons 1970). In part their scarcity may be attributable to a limited amount of excavation and the relatively short time involved; it almost certainly is related to the widespread adoption of European brass kettles throughout this region (Dilliplane 1980).

The limited production of ceramic vessels by native women during the seventeenth century does not necessarily reflect their insignificance in the overall cultural system. At the same time ceramic vessels are being replaced by European kettles, they are constructed in a wider variety of sizes and forms, and decorated with a greater degree of elaboration than ever before. Seventeenth-century ceramic vessels typically have elaborately decorated collars and castellations, often adorned with appliqued effigies (Fowler 1966; Rouse 1947; Smith 1950; Snow 1980). While archaeologists have long recognized this dramatic increase in decorative complexity, little effort has been made to explain it. Two main classes of vessels were produced during this period: large cooking vessels, recovered archaeologically from habitation contexts, and miniature vessels recovered exclusively from burials (Goodby 1994:198–201). This latter class of vessel appears to lack pre-contact antecedents, and suggests that ceramic vessels took on a new ritual significance during the seventeenth century.

In the archaeology of the seventeenth century in southeastern New England ceramics have been used to identify both the tribal affiliation of individual potters (e.g., Fowler 1974) and ceramic traditions associated with historically known tribal groups. The ceramic assemblage from the Mohegan site of Fort Shantok became the basis for a proposed "Shantok Tradition," assumed (but never proven) to be the distinctive product of Pequot-Mohegan potters (Lavin 1986; Rouse 1947; Solecki 1950, 1957; Williams 1972). Due in part to limited samples, ceramic traditions associated with the adjacent Narragansett and Wampanoag peoples were never defined. The relatively few seventeenth-century vessels known from the Narragansett and Wampanoag areas bore an uncanny resemblance to Shantok vessels (Fowler 1974:16; Robbins 1959:64; Simmons 1970; Williams 1972:350), a fact attributed by some to a diaspora of Pequot refugees following the Pequot War (Fowler 1974; Solecki 1950).

Shantok ceramics (Fig. 7.3) are described as primarily shell-tempered and relatively thin, with a "soapy" texture, varying from dark gray to buff in color (Rouse 1947:15–16). Shantok pots typically have "round" (globular) bases, shoulders, constricted necks, and collars. Collars are frequently "lobate," in that the base of each is marked off by prominent, roughly triangular lobes. Rims are castellated, with *castellations* projecting "abruptly upwards and outwards from the rim" (Rouse 1947:16). Decoration, while elaborate, was limited to collars, lobes, and castella-

Figure 7.3. Castellated rim sherd from Swansea, Massachusetts. (Courtesy of the Robbins Museum of Archaeology of the Massachusetts Archaeological Society.)

tions, and included incising, punctation, and modeling. Prominently displayed on ceramic vessels were:

> Small lugs modeled in the form of human or animal heads are attached to several rim points. . . . A number of nodes also occur either on castellations or beneath them at the bases of the collars. These are more or less oval in shape. Some are split vertically, and others decorated with cross hatchings. The former have a phallic appearance, while the latter resemble ears of corn. (Rouse 1947:16)

The origins of the Shantok Tradition were, according to Rouse, "obscure" (1947:22), but he argued that its limited distribution corresponded to the limits of the Mohegan-Pequot peoples. Recently, McBride has argued that Shantok ceramics were the distinct product of the Mohegan people, and were manufactured only after the defeat of the Pequot by the Mohegan, Narragansett, and English in 1637 (McBride 1990:99). If so, then we would expect the distribution of Shantok ceramics to be concentrated in eastern Connecticut, and thus to be, in some way, distinct from the ceramics of the Narragansett and Wampanoag.

CERAMIC ANALYSIS

Methodology

A total of 43 ceramic vessel lots was utilized in this study. Twenty-four were from the Narragansett tribal area, and 19 were from the Wampanoag area. Vessels from habitation and mortuary contexts were included in the sample from both areas. This sample includes nearly every known vessel lot from this area and time period. As noted above, ceramic vessels were not produced in abundance during this time period, and all contemporaneous samples from New England are comparatively limited. The methodology employed for this analysis was modeled on that employed by previous studies utilizing attribute analysis in New England (Dincauze 1975; Goodby 1995; Hamilton and Yesner 1985; Kenyon 1983, 1985; Petersen 1980, 1985; Petersen and Power 1985).

The primary unit of analysis in this study was the vessel lot. A vessel lot is defined as all sherds assignable to a single vessel. Individual vessel lots can thus range from a single sherd to hundreds of sherds to whole vessels. The use of vessel lots as a basic unit of analysis has become a standard practice in the Northeast. An alternative approach, which uses individual sherds as the unit of analysis, introduces considerable distortion into the sample due to the differential breakage and recovery of sherds from individual vessels (Dincauze 1975; Kenyon 1983:61; Petersen 1985). Sorting of vessel lots was undertaken for each site-specific assemblage, except for those cases where only whole vessels were present and this approach was unnecessary. When vessel lot sorting was required, rim sherds were used to sort vessel lots, as they include a wider range of mutually exclusive attributes, including decoration, thickness, lip form, and profile, in addition to temper type and size.

Attribute analysis is an explicit attempt to describe, systematically and objectively, a wide range of attributes present in any ceramic assemblage, without making a priori assumptions about their relative importance. A range of technological, morphological, and decorative attributes was recorded for each vessel lot. Technological attributes included temper type, temper size, percentage of temper relative to paste, temper sorting, surface treatment on both interior and exterior surfaces, rim sherd thickness, body sherd thickness, and manufacture technique. Morphological and decorative attributes included height, orifice diameter, rim form, lip form, number of castellations, presence and type of effigies, lip decoration, and decorative technique. As overall decoration exhibited considerable variability within each tribal area, vessel lots were not classified according to motifs. Contextual attributes, such as associated artifacts, context of recovery, presence of carbonized remains on sherd surfaces, and the age and sex of associated skeletons were also recorded.

Results of Analysis

Ceramic assemblages from the Wampanoag and Narragansett tribal areas were found to be technologically indistinguishable from one another (Tables 7.1 and 7.2). The use of particular decorative techniques and effigies was also identical in

Figure 7.4. Castellated rim sherd from the Titicut site, Bridgewater, Massachusetts. (Courtesy of the R. S. Peabody Museum of Archaeology, Phillips Andover Academy.)

both areas. Vessel form and overall decorative motif exhibited considerable variation, with no one form or motif seemingly exclusive to either tribal area. Neither technological nor decorative style exhibited a spatial distribution corresponding to the boundary between these two groups. Interior and exterior surfaces are smoothed, and fine, well-sorted shell temper predominated in both areas (Fig. 7.4). Vessels from both areas are nearly identical in average sherd thickness and in the predominant use of coiling as a manufacture technique. They typically have globular or semiglobular bodies, constricted necks, and distinct collars (Fig. 7.2). Lips are either square or rounded. Castellations typically appear on collars, numbering between one and four (Figs. 7.3, 7.4, 7.5). Decoratively, they are alike in the predominance of complex, incised and often lobate collars, and the use of three basic types of effigies (corn-ear, human face, phallus) on the peaks of castellations. Unlike their Late Woodland counterparts, however, seventeenth-century vessels were produced with more uniformity and skill. Surface treatment was more uniform, temper was finer and better sorted, vessel walls were thinner, and less evidence of breakage along coil junctures was noted.

All of these characteristics are present in the Shantok Tradition as well (Rouse 1947; Williams 1972). The high degree of decorative and morphological variability described for the Fort Shantok assemblage was also characteristic of the Wampanoag and Narragansett assemblages. Potters in all three areas made similar technical choices regarding the use of temper, vessel construction, surface treatment, and vessel form. This patterning suggests that a single technological style ex-

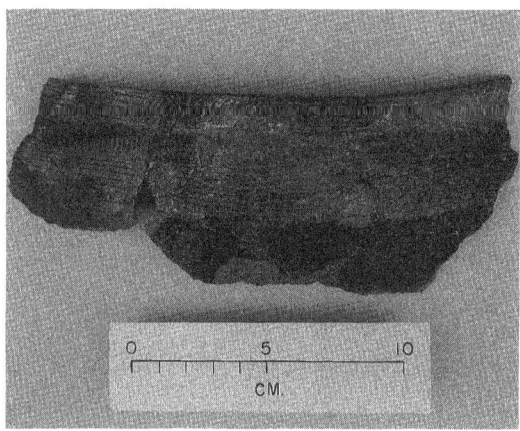

Figure 7.5. Incised rim sherd from the Titicut site, Bridgewater, Massachusetts. (Courtesy of the R. S. Peabody Museum of Archaeology, Phillips Andover Academy.)

isted throughout southeastern New England, much in the way it had during the preceding Late Woodland period. Thus, it seems that the original association of "Shantok" ceramics with the Pequot-Mohegan was premature, as comparable vessels are equally prominent in Narragansett and Wampanoag assemblages.

The lack of correspondence between all aspects of style and seventeenth-century social boundaries provides the basis for a discussion of the complexities involved in recognizing social boundaries in archaeological contexts. As Conkey (1990:11), Hodder (1979a, 1982), Lemonnier (1986:180), Stanislawski (1978), and others have observed, there is no necessary reason why social boundaries should be marked by any one category of material culture, or by material culture at all, even when such boundaries exist and are important elements structuring the interaction and relationships between distinct peoples. Ethnoarchaeological studies suggest that ceramic style may or may not have any relationship to social interaction, and when there is a relationship, the nature of that relationship cannot be understood by applying universal (i.e., decontextualized) assumptions. Recent ethnographic and ethnoarchaeological research on style supports a particularist approach that assumes little about the relationship of social patterns and style other than its complexity and dependence on the constraints of the specific social, cultural, and historical context (Braithwaite 1982; Conkey 1990; Hodder 1979a, 1982, 1990; Longacre 1981). Simply noting the correlation (or, in this case, the lack of correlation) between material culture variability and social boundaries does little to explain why such relationships do or do not exist: to understand this patterning, "such correlations and such social processes assumed to lie behind them . . . must be supported by an independent archaeological argument" (Conkey 1990:11).

In this study, the "independent argument" for relatedness is derived from a detailed consideration of the larger cultural and historical context of the seventeenth century grounded in archaeological and ethnohistoric data and constructed in relation to a developing understanding of the nature and cause of social boundary formation among indigenous peoples undergoing colonization.

Technological Patterning, Decorative Style, and European Colonialism

The technological style of seventeenth-century ceramics from southeastern New England reflects an older cultural unity upon which the social boundaries of the seventeenth century were imposed. Ceramic decoration, in contrast, actively expresses the response of individual potters to these boundaries. Women did not choose to use their increasingly elaborate decorations to mark affiliation with a particular social or political entity. In doing so, they were actively expressing a preference for the traditional ties over the recently intensified separations, and so were actively resisting the effects of colonialism. Ceramics were decorated in ways that blurred social boundaries, because the recognition and acceptance of these boundaries was strongly contested within native communities. This practice may have been, in part, an act of resistance against the increasing divisions among native societies imposed by the divide-and-conquer politics of the English colonialists and the ambitions of male sachems.

Clearly, there was an important debate ongoing during the seventeenth century about the nature of relations among native peoples. The discrepancy between the inter-tribal and intra-tribal political tensions and occasional inter-tribal warfare of the seventeenth century, and the "pan-tribal" commonalities of ceramic technology and design, suggests that ceramic style was used by women in a debate over the importance of unity among native peoples during this period. As such, these women were commenting on the ongoing struggle and emerging divisions within native societies, arguing through their expressions of unity in ceramic design that unity among native peoples was a desired social end. This sentiment was not limited to female potters, as a quote from the seventeenth-century Narragansett sachem Miantonomi illustrates. Speaking to the Montauk peoples of eastern Long Island in 1642 in favor of a united native stance against the English, he said:

> For we are all Indians as the English are [all one people], and say brother to one another; so must we be one as they are, otherwise we shall all be gone shortly, for you know our fathers had plenty of deer and skins, our plains were full of deer, as also our woods, and of turkies, and our coves full of fish and fowl. But these English have gotten our land, they with scythes cut down the grass, and with axes fell the trees; their cows and horses eat the grass, and their hogs spoil our clam banks, and we shall be starved. (Simmons 1989:42)

Unfortunately for the Narragansett and other native peoples of southern New England, while this sentiment was certainly an important part of the existing political consciousness of the day, it was not consistently held or acted upon. The very next year, in fact, Miantonomi was murdered by Uncas as part of an English-Mohegan conspiracy following a renewal of hostilities between the Mohegan and the Narragansett.

While the notion that the Shantoklike characteristics of vessels from Rhode Island and Massachusetts are the product of a Pequot "diaspora" is overly simplistic, the historical timing of the dispersal of the Pequot and the widespread appearance of Shantoklike ceramics does suggest a relationship between the two events. It is unclear, however, why peoples in areas adjacent to the Pequot-Mohegan would adopt their ceramic styles following the Pequot War, or why the products of refugee Pequot potters would come to dominate the archaeological record. A more likely connection may be found in the effects of the Pequot War on aboriginal societies and on their relations with the English. Roger Williams noted that the Narragansett were appalled by the brutality of the English exhibited in the attack on the fort at Mystic, which contrasted with traditional native warfare in which casualties were consistently low (LaFantasie 1988). The slaughter at the Pequot fort in Mystic may have compelled many native peoples to seek stronger bonds with each other in the face of continuing English aggression and political manipulation. A renewed emphasis on unity may be directly related to the experience of the Pequot, but may not have resulted in the dispersal of potters and pottery styles suggested by some scholars. Instead, potters in the aftermath of the Pequot War were careful to not create boundaries and divisions with ceramic styles but instead sought to express varying degrees of unity.

Two of the most obvious characteristics of seventeenth-century ceramics are their unprecedented degree of decorative elaboration and their high technical quality. This vitality in a traditional medium is paralleled by the intensification of traditional mortuary rituals reflected in Narragansett cemeteries (Robinson et al. 1985; Rubertone 1993:29). This florescence develops in spite of the fact of ongoing and intense acculturative pressures whose effects were most pronounced in the realm of material culture. The elaboration of ceramics can be seen as an overt act of resistance against the changes in traditional material culture, and, more broadly, against the changes taking place in society at large. The notion of resistance provides a credible explanation for this phenomena, is reflected in other aspects of material culture (Handsman 1988, 1990; Rubertone 1989, 1993) and explains both the fact of elaboration and the pan-tribal nature of decorative style.

In the context of the seventeenth century, the very production of pottery can be viewed as a political act, used to express and defend traditional native identity and culture. Such expressions were themselves forms of resistance to the process of acculturation. Flaked-stone tools are conspicuously absent from seventeenth century sites, reflecting the rapid adoption of arrowheads made from scraps of brass kettles, iron knives and axes, and other tools of European manufacture. Ceramic vessels, in contrast, continued to be manufactured until (at least) King Philip's War, despite the widespread availability (and use) of more durable European kettles. In the face of rapid acculturation and the loss of many elements of traditional material culture, some native women persisted in the manufacture of ceramic vessels, creating them with considerable skill and decorative elaboration. Other types of traditional material culture associated with women were similarly retained, as Roger Williams (1866:124) noted: "The Indian Women to this day, (notwithstanding our Howes) doe use their naturall Howes of shell and Wood." Traditional basketry and textiles likewise continue in use through the seventeenth century as seen at the RI-1000 site (Robinson and Gustafson 1982), and basketry continues as a traditional form of material culture produced by women up to the present (McMullen 1987). Ceramic pots continued to be made at least until King Philip's War (1675–76); their apparent disappearance after this time coincides with the loss of the last vestiges of native autonomy, and, in a sense, with an end to the debate over how to best preserve that autonomy.

TECHNOLOGICAL PATTERNING AND SOCIAL BOUNDARIES

In sum, the spatial distribution of technological and expressive style in seventeenth-century ceramic vessels in this region has no relationship to historically documented social boundaries. This is a reflection of the fact that these boundaries were of recent origin, that their formation was a response in large part to the pressures of European colonialism, and that the recognition and acceptance of these boundaries by native people (and thus their expression in material culture) was strongly contested (Robinson 1990). Rather than interpreting all aspects of style in relation to ethnic "groups," in this instance the concepts of gender and faction are more useful (Brumfiel 1992). By refusing to mark the emerging social boundaries of the seventeenth century, native women (or that subset of women still manufacturing ceramic vessels) used style in a manner that crossed increasingly tense social boundaries, leaving for the archaeologist the puzzling phenomena of contradiction between the spatial distributions of material culture and historically known social groups.

The concept of technological style is still useful—and even essential—in interpreting the cultural and behavioral meaning of ceramic variability in this context. Narragansett, Wampanoag, and Pequot-Mohegan women *were* making technical choices within a distinct cultural tradition (Sackett 1990:33). This tradition is the centuries-old pattern of ceramic manufacture in this region, within which a common technological style reflects the strongly similar and closely related cultures of the Woodland period. The common technological style seen in seventeenth-century ceramics is a direct outgrowth of the common patterns of temper, surface treatment, mode of construction, and morphology that extend across southern New England for three-thousand years prior to European contact. In choosing not to mark social boundaries with the decorative elements of ceramic vessels, potters were expressing an attachment to the older, broader, and inclusive boundaries reflected in the technological aspects of style. Rather than merely providing a cautionary tale about the pitfalls of assuming a direct correlation between material culture variability and ethnic boundaries, this example encourages us to ask more complex questions about boundaries and individuals' behavioral responses to them.

In discussing his concept of isochrestic variation, Sackett uses the analogy of a World War I battlefield. He notes that the archaeologist of such a battlefield site should be able to readily identify the distinct markers of the opposing ethnic groups, variation that occurs simultaneously with structures (such as trenches and command posts) common to both sides (Sackett 1990:38). However, in a circumstance of miraculous preservation, the archaeologist of such a battlefield might also find curious commonalities in the Marxist and anarchist literature that was read by the soldiers of both sides, literature that emphasized the class identity of soldiers as workers, an identity that transcended ethnic boundaries and, to the extent it was adopted by the combatants, produced behavioral consequences in fraternization across boundaries. It is easy to see in modern contexts when social boundaries exist they are not always givens, immutable and accepted uncritically by the individuals they bound. They can (and could) be accepted, transformed, or rejected outright by the conscious action of individual actors, particularly in prestate societies where boundaries were often fluid, permeable, and loosely defined. In all human societies, ordinary people may question and sometimes reject the social boundaries imposed by the powerful, muddying the relationship between behavioral variability and ethnic or social groups. This should not surprise us nor discourage us as archaeologists.

Identity is a quality that individuals and even entire groups can transform with considerable flexibility, in ways that create social boundaries, compromise existing boundaries, or eliminate boundaries altogether. In the search for social boundaries in the archaeological record we may see, in single places and times, patterns that clearly demarcate groups (e.g., Petersen and Hamilton 1984) and patterns that deny them—and we should not assume that only the former are useful or significant. This is particularly important in studying indigenous peoples undergoing colonization. While the search for isochrestic variation in technological style may be useful for detecting boundaries among traditional, nonhierarchical societies, their use should be combined with other lines of evidence (archaeological, linguistic, historical, ethnographic) to interpret the meaning of material culture patterning. Considering technological patterning in relation to the broad spectrum of archaeological and other data will better enable archaeologists to interpret the ambiguities and contradictions that often lurk beneath social-political boundaries. In southeastern New England, neither technological patterning nor decorative style corresponds to the known social boundaries of the seventeenth century, suggesting that these boundaries were neither very old nor accepted unequivocally by everyone. As anthropologists, we want to know not only how to detect boundaries, but how people behaved in response to them.

CONCLUSION

The interpretation of ceramic style presented in this paper does not challenge the importance of technological patterning in the search for social boundaries, nor does it privilege it. Instead, it suggests adding another level of behavioral variability to the analysis of style: behavior resulting from the active role of the individual, consciously and deliberately manipulating style to either support or reject the centrality of social boundaries in social life (Hodder 1986). The literature on isochrestic variation, technological style, and material culture patterning generally seems to privilege the broad social or ethnic group as the primary unit of analysis, assuming that individuals within groups adhere to patterns set by tradition and common cultural rules for the production of material culture. The boundaries of these groups, either "detected" archaeologically or assumed a priori, become the containers both for particular cultures and for specific archaeological interpretations. In doing so, the significance or even existence of individual action in the "performance" of material culture production is minimized as individual behavior is interpreted as simple submission to group norms, and our ability to detect and interpret conflict, dissent, and resistance is effectively abandoned. This has resulted in archaeological interpretations that assume, rather than question, behavioral and ideological conformity within bounded social groups.

None of these comments should be taken to mean social groups and social boundaries did not exist in some form, that they were not important, or that we cannot find material culture patterns that inform us about their distributions temporally and spatially. However, in the search for social boundaries, it is the interplay of "active" and "passive," instrumental and adjunct styles situated in particular, unique contexts that is most revealing, not only of the boundaries themselves, but the extent to which they were recognized and accepted by those whose social universe they divided. Given the highly contextualized nature of most stylistic behavior, and the innumerable contingencies involved in detecting such behavior in material culture, there appears to be no general principle, or set of principles, that will allow us to explain human behavior in relation to social boundaries in every context. In searching for and understanding social boundaries, we are engaged in an interpretive quest at best. Our interpretations should include the recognition that ambiguity and contradiction are central features of both social life and the archaeological record. Seeking ambiguity and contradiction while we search for boundaries will continue to inform us about how, where, and why social boundaries shape the lives of human beings.

ACKNOWLEDGMENTS

The author gratefully acknowledges the following individuals and institutions for providing access to data discussed in this paper: Carol Barnes and Pierre Morenon, Rhode Island College; Paul Robinson, Rhode Island Historic Preservation Commission; James Bradley and Malinda Blustain, R. S. Peabody Museum of Archaeology; Nanepashemet, Plimoth Plantation; Mr. William Taylor, Middleboro, Massachusetts; Ruth Warfield, Tom Lux, and Curtis Hoffman, Robbins Museum of Archaeology; Ned Dwyer, Rhode Island School of Design; Mary Minor, Jamestown Historical Society; Deborah Cox, Alan Leveillee, and Duncan Ritchie, The Public Archaeology Laboratory, Inc.

This paper is an outgrowth of research conducted for the author's doctoral dissertation, completed in 1994 in the Department of Anthropology, Brown University. The author gratefully acknowledges the inspiration and constructive criticism provided by Patricia Rubertone. Douglas Anderson and William Beeman (Brown University), Russell Handsman (University of Rhode Island), and Paul Robinson are thanked for their helpful comments and criticism. Michelle Hegmon is thanked for her comments on earlier drafts of this paper. Miriam Stark is thanked for including the author in the original symposium at the 1995 annual meeting of the Society for American Archaeology, for considerable editorial guidance, and for her good humor and seemingly limitless patience. All errors of fact or interpretation are the sole responsibility of the author.

Research discussed in this paper was funded in part by the Graduate School of Brown University and the Department of Sociology and Anthropology, Wheaton College, Norton, Massachusetts. The author was inspired by the humanistic approach to the study of southern New England's original inhabitants embodied in the work and thought of the late Nanepashemet. This paper is offered in his memory.

NOTE

1. The single term "Pequot Mohegan" occurs frequently in the literature (e.g., Salwen 1978; Starna 1990), while elsewhere the distinction between the Pequot and Mohegan is emphasized (e.g., McBride 1990). Seventeenth-century accounts indicate that the "Mohegan" were at one time part of a larger "Pequot" political alliance (Salwen 1978:172), but split off when the Mohegan sachem Uncas rebelled against the Pequot sachems, with whom he claimed to share a common ancestry (Weinstein-Farson 1991:10–11). The Mohegan continued as a distinct community following the defeat of the Pequot sachems in the Pequot War, and were joined by many Pequot refugees (Hauptman 1990:76).

8

Coursed Adobe Architecture, Style, and Social Boundaries in the American Southwest

CATHERINE CAMERON

The identification of social boundaries and movement of people across those boundaries has long been a primary concern for archaeologists in the American Southwest. Southwestern archaeologists have, however, sometimes made simplistic assumptions about the relationships between material culture and cultural affiliation. The idea of technological style (although in some ways implicit in traditional definitions of cultural boundaries), represents a new approach to the identification of social groups in the archaeological record. In this chapter I evaluate technological style in traditional explanations of architectural change in the Southwest. Specifically, I explore an interesting archaeological phenomenon: in at least three areas of the Southwest, at approximately the same time, coursed adobe architecture became a common building material. In each case, technological style has been used (although not always explicitly) to explain the change. I evaluate other stylistic determinants or themes against the proposed archaeological explanations.

The present study evolved out of an attempt to better understand the best documented migration in Southwestern prehistory: the late thirteenth-century migration from the northern San Juan region to the northern Rio Grande region (Fig. 8.1). By A.D. 1280, the entire northern San Juan region was abandoned (as well as the northern portions of the adjacent Chaco and Kayenta regions). At the same time, there was a dramatic increase in population in the northern Rio Grande. Ceramic types found in both areas are remarkably similar. The major problem confronting the northern San Juan-northern Rio Grande migration is architecture. There are significant differences in architectural elements between the

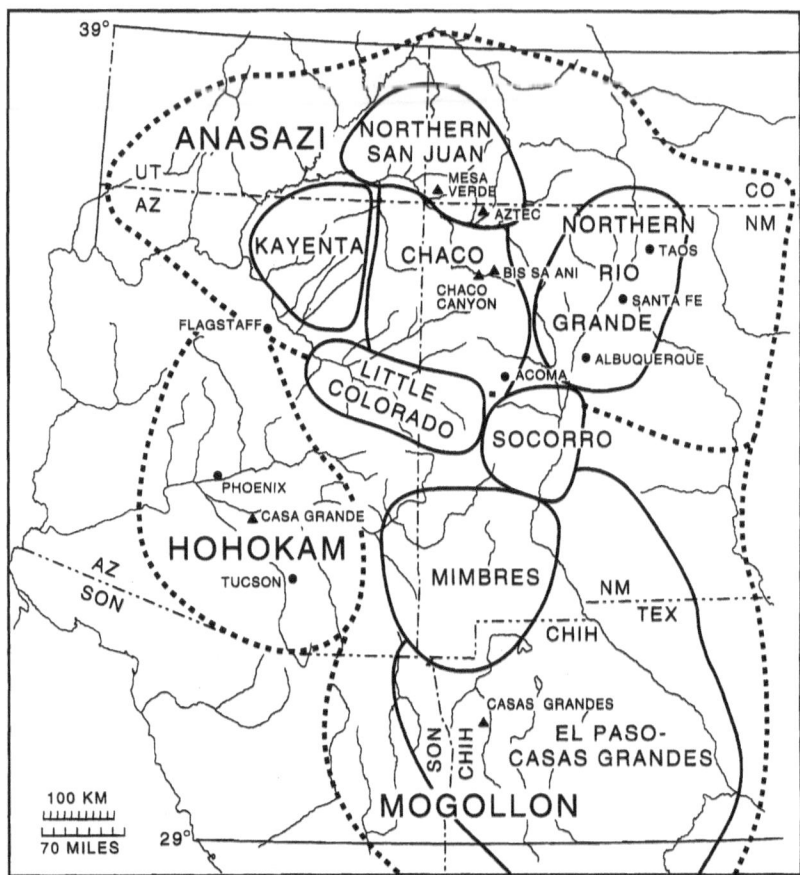

Figure 8.1. The American Southwest, showing prehistoric culture areas (dashed lines) and archaeological regions (solid lines) described in text.

two areas. Buildings in the northern San Juan region are built almost exclusively of well-coursed sandstone masonry, whereas buildings in the Rio Grande are built of coursed adobe or of more crudely constructed masonry.

Understanding the northern San Juan-northern Rio Grande migration will require new methods of identifying social boundaries in the archaeological record. This is an initial attempt to evaluate some of the archaeological concepts that link material culture and social boundaries in the consideration of a specific architectural element: wall fabric. In each of the three regions of the Southwest examined here, northern Rio Grande, Hohokam, and Mimbres, archaeologists have proposed migration to explain dramatic cultural changes. In each case, changes in architecture—especially the use of the coursed adobe technique—figure prominently in either confirming or denying that migration occurred. In all three cases, the assumption made by archaeologists is that an historical architectural tradition—a technological style—was either developed by local residents or introduced by immigrants. Here I offer a consideration of other, not necessarily exclusive, determinants of the style.

THE IDENTIFICATION OF PREHISTORIC SOCIAL BOUNDARIES IN THE AMERICAN SOUTHWEST

Social boundaries in the Southwest have been defined primarily by distributions of ceramic types and architectural styles. By the mid-1930s, three broad cultural groups had been defined: the Anasazi on the Colorado Plateau, the Hohokam in the southern deserts of Arizona, and the Mogollon in the mountainous area between the other two cultures (Fig. 8.1). These divisions were based primarily on differences in ceramics and architecture: stone or adobe pueblos and black-on-white pottery in the Anasazi area; pithouses or adobe compounds and red-on-buff pottery in the Hohokam area; and a combination of pithouses and pueblos and brown ware pottery in the Mogollon area. All three "culture areas" were further subdivided into "regions" that were also defined on the basis of subdivisions of ceramic types and variations in architectural traits.

For decades, Southwestern archaeologists have used these culture areas and regions, defined by stylistic variability in artifact types, as if they directly represented social groups; there were few attempts to define the sort of sociocultural group these archaeological manifestations might represent. Identifying spatial and temporal groupings of artifacts has facilitated archaeological comparison through time and across space, as well as enabling study of such important prehistoric behavior as trade between regions. But such grouping requires that the archaeologist ignore or minimize evident spatial and temporal gradients among stylistic elements. Furthermore, archaeologists imbue the groups with a cultural reality that may not be inherent in the archaeological data (Dean 1988). The idea that shared material culture means shared cultural identity has recently come under question (Speth 1988; Upham et al. 1994). Even in contemporary societies, ethnic boundaries are difficult to characterize and are not generally sharp nor stable (Barth 1969a; David et al. 1991; Levine and Campbell 1972).

One of the biggest challenges to the culture area concept used by archaeologists has been the identification of population movement. If a group of people can be identified by the pottery, architecture, and other artifacts they make, then, when they move into a new region, their material culture should contrast dramatically with the material culture of an indigenous population. We know that population movement happened (Haury 1958), yet such "site unit intrusions" (Willey and Lathrap 1956) are extraordinarily rare in the archaeological record. Cordell (1995) has argued that because immigrant populations tend to assimilate with indigenous populations they may not retain a recognizable complement of cultural elements. Alternatively, in many cases, the remains most accessible to archaeologists (ceramics and architecture) may not have been distinct between immigrant and indigenous populations (Hegmon, this volume). The difficulty for the archaeologist is to identify social boundaries where the material evidence for boundaries may be ambiguous or where material culture differences may never, in fact, have been sharply defined.

ARCHITECTURE IN SOCIAL CONTEXT

Vernacular buildings and builders interact; buildings provide essential shelter and at the same time reflect social values and ideals (Rapoport 1969). As a result, architecture is perhaps one of the best classes of material culture in which archaeologists can seek clues to social identity. Architecture functioned as a spatial focus for prehistoric activities and as a container for the artifacts used in those activities. For example, both the size and organization of the household and the dynamics of community development (establishment, growth, decline, and abandonment) can best be explored through studies of domestic architecture (Adams 1983; Cameron 1991; Dean 1969; Reid 1973).

Architecture not only reflects social organization and values, it also can play a key role in shaping social values. For example, in the Southwest, the massed, multistoried structures that characterized both prehistoric and historic Puebloan architecture began as large, symbolic, public structures (Chacoan Great Houses; Lekson and Cameron 1995). In the A.D. 1200s, when these large buildings began to be used domestically, their close living spaces may have helped shape the tightly-knit Puebloan communities that we know today (Cameron 1995; Lekson and Cameron 1995).

Architecture has been interpreted from a variety of perspectives. Shelter is a basic human need. Domestic architecture is often perceived as a direct response to the need for protection from the environment and of the materials available in that environment with which to construct shelter (Fitch and Branch 1960). In the Southwest, the design of the multistoried structures that characterize both prehistoric and historic Puebloan architecture has long been explained by arguments of thermal efficiency (Fewkes 1906; Knowles 1974; Taylor 1983).

Architecture is far more than just shelter, however. Architecture structures daily use of space by breaking space into compartments within and around which other activities take place. By defining patterns of encounter and avoidance, architecture directly (as well as symbolically) reflects social relations (Hillier and Hanson 1984; Rapoport 1969). For these reasons, the examination of style in architecture requires special consideration not accorded other artifact types when seeking clues to social boundaries. Buildings are stationary; typically, they are not moved from place to place so they have less chance to be viewed by individuals outside the community. As a result, they might be less likely (than, for example, traded items or clothing) to be used in *active* cultural signaling of group membership; people within the community already know the messages they might send (Wobst 1977). On the other hand, because the use of space encoded in architecture is so loaded with cultural information, architecture will signal a variety of messages—about ideals, values, and relationships—either consciously or unconsciously to both residents and visitors.

Walls, of course, are a key element of any architectural construction and are especially important to archaeologists because they are generally all that is left of prehistoric architecture. Roofs, rooflines, and exterior (and, often, interior) decoration have usually weathered away, leaving walls or wall stubs as the only element left to analyze. The selection of wall fabric certainly has environmental constraints, but most environments also offer many choices of materials. Social information may be found in the choices that are made.

ADOBE CONSTRUCTION IN THE SOUTHWEST

The coursed adobe technique, the subject of this chapter, was only one of a number of mud wall construction techniques in use prehistorically (see Moquin 1992 for a brief summary). "Jacal," a wood and brush wattlelike superstructure covered with mud, was common throughout the Southwest in prehistoric and historic time periods. In spite of scattered early reports of prehistoric adobe bricks in the northern Southwest (Fewkes 1911; Morris 1944), until recently most archaeologists assumed that brick was a Spanish introduction (Johnson 1992; Judd 1916). Recent research and reexamination of old reports have shown that adobe bricks were used prehistorically (Gann 1992; Johnson 1992; Moquin 1992). At present, all reported occurrences of the prehistoric use of adobe brick are north of the Mogollon Rim, primarily associated with Pueblo cultural manifestations dating after the twelfth century (Johnson 1992). A variety of techniques were used to form adobe bricks, including hand-molding (Fewkes 1910, 1911; McGimsey 1980; Morris 1944), form-molding (Gann 1992; Johnson 1992), and "turtle-backs," (hand-molded, lenticular bricks; Kidder and Guernsey 1919; McKenna and Truell 1986). Another widely used technique consisted of adobe walls randomly interspersed with stone spalls, cobbles, or boulders (Crown 1991a:151; Di Peso 1974, vol. 4:216; Hayes and Lancaster 1975).

Coursed adobe was one of the most common mud-wall techniques in the Southwest. Unlike jacal and turtle-backs, coursed adobe was functionally equivalent to stone construction for load-bearing walls. Like masonry buildings, coursed adobe structures could be multistoried, were often terraced, divided space into functionally discrete cells that could be entered through ceiling hatches or doors, and were lighted and ventilated with wall fenestrations.

Coursed adobe technique (also called "English cob"; Fig. 8.2) involves using a stiff mud that is built up in courses (Wilcox and Shenk 1977). Stubbs and Stallings (1953:26) provided an early description of coursed adobe at Pindi Pueblo in the northern Rio Grande Valley near Santa Fe.

> The walls were built by taking large handfuls of plastic clay and patting them into place upon the foundation. Each lot added was molded and smoothed with the hands on to the mass below. When the adobe had been built up to the desired thickness and the height allowed by the consistency of the mass, the course was left to dry sufficiently to support the next layer. The courses average from 15 to 18 inches in height; some are only a few inches and others may be as much as 2 feet. . . . The top of each course has a general convex surface, still showing hand-prints of the hand molding process of construction. . . . In drying, shrinkage of the adobe in the long wall courses had a tendency to crack them vertically into long blocks. This gives the appearance of having been made in sections, but examination shows otherwise. The courses are usually as long as the wall of which they are a part.

There is dispute among Southwestern archaeologists about the methods used to manufacture coursed adobe walls. At least at some sites, archaeologists have suggested that wall construction actually involved a "poured" or "puddled" technique instead of the hand-built method described by Stubbs and Stallings. In the puddled method, thick mud is poured into movable forms and then allowed to dry. Puddled adobe was first identified at the site of Los Muertos in the 1880s by Frank Hamilton Cushing (Haury 1945, cited in Wilcox and Sternberg 1981:19). Later, Cushing (1889) suggested that the technique was also used at the site of Casa Grande, near Phoenix, Arizona (Steen 1965). Di Peso (1974) believed that the puddled adobe technique was used at Casas Grandes (also called Paquimé) in northern Chihuahua. Wilcox and Shenk (1977), however, have argued persuasively that the poured

Figure 8.2. Coursed adobe walls at the fourteenth-century site of Arroyo Hondo Pueblo near Santa Fe, New Mexico. (Courtesy of the School of American Research, Santa Fe.)

adobe technique was not used at either Casa Grande, Arizona, or Casas Grandes, Mexico, and suggest that the hand-built coursed adobe technique may have been the predominant technique used at all adobe walled sites in the Southwest (see also Judd 1916; Kidder et al. 1949).

Coursed adobe in the prehistoric Southwest apparently had a wide distribution, from western Utah and eastern Nevada (Judd 1926; Madsen 1989; Metcalfe and Heath 1990; Shutler 1961; Wormington 1955) to northern Mexico (Di Peso 1974; Doolittle 1988; Kelley 1995). The question of most interest for this study is the timing and nature of the introduction of the technique into the Southwest. Early use of coursed adobe technique has been reported in the Taos area of the northern Rio Grande region during the Valdez phase (Blumenschein 1958; Loose 1974, Luebben 1968; Moquin 1992) and Virgin and Sevier-Fremont culture areas of southwestern Utah and southeastern Nevada (north and west of the Kayenta and northern San Juan regions on Figure 8.1; Shutler 1961; Wormington 1955). In both cases, sites where the technique was used are poorly dated, but sites in these areas seem to have been occupied between A.D. 900 and 1200. Much of the Virgin and Sevier-Fremont culture areas were abandoned by A.D. 1150 (Lyneis 1996; Wormington 1955) so the technique must have been in use there before that time.

Some Southwestern archaeologists believe that the coursed adobe technique was introduced from Mesoamerica. Breternitz and Marshall (1982:435–436) note that the coursed adobe construction in western Utah and eastern Nevada, although poorly dated, appears to occur earlier than it does in the Anasazi area (where it dates primarily to the Pueblo II period) and may have been introduced into these western areas from northern Mexico through cultures in western Arizona. Breternitz and Marshall (1982:436–437) do not connect later use of the coursed adobe technique in the Anasazi area or elsewhere with its earlier use in western Utah and eastern Nevada, but seems to suggest an independent origin, perhaps from Casas Grandes. Di Peso (1974, vol. 4:216–217), however, suggests a Peruvian origin for the "puddled" adobe technique he found at Casas Grandes with introduction into the Southwest through western Mexico. Hosler's recent (1994) demonstration of the connection between metallurgy in western Mexico and metallurgy in coastal South America strengthens Di Peso's suggestion.

The suggestion that the coursed adobe technique was introduced from Mesoamerica is questionable because the technique was apparently rare beyond northern Mexico. In Mesoamerica and extending into northern Mexico, adobe bricks seem to be a common mud wall construction technique (Diehl 1983; Thomsen 1960; Tozzer 1921; Yampolsky and Sayer 1993) although "solid clay walls" are occasionally reported (e.g., Healan 1974, 1989:61; these may be melted adobe brick walls, however). In Zacatecas, in north-central Mexico, domestic architecture at the sites of La Quemada and Alta Vista were of adobe brick (B. Nelson, pers. comm. 1995). Further north in the Santa Maria Basin of western Chihuahua, Mexico, which is about 150 km south of Casas Grandes and 300 km south of the New Mexico border, coursed adobe walls have been found at least as early as A.D. 1200 and perhaps earlier (Kelley 1995). The use of coursed adobe at approximately the same time has been reported further west in the Valley of Sonora, Mexico (Doolittle 1988). Of course, coursed adobe walls were used in the construction of Casas Grandes, which lies approximately 150 km south of the New Mexico border.

Style and Wall Fabric

Almost three decades ago, archaeologists began a lengthy discussion of the types and functions of style in the archaeological record (Binford 1973; Bordes and de Sonneville-Bordes 1970; Conkey and Hastorf 1990; Hegmon 1992, and this volume; Lechtman 1977; Lemonnier 1986; Sackett 1973, 1990; Wiessner 1983; Wobst 1977). This discussion has resulted in recognition that the relationship between social groups and material culture is much more complex than traditional culture-area designations would imply. Of the numerous classifications of style that have developed during this discussion, two of the most useful may be technological style and "emblemic" style.

"Emblemic" style has been defined by Wiessner (1983; also Sackett's [1990:36] "active" style or Wobst's [1977] "information exchange" model) as technological steps that express social values used to reinforce social themes within a group or to serve to demarcate social characteristics in relation to other groups. In other words, emblemic style suggests that the maker of an artifact is consciously sending a message—often about the makers' group affiliation—to anyone who views that artifact. Unfortunately for archaeologists, the use of emblemic style is contextually variable, depending on the relationships of the maker with the potential viewer of the object, economic and socio-political conditions, and type of object that will carry the message, among other variables. The greatest difficulty for archaeologists lies in identifying intentionality in the past.

The concept of technological style (the focus of this volume) recognizes that artifact manufacture follows a historically derived cultural template that defines "the way things should be done." Technological style may be stable over long periods of time and generally expresses social information unconsciously (Childs 1991; Lechtman 1977; Sackett 1990), but it may also be the result of conscious choices about how manufacturing steps should be accomplished (Gosselain, this volume). The intent of technological style is *not*, however, overt signaling of group membership. For this reason, technological style exhibited in material culture should not be subject to the problems of context and intentionality that make emblemic style so difficult for archaeologists to use in identifying social boundaries. Although current studies of technological style use the concept to develop new methods to identify prehistoric social boundaries, the idea that social or ethnic groups have particular ways of manufacturing things or accomplishing tasks is, of course, inherent in the traditional culture area concept.

The selection of wall fabric is probably most often attributed to an environmental determinant. In other words, construction will be accomplished with the most functionally appropriate, readily available material. Fitch and Branch (1960) is the classic presentation of this environmental perspective, although virtually all contemporary scholars of vernacular architecture see the environment as a conditioning rather than a limiting factor in construction materials and methods adopted (e.g., Nabokov and Easton 1989).

Emblemic style, technological style, and the environment are each useful concepts for exploring how the selection of wall fabric in the Southwest might be used to identify social boundaries. First, emblemic style: wall construction materials serve as an active reminder of group boundaries/membership. Second, technological style: a builder will build walls in a particular way simply because that is the way his or her parents and grandparents built walls. Third, the environment will condition which materials will be most readily available and appropriate. In the present analysis, the strengths, ambiguities, and weaknesses of each of these three determinants of the selection of wall construction materials will be considered.

Bis sa ani

An intriguing case from the northern Southwest illustrates how these three determinants of wall construction materials—emblemic style, technological style, and environment—might be used in explaining architectural patterns in the Southwest. Bis sa ani is a Chacoan great house, part of the Chacoan regional system (Breternitz et al. 1982). The extent and nature of this regional system is uncertain, but its center was clearly at Chaco Canyon in northwest New Mexico, and this center was connected by wide, straight roads to great houses scattered across the northern Southwest at distances as great as 150 miles from Chaco Canyon (Lekson et al. 1988). Great houses were built with massive, multistoried masonry walls using a core and veneer method with elaborate facing techniques that produced beautiful walls that could (and did) stand for centuries (Lekson 1986).

Bis sa ani was built near the end of the Chacoan era, about A.D. 1130, just outside Chaco Canyon on a clay pinnacle in the middle of the Escavada Wash valley. The Bis sa ani great house consisted of two structures. One was built of typical Chaco-style sandstone masonry and the other of massive coursed adobe (Fig. 8.3).

Bis sa ani is one of only two great houses to show the use of coursed adobe in the widespread Chacoan regional system. Another has been reported at the Aztec Ruins great house complex, about 60 miles north of Bis sa ani (Stein and McKenna 1988:21). A few of the small, residential "community" structures built around Bis sa ani were also of coursed adobe (Dykeman 1982). At A.D. 1130, the coursed adobe structures at Bis sa ani predate almost all coursed adobe architecture in the Southwest, except perhaps at Taos and the southwestern Utah-southeastern Nevada regions discussed above. Aside from the coursed adobe used at Bis sa ani, only a very few prehistoric coursed adobe structures are known for the entire Four Corners area throughout its prehistoric occupation.[1] Ceramics found at the Bis sa ani great house and at the coursed adobe community structures were of styles characteristic of the Socorro region, almost 150 km south of Chaco Canyon (Fig. 8.1).

Chacoan great houses were public architecture and most archaeologists interpret them as heavily symbolic structures: public symbols of "Chacoaness" (e.g., Fowler and Stein 1992). The coursed adobe great house at Bis sa ani may have reflected "Chacoaness," but the selection of coursed adobe for construction might also have exhibited emblemic style in another way, by signaling a subset of Chaco

Figure 8.3. Bis sa ani Pueblo, a Chacoan great house constructed during the early twelfth century near Chaco Canyon, New Mexico. Coursed adobe walls are visible in the center of the photograph. (Dick Meleski, photographer, University of New Mexico Archives Collection.)

membership. In other words, the builders of the coursed adobe structure may have been actively proclaiming, through the use of this particular wall construction technique, that they were different and distinct from other Chacoan people.

One difficulty with this interpretation would be if the masonry structures at Bis sa ani had been plastered, making their method of construction invisible to the viewer. Walls at Chacoan great houses were plastered (Lekson 1986:29), perhaps most of the time. At Bis sa ani, however, at least one exterior masonry wall was not plastered (Marshall 1982:264), whereas an exterior wall of the adobe structure was plastered red-tan and white (Marshall 1982:187). The difference in construction methods would have been visible.

Alternatively, Bis sa ani's builders may have been immigrants from the Socorro region—as the ceramics indicate. If so, they may not have actually intended to signal their separateness. When it came to building a structure, they may have simply preferred to build using technologically familiar materials, a technological style. The fact that the small, residential "community" structures built around Bis sa ani were also of coursed adobe and contained Socorro ceramics, strengthens the idea that a familiar technological style of construction was used by immigrants.

There is a potential chronological problem, however, with the suggestion that southern immigrants to Bis sa ani brought their own technological style with them. The use of coursed adobe was apparently not a long-standing technique in the Socorro area where the immigrants were supposed to have originated. This area was sparsely occupied prior to A.D. 1000 and coursed adobe may not have been used prior to A.D. 1200 (Wimberley and Eidenbach 1980:227).

Alternatively, environmental factors can be evoked. The builders of the coursed adobe structure at Bis sa ani may not have been immigrants. The highly skilled Chacoan masons, having located their building on a clay pinnacle (high points were often selected for Chacoan great house construction), may have simply become tired of hauling in stones and decide to build in coursed adobe—letting the available environment triumph over any other stylistic considerations. If this were the case, it is unclear where they learned this technique of construction. Furthermore, the adobe structure appears to have been the first built at the site (although all structures at Bis sa ani appear to have been built within a short period of time—10 years).

In the Bis sa ani case all three determinants—emblemic style, technological style, and the environment—may be at work. The importance of the Bis sa ani case is that it demonstrates some of the problems in using these determinants to explain social boundaries and population movements. In the following sections, I continue to explore the strengths, ambiguities, and weaknesses of these determinants of wall construction by examining the use of coursed adobe in three other areas of the Southwest where it becomes predominant. In each of these areas, a coursed adobe *technological style* has been proposed to either confirm or deny that migration has taken place. An environmental determinant underlies most of these explanations: technological styles are developed using the most functionally appropriate, readily available building materials. The final section of the paper takes a broad look at the potential explanatory power of each of the three determinants, especially emblemic style, for exploring what seems to be the sudden, widespread introduction or development of a new method of wall construction.

Coursed Adobe in the Northern Rio Grande

In the northern Rio Grande region (Fig. 8.1), technological style has been used in competing explanations to both confirm and deny significant immigration from the northern San Juan region in the late thirteenth century. The facts in the case are not in dispute, but they permit alternate interpretations. Most of the region experienced a significant increase in population between A.D. 1250 and 1300

Coursed Adobe Architecture in the American Southwest 195

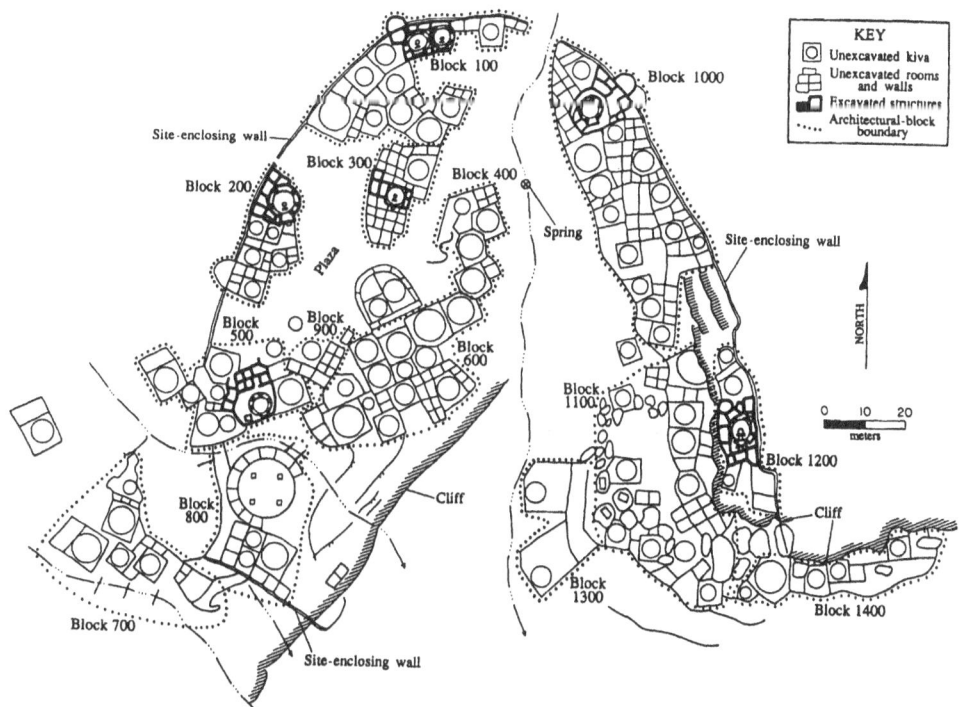

Figure 8.4. Sand Canyon Pueblo, located in the northern San Juan region near Cortez, Colorado, was built in the late thirteenth century near the head of a small canyon. (Courtesy of Crow Canyon Archaeological Center, Bruce Bradley 1992.)

(Collins 1975; Crown et al. 1996). Ceramic types found at many of these sites are almost identical to types found in the northern San Juan region. At about the same time, the northern San Juan region was almost completely abandoned. The evidence for immigration of people from the northern San Juan region seems overwhelming except that post-migration Rio Grande architecture differs markedly from northern San Juan architecture.

Site layout in the northern San Juan region consisted of blocks of rooms laid out in parallel rows or strung out following the topography of canyon heads or rock shelters (Fig. 8.4). Site layout in the northern Rio Grande region was plaza-oriented (Fig. 8.5). Structures called "kivas" (whose function is under debate, Lekson 1988, 1989) are round in the northern San Juan region, but may be round or square in the northern Rio Grande, and internal kiva features in each area are different, distinctive, and differently oriented. Finally, villages in the northern San Juan region are built almost exclusively of carefully coursed sandstone masonry

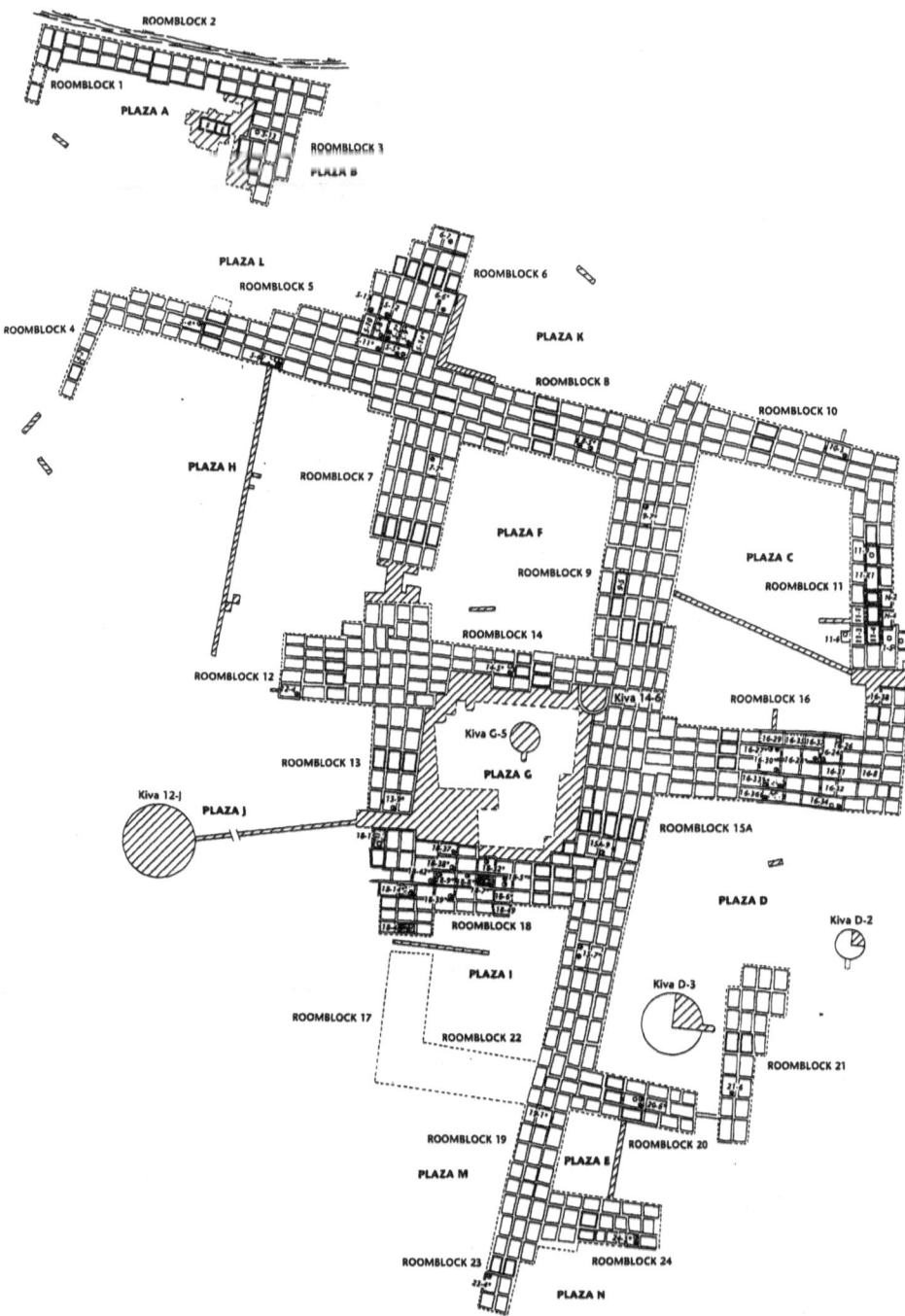

Figure 8.5. Plan view of the initial occupation of Arroyo Hondo Pueblo, a fourteenth-century site located in the northern Rio Grande region near Santa Fe, New Mexico. (Courtesy of the School of American Research.)

Figure 8.6. Stone masonry typical of the northern San Juan region at Twin Towers, a thirteenth-century structure at Hovenweep National Monument, southeastern Utah. (Courtesy of Stephen H. Lekson, photographer.)

(Fig. 8.6). Sites in the Rio Grande are made of coursed adobe or quite different stone masonry, generally cruder than that in the northern San Juan (Fig. 8.7).

Rio Grande archaeologists have tended to see a strong architectural technological style in their region (although they haven't used the term "technological style"). In fact, significant continuity in building techniques from early Rio Grande periods through the post-migration period has been used by some scholars to deny that significant immigration from the northern San Juan occurred (e.g., Steen 1983:170; see Cordell 1989:315–317 for a summary of these views). Even archaeologists who acknowledge that the migration occurred use technological style implicitly to explain the difference in architecture between the two regions. For these archaeologists, architectural differences are explained by assuming that the immigrant population joined with and was assimilated by the local indigenous population. In other words, people from the northern San Juan migrated into the Rio Grande as small family groups who then integrated with existing Rio Grande communities with whom they already had long-standing ties of trade or kinship (Cordell 1979a:103, 1979b:150–151; 1989:317; Wendorf and Reed 1955:164 for early

Figure 8.7. Stone masonry at Rowe Ruin, a fourteenth-century site located near Santa Fe, New Mexico, in the northern Rio Grande region. (Courtesy of Linda S. Cordell.)

immigrations; see also Kohler 1993:296 for emigration from the northern San Juan region in small groups). In this scenario, coursed adobe would be a local technological style adopted by immigrants.

Both the proposal that immigrants assimilated with a local population and adopted the local architectural technological style (or perhaps moved into existing built space) and the denial of migration based on architectural differences assume that coursed adobe architecture was developed from earlier mud-based architectural techniques in the Rio Grande (see Steen 1983:170). There is reason, however, to question whether the coursed adobe technique was a long-standing technological style in the northern Rio Grande. The area was sparsely occupied prior to A.D. 1200. Much of that limited occupation was in pit structures or jacal structures; above-ground structures of either masonry or adobe are rare (Cordell 1989:309–310).[2]

In the Taos area coursed adobe walls do occur in apparently early contexts. Coursed adobe was used to line the walls of some Valdez phase pit structures in the Taos area and a few above-ground coursed adobe structures were associated with these pit structures (Blumenschein 1958; Loose 1974; Luebben 1968). The Valdez phase has been dated between A.D. 900 and 1200 but the place of the coursed adobe structures within that span has not been firmly established. Large villages of coursed adobe or masonry did not actually become common until well after A.D. 1300, about the same time as (or later than) immigration from the northern San Juan took place. It is not clear whether the coursed adobe technology was developed by indigenous populations within the northern Rio Grande, whether it evolved among intermingling groups after significant immigration, or whether it arrived with immigrating populations from areas other than the northern San Juan (see below).

Both coursed adobe and stone are used in wall construction in the northern Rio Grande and a few archaeologists have tried to link the use of stone to the immigration of people from the northern San Juan region (Wendorf and Reed 1955). Unfortunately, in addition to differences in the quality of stonework between the two areas, there is no consistent chronological transition between the two building techniques (Cordell 1989:315; see also Lambert 1954:13). Other archaeologists have explained the occurrence of the two construction materials as the result of environmental determinants. The assumption is that Rio Grande builders used stone when it was available and when it was not, they used mud (Creamer et al. 1993:14–15). Builders in the northern San Juan had abundant sandstone so of course they built masonry structures. The value of one technique over the other in terms of materials, labor costs, or technological efficiency has not been considered.

The Hohokam Region

As in the northern Rio Grande, technological style has also been used implicitly to explain architectural change in the Hohokam area of southern Arizona (Fig. 8.1). Before A.D. 1150, Hohokam domestic architecture consisted of shallow brush-and-mud pit structures (or, more properly, "houses-in-pits") that were arranged in groups of three or four around a small, common courtyard (Fish and Fish 1991). Hohokam communities included many of these courtyard groups as well as public architecture such as ballcourts and platform mounds.

After A.D. 1150 throughout the Hohokam area, above-ground "compounds" largely replaced pit structures (Doyel 1991). Compounds consist of a rectangular enclosing wall with rooms built along the interior of the wall (Fig. 8.8). The walls of the compound and interior rooms were built of coursed adobe or, at some buildings, adobe built up over and around upright wooden posts (the latter technique is called "post-reinforced adobe"). Coursed adobe is used also for a new public architectural form, the great house, which formed the focus of one or perhaps a few Hohokam communities (Wilcox and Shenk 1977). Platform mounds, an earlier public architectural form, also began to be enclosed with coursed adobe walls after A.D. 1150 (Crown 1991a; Doyel 1991).

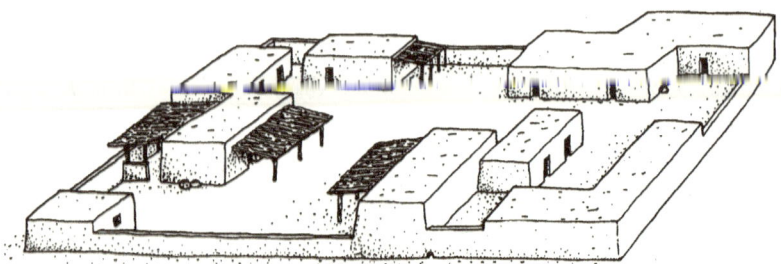

Figure 8.8. Artist's reconstruction of a coursed adobe Hohokam compound. (Drawing by Ben Mixon, redrafted in Doyel 1991:254. Courtesy of Ben Mixon and David Doyel.)

The Hohokam area after A.D. 1150 was, therefore, marked by a dramatic change in domestic architecture, as well as changes in public architecture. When the Hohokam culture was first defined in the 1920s, this dramatic architectural change, which was accompanied by a change in ceramic types and other elements of material culture, was interpreted as marking the immigration of a new group of Puebloan people called Salado (Gladwin et al. 1937; Haury 1945, 1976; Stark et al. 1995 and this volume).

During the 1960s, most Southwestern archaeologists abandoned migration as an explanation for culture change (Cameron 1995). Beginning in the 1960s and continuing to the present day, one of the major thrusts of Hohokam research has been the argument that the Classic period was the result of an in-place cultural development, not immigration (Doyel 1980). Unlike the Rio Grande, this effort has been largely successful and most archaeologists now see Hohokam Classic culture as an in-place development from earlier populations, based in part on perceived continuities in architectural traits (Crown 1991a; Doyel 1991; Sires 1987). According to this explanation, architectural changes apparent in the prehistoric record reflect in-place evolution of architectural methods, an explanation that fits both the concepts of technological style and an effective use of the environment: builders used locally available materials and evolved specific architectural techniques that were passed from generation to generation.

The Mimbres Region

In the Mimbres region of southwestern New Mexico (Fig. 8.1) immigration of peoples with a particular architectural technological style has been used to explain prehistoric architectural change. From A.D. 1000 to 1150, architecture in the Mimbres area consisted of compact, stone masonry room blocks, with villages containing as many as 200 rooms. In fact, when they began to be built about A.D. 1000, these were the first true "pueblos" in the sense of multiroomed, contiguous village structures (Lekson 1990). The walls of Mimbres pueblos were built primarily of river cobbles set in abundant mud mortar. This construction technique was not as durable as the sandstone masonry used by prehistoric people in the Anasazi area and today Mimbres people are known far more for the beauty of their decorated ceramics than for their architecture.

After A.D. 1150, the Mimbres pattern ends; the standard explanation is collapse and depopulation or cultural replacement (LeBlanc 1983; Nelson and LeBlanc 1986).[3] The distinctive Mimbres black-on-white pottery was no longer made, nor were there large pueblos built of river cobble masonry. After A.D. 1150 in the lower Mimbres Valley, as well as a much wider area of southern New Mexico and southern Arizona (the El Paso-Casas Grandes area, Fig. 8.1), moderate-sized, (or occasionally large) villages were built of coursed adobe. These sites are architecturally similar over a wide area of southern New Mexico and northern Mexico although they have been divided into various cultural units (i.e., the Animas phase, the Black Mountain phase, and the El Paso phase) based on differences in proportions of various ceramic types (LeBlanc 1989:194; Schaafsma 1979; see also Phillips 1989). These coursed adobe pueblos have been attributed to the influx of new people with new ceramic styles and new architectural techniques who may have assimilated the remaining Mimbreños. In other words, a technological style, developed elsewhere and imported by immigrants, explains architectural change.

IMPLICATIONS OF COURSED ADOBE ARCHITECTURE

To summarize the evidence presented above, the coursed adobe technique of wall construction occurs across the southern Southwest after about A.D. 1150. In the Hohokam area of southern Arizona, as well as much of southern New Mexico, coursed adobe above-ground structures replace pit structures or houses-in-pits as the common domestic architectural form. In the Mimbres area, coursed adobe pueblos replace river cobble masonry pueblos after A.D. 1150. In the northern Rio Grande, the situation is more complex. Before A.D. 1150, pit structures apparently were the most common architectural form. After A.D. 1250, large pueblos with either coursed adobe or masonry walls occur and become common after A.D. 1300, but there is apparently no consistent chronological transition between the two building techniques.

The sudden occurrence of coursed adobe architecture after about A.D. 1150 throughout much of the southern Southwest and in the northern Rio Grande has been explained in three different regions as the result of a technological style that

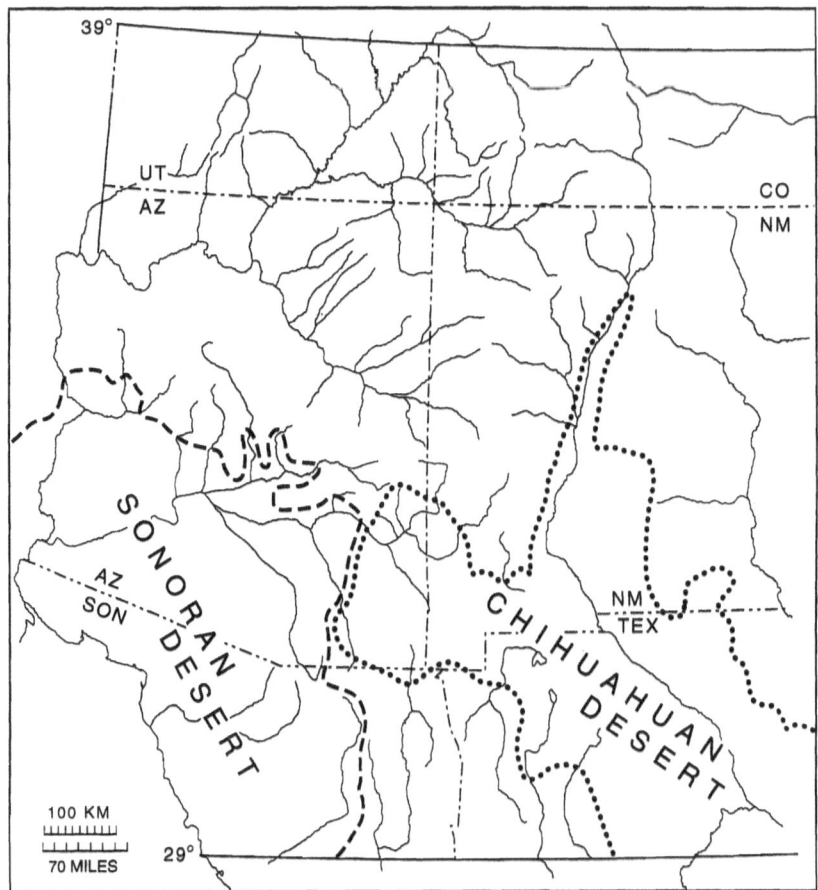

Figure 8.9. Southwestern deserts (after Brown and Lowe 1980).

was either developed by local residents or introduced by immigrants. The role of the environment in providing appropriate materials is implicit in most of these explanations.

It is useful to examine environmental determinants more broadly, however. The distribution of coursed adobe architecture in the greater Southwest (southern Arizona, southern New Mexico, and northern Mexico) is approximately coterminous with the distribution of the upper Chihuahuan and upper Sonoran Desert biotic zones (Fig. 8.9; Brown 1982). These environmental zones are characterized by very low rainfall, which would be especially appropriate for long-term maintenance of adobe structures; in fact, after A.D. 1150, we might refer to this entire area as "the Southern Adobe Belt." Significantly, the Chihuahuan biotic zone makes a 250-

mile spikelike bulge into the northern Rio Grande region in places where adobe structures were found (Fig. 8.9). The co-occurrence of coursed adobe and desert biotic zones, while close, is not exact. For example, structures in the Taos area of the northern Rio Grande region, about 90 miles north of Santa Fe, New Mexico, are of coursed adobe, yet rainfall there is relatively high. Coursed adobe is also found in a band from the Little Colorado area to Acoma, which is outside both of these southern desert areas (Fig. 8.1).

If coursed adobe was the environmentally appropriate and most readily available material over a wide area of the southern Southwest and the northern Rio Grande, it seems likely that the technique was developed by people in those areas —a technological style. But this does not seem to be the case. Coursed adobe construction seems to appear suddenly between about A.D. 1150 and 1200 in a number of areas of the Southwest. Perhaps even more remarkable is the isolated occurrence of the coursed adobe great house at Bis sa ani perhaps two decades before it becomes widespread elsewhere. The technique may have been in use prior to A.D. 1150 in a few scattered places: southeastern Nevada and western Utah (Judd 1926; Wormington 1955), parts of northern Mexico (Kelley 1995), or the Taos area, although dating in these areas is not secure. This seeming sudden occurrence of a new construction method does not fit well with the concept of the in-place development of a construction technique that made use of appropriate, readily available materials.

If the coursed adobe technique was not a technological style developed in the Southern Adobe Belt or the northern Rio Grande, then where was it developed? This question is largely beyond the scope of this paper, but certain observations can be made. The idea that the coursed adobe technique was introduced from Mesoamerica seems unlikely, as the technique is uncommon there; instead, adobe bricks seem to be far more commonly used in mud wall construction (see above). Di Peso's suggestion that the coursed adobe technique was introduced from Peru through western Mexico will require far more study of construction techniques in western Mexico and beyond. The timing of the use of coursed adobe in western Utah and eastern Nevada should also be examined, as well as the exact techniques of construction.

The third determinant, emblemic style, can be considered with regard to the distribution of coursed adobe architecture and ideology in the Southwest. E. Charles Adams (1991) has suggested that the kachina religion, an ideological construct still found in modern pueblos, developed in the Little Colorado area of east-central Arizona (Fig. 8.1) in the late 1200s and spread from there to the Rio Grande. The kachina religion, which involves public participation in ceremonial dances often in enclosed plazas, may have been an important mechanism of social organization that permitted the aggregation of many people in large villages (Adams 1991).

Adams believes a "group of elements" became associated with the kachina religion in the Little Colorado area, including enclosed plazas, rectangular kivas, Fourmile style pottery, and a distinctive iconography found in rock art and other media. He argues that some of these elements, including enclosed plazas and iconographic features had their ultimate origin in northern Mexico, but he believes that the kachina religion itself developed in the Little Colorado area. Then, he suggests, during the 1300s or even 1400s, the kachina religion was introduced to the Rio Grande area. Adams (1991:124–125) notes that coursed adobe walls are found in the Little Colorado area in the thirteenth and fourteenth centuries and implies that the technique is the result of architectural influences from southern Arizona.

Polly Schaafsma (1992) offers an alternative reconstruction of the origin of the kachina religion. She suggests that the kachina religion was present in the Mimbres area in the early 1100s (Schaafsma and Schaafsma 1974). Aggregated pueblos with interior plazas are found in both the old Mimbres area and in adjacent areas of south-central New Mexico especially after A.D. 1150. Schaafsma suggests that large, coursed adobe-walled, planned villages are found first in the southern deserts of New Mexico and northern Mexico as the manifestations that archaeologists have called the El Paso phase, the Black Mountain phase, and the Animas phase (El Paso-Casas Grandes area, Fig. 8.1). Schaafsma further suggests that the development of large villages in the northern Rio Grande after A.D. 1300 was in part the result of the introduction of the kachina religion and other aspects of social organization from these southern origins. She cites Kelley (1984) and Sebastian and Levine (1989) in suggesting population movement into the Rio Grande region from the Tularosa Basin and the Sierra Blanca area of southeastern New Mexico (Schaafsma 1992:88).

The distribution of the kachina religion as defined by both Schaafsma and Adams is largely coterminous with the distribution of coursed adobe architecture.[4] If Schaafsma is right, the kachina religion is found, possibly in the mid to late 1100s, in southern New Mexico in association with large, planned, coursed adobe villages. Coursed adobe is found later in the northern Rio Grande and in a band from the Little Colorado area to Acoma, both areas where the kachina religion occurred historically (Fig. 8.1). Could coursed adobe have been part of the material manifestations of this new ideological movement? Might coursed adobe architecture have functioned as emblemic style?

There are a number of problems with this suggestion. If Adams is correct about the origin of the kachina religion in the Little Colorado area and if (as suggested here) coursed adobe walls are somehow associated with the religion, then wall fabric there should be primarily coursed adobe. In fact, wall fabric in the Little Colorado area is remarkably variable. Not only are coursed adobe and stone masonry found, as in the northern Rio Grande, but the use of adobe brick has been demonstrated at some sites (Gann 1992; Johnson 1992). Adams (1991) suggests that diverse groups aggregated in the Little Colorado area and the kachina religion resulted from the need to integrate and bind these groups together. Variability in wall fabric may reflect diverse populations.

If Schaafsma is correct about the early origins of the kachina religion in the Mimbres and El Paso-Casas Grandes areas (Fig. 8.1) and if (as suggested here), coursed adobe walls are somehow associated with the religion, then the kachina religion might be expected wherever there are coursed adobe walls. Yet, no indication of the kachina religion has been found in the Hohokam area where coursed adobe architecture appeared at approximately the same time it appeared in the El Paso-Casas Grandes area. But there does appear to have been a significant religious-ceremonial transformation in the Hohokam area at this time, signaled especially by the replacement of ballcourts with platform mounds as the primary form of public architecture. Interestingly, coursed adobe walls begin to be used to enclose platform mounds—suggesting, possibly, emblemic function for the coursed adobe technique of wall construction.

CONCLUSIONS

This chapter examined the ways that the appearance of coursed adobe architecture has been interpreted in three different regions where it occurs in the Southwest. The study began as an effort to understand the late twelfth and thirteenth-century migration from the northern San Juan region to the northern Rio Grande. In both of the other regions examined (Hohokam and Mimbres), migration had been proposed, at some point, to explain the occurrence of coursed adobe architecture. In the northern Rio Grande, the coursed adobe technique had been considered a long-standing technology and it was used as evidence that the northern San Juan-Rio Grande migration did not take place or was of insignificant magnitude. These regional explanations for the occurrence of coursed adobe have been examined, and then the occurrence of the technique in the Southwest was examined broadly using three stylistic determinants: the environment, technological style, and emblemic style.

Although this study is exploratory, it seems likely that all three determinants functioned together to govern the selection of wall construction materials in the Southwest. Both the environment and technological style should have conditioned the development of the technique. While it seems unlikely that coursed adobe was simply a construction method developed where there was insufficient stone available, the environment may have determined where coursed adobe provided an effective construction technique. It might then have been selected as a traditional method of construction—a technological style.

The linkage of coursed adobe architecture with emblemic style and ethnic signaling or ideology is far more difficult to evaluate because the intentions of the builder are archaeologically problematic. The very early association of coursed adobe with a Chacoan outlier, a structure that seems to have been highly charged with social meaning, is intriguing, as is the relationship between the kachina religion and coursed adobe, suggested above.

The sudden appearance of the coursed adobe construction technique across a large part of the southern Southwest seems likely to have signaled something more significant than simply the concurrent development (or even spread) of a new and useful building technique. The role of both population movement and of ideology in explaining the distribution of coursed adobe construction should be explored in greater depth. If prehistoric people were either intentionally (emblemic style) or unintentionally (technological style) signaling membership in particular social or ethnic groups through the use of a particular wall fabric, the coursed adobe technique holds great promise for the exploration of social boundaries in the Southwest. This building technique may then provide useful information about movements across these social boundaries.

ACKNOWLEDGMENTS

Abundant thanks are due Miriam Stark who invited me to contribute a paper, first to her symposium on technical choices, organized for the 1995 Society for American Archaeology meetings in Minneapolis, and then to this volume. This paper owes much to her work as organizer and editor. Thanks go to Steve Lekson for many discussions on adobe architecture and comments on several versions of the paper. Much gratitude is also owed to a number of colleagues who either discussed adobe architecture with me or read and commented on versions of this paper: Michael Adler, Linda Cordell, David Doyel, Michelle Hegmon, Jane Kelley, Timothy Kohler, Fred Lange, Michael Marshall, Ben Nelson, James Sackett, Payson Sheets, Miriam Stark, and Barbara Voorhies. Many thanks to all and no blame to any for shortcomings of this paper.

NOTES

1. See Morris (1915, 1944) and Stein and McKenna (1988) for other examples of coursed adobe architecture in the Four Corners area. All cited examples are located around Aztec Ruin, a

Chacoan great house in northwestern New Mexico, and at least one (Morris 1915) clearly post-dates the Chacoan era.
2. Both Pueblo Arroyo Negro (LA 114) and LA 835 may be early examples of the use of coursed adobe architecture in above-ground structures in the northern Rio Grande, possibly dating as early as the mid-eleventh century. The dating of the coursed adobe portions of these sites is poorly understood, however, and may be later than the earliest dates obtained from each site (Robinson et al. 1972; Smiley et al. 1953).
3. Recently, several archaeologists have identified significant continuities in the Mimbres and post-Mimbres cultures leading them to question the proposed immigration of new peoples into the Mimbres area (Lekson 1992:133).
4. The kachina religion has apparently never been part of religious observances in the Taos area of northern New Mexico, yet this area may have the earliest examples of the use of coursed adobe in the Rio Grande (Valdez phase pit structures).

9

Social Boundaries and Technical Choices in Tonto Basin Prehistory

MIRIAM T. STARK, MARK D. ELSON, AND
JEFFERY J. CLARK

The identification of distinct social groups has been a long-standing concern throughout the history of Southwestern archaeology. Boundaries in material culture patterning have commonly been described at scales that range from the "culture area" and "province" levels to those of regional "alliances" and "macrosystems." The culture areas known as Mogollon, Anasazi, and Hohokam, for example, are familiar to any archaeologist who has spent even a little time studying Southwest prehistory. Archaeologists have generally identified boundaries for these areas by trait distributions of key artifact types, selected behavioral practices, environmental adaptations, and architectural traditions (see also Cameron, this volume).

Cross-cultural research, however, has made it increasingly clear that many (if not most) of these archaeological social boundaries exceed the scale of social boundaries in traditional small-scale societies (also see MacEachern, this volume). Reliance on particular types of data in Southwestern research, particularly on decorated ceramics and monumental architecture, has limited the range of social scales that are observable in the archaeological record. A conceptual paradigm, beginning in the earliest days of Southwestern research, emphasizing homogeneity and stressing the similarities within large, environmentally defined areas, has further influenced the nature of investigation.

A critical—if often overlooked scale of analysis—lies at the local level. As used in this study, "local systems" refers to well-bounded, small-scale systems whose participants engaged in regular face-to-face interaction (see Adler 1992; Gregory 1995). Previous archaeologists have observed spatial discontinuities in ceramic patterning at subregional scales in the Southwest (early examples include Colton 1953; Haury 1945:95–96; and Wheat 1955). However, few conceptual and methodological tools are available for examining social boundaries within this smaller framework. The idea of local traditions—as these are manifested in technological variability—is not unfamiliar to archaeologists.

We concentrate our research on the Tonto Basin in east-central Arizona (Fig. 9.1), an area that is traditionally viewed as the heartland of the Salado culture. What our study provides are a conceptual framework and analytical methods for examining this type of material culture patterning. This conceptual framework emphasizes how differences in a wide range of technical choices made during the production process effect the final appearance of the manufactured good. We attempt, following Lemonnier (1986:180), to view goods as the result of these choices and to place them into broader technical systems in which they participated in the prehistoric past.

We use architectural and ceramic data from the Roosevelt Community Development Study to examine technical choices in the prehistoric Tonto Basin of central Arizona. We examine changes in social boundaries at two critical points in the developmental sequence of a small local system located in the eastern portion of the basin. The first is the Colonial period (beginning ca. A.D. 750), in which a technological approach illustrates processes of population movement into the study region. The second is the early Classic period (beginning ca. A.D. 1250) where technological attributes reflect patterns of co-residence between groups with different enculturative backgrounds that participated in the same local system. Use of this technological framework for the Colonial period confirms a previous model based on stylistic variation. Yet this same focus on technological variation in the Classic period also identifies previously undetected dimensions of social interaction.

SOCIAL BOUNDARIES IN THE ARCHAEOLOGICAL RECORD

The study of stylistic behavior and prehistoric social boundaries, popularized first during the New Archaeology, has continued to be important in Southwestern archaeology (see Hegmon 1992 and S. Plog 1983 for reviews). Previous research has been exhaustive in its regional scope but somewhat myopic in its theoretical perspective. Most research concentrates on decorated ceramics, with some attention to stylistic variability in lithic artifacts and monumental or large-scale integrative architecture. Most studies employ an approach (following Wobst 1977) that involves the use of goods as tools in social strategies. In this framework, generally called the information exchange model, material culture is seen as a mechanism

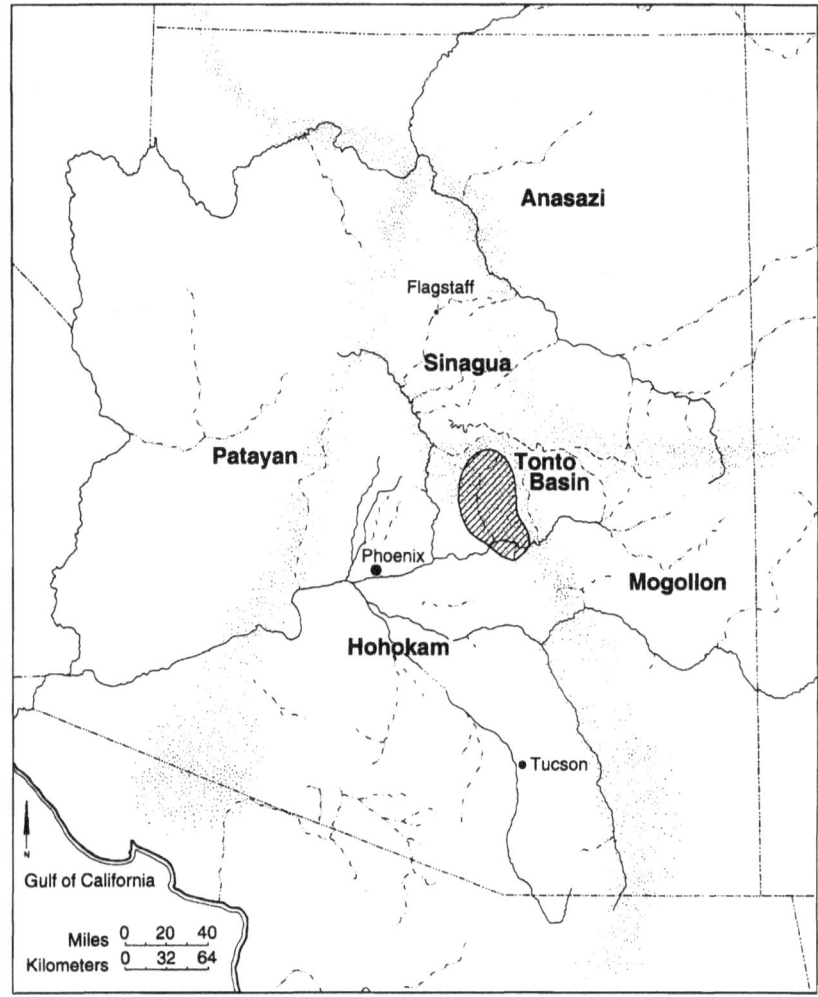

Figure 9.1. The Tonto Basin in central Arizona with surrounding prehistoric culture areas, as traditionally defined.

for the conscious conveyance of information, although the information content may vary depending on the viewer. This restrictive definition of material style has focused research on a narrow range of objects that obviously appear (to the archaeologist) to contain stylistic meaning—painted pottery, carved shell, personal ornaments, and projectile points, for example—while short-shrifting undecorated everyday goods. Studies of utilitarian ceramics (Dietler and Herbich 1989, 1994a, this volume; Sterner 1989) and of boundary regions with multiple ethnic groups (Hodder 1979a) have identified limitations of this model.

The approach that we use in this study differs from the information exchange model in its focus and theoretical framework and draws from an "anthropology of technology" (see Pfaffenberger 1992). In such an approach, all goods (not simply those with decoration) convey information about behavior. This technological patterning both embodies and generates meaning in different cultural traditions. Spatial discontinuities in technological traditions—not simply in stylistic patterns—should reflect social boundaries in the material record. Most research in the prehistoric Southwest focuses narrowly on stylistic dimensions of material culture: on platform mounds and great kivas but not on domestic architecture, and on polychromes but not on utilitarian ceramics. Here we suggest some systematic methods for studying technological variability in these mundane goods of everyday life. We believe that these goods, precisely because they are mundane, are more indicative of prehistoric social boundaries than goods that are consciously manipulated for conveying information, the content of which the archaeologist can only approximate.

IDENTIFYING TECHNOLOGICAL BOUNDARIES IN THE ARCHAEOLOGICAL RECORD

Objects of material culture are created through technological processes, each of which consists of behaviors and techniques. Artisans make choices in their selection of certain behaviors and techniques, and they can execute some steps in the technological process in numerous ways. This latitude in decision-making during the technological process creates myriad different approaches to accomplishing the same goal; it also generates variability in the manufactured goods. We may disagree on the nature and scale of social groups in the prehistoric past, but most archaeologists agree that we can observe technological variability in the archaeological record. One way to do so is through the systematic analysis of how goods are made; we gloss the sum of these technical choices here as technological style.

The term technological style refers to the "formal integration of the behaviors performed during the manufacture and use of material culture, which, in its entirety, expresses social information" (Childs 1991:332). It represents the outcome of repetitive and mundane activities associated with everyday life, and is often understood as "the way things are always done" (Wiessner 1984:161, 195). Alternatives selected by artisans in their choice of materials, in the configuration of their products, and in their decorations reflect a thoroughly internalized understanding of the manufacturing tradition that they generally pass on from one generation to the next (Gosselain 1992b; Lechtman 1977:15; Mahias 1993; Sackett 1986:268–269, 1990:33, 37).

Technological styles inform on the enculturative background of producers, since they represent the sum of the technical process: raw materials, sources of energy, tools and scheduling (Lemonnier 1993a:4, van der Leeuw 1993). The particular usage of a good also shapes its technological style. These choices, rather than simply the raw materials and production tools, are crucial in determining the outcome of a product. Technological styles thus reflect conscious and unconscious elements of technical choices, and are reasonably stable through time.

Although most goods express variation as technological styles, socially meaningful trends in these styles are perhaps most easily observed in utilitarian goods. Utilitarian goods may be more sensitive to cultural boundaries than are nonutilitarian goods, which vary in their degree of closure depending on numerous variables (Barth 1969a; Hodder 1979a, 1985). A wide body of ethnographic and ethnoarchaeological data, for example, suggests that social boundaries for decorated ceramics and projectile points are often permeable. These artifact classes, although commonly used, are generally poor indicators of cultural identity (Collett 1987; DeCorse 1989; Larick 1987; Lyons 1987; Sterner 1989). One reason for this problem may lie in contrasting contexts of production and distribution: at least some types of goods reveal more about the identity of their producers than about their consumers (Dietler and Herbich 1994a, this volume).

Media on which technological style is most easily studied include ground stone tools, chipped stone tools, and utilitarian pottery (Dean 1988; Sackett 1985a; Sterner 1989). Utilitarian pottery is a particularly sensitive medium given the wealth of recent attention by ethnoarchaeologists (e.g., Kramer 1985; Longacre 1991a, 1991b; Stark 1995c). Architecture, the most complex and least portable of all artifacts, also contains abundant information on technological style. Meaningful architectural variables include construction techniques, construction materials, and the organization of space in domestic contexts (Baker 1980; Baldwin 1987; Cameron, this volume; David et al. 1991; Ferguson 1992; Hillier and Hanson 1984; McGuire 1982).

TECHNOLOGICAL STYLES AND SOCIAL BOUNDARIES

The cultural conservatism inherent in technological styles is one of its greatest strengths in studies of social boundaries in the archaeological record. Technological style is more resistant to change than is stylistic variation via decorative parameters, because stylistic variation does not significantly alter the manufacturing process (Gosselain 1992b:582–583; P. Rice 1984:252; Wiessner 1985). Both the types of goods that are used (Welsch et al. 1992; Welsch and Terrell 1994; Wiessner 1983), and variation in the styles of widely used goods may reflect social boundaries (Sackett 1986:270).

Examination of a specific class of material culture provides clues regarding a particular technological style. Spatial variability should be evident, then, at two levels: in the technological style that shapes each artifact class, and in the suite of technological styles that comprise a culture's technical system. In the present case, we devote substantial attention to understanding variability in the operational sequence represented in domestic architecture and utilitarian ceramics. Several previous ceramic studies have adopted similar approaches (Gosselain 1992b; Mahias 1993; van der Leeuw 1993), but less attention has been devoted to this topic in architectural studies (but see Cameron, this volume; Lemonnier 1986). Because ceramics and architecture are the two data classes used in this study, specific attributes of their technological styles are briefly discussed below (see also Stark, Clark, and Elson 1995).

SOCIAL BOUNDARIES IN THE PREHISTORIC TONTO BASIN

Data presented here were recovered primarily from investigations undertaken by Desert Archaeology, Inc., on the Roosevelt Community Development Study (RCD), which involved the investigation of 27 prehistoric sites in the eastern portion of the lower Tonto Basin (Fig. 9.2). The RCD Study was one of three related archaeological projects undertaken as part of the Bureau of Reclamation's Central Arizona Project before the raising of the dam (and water level) of Roosevelt Lake (see Doelle and Craig 1992; Pedrick 1992). Researchers from these projects have now excavated portions of nearly 150 archaeological sites, and approximately 400 additional sites are known through survey data. This has enabled us to define the eastern Tonto Basin local system, which includes all sites along a 6-km stretch of the Salt River, and is separated from the next local system downstream by a 3-km area of limited prehistoric settlement (Fig. 9.2; see Gregory 1995; Stark, Clark, and Elson 1995). This provides one of the strongest regional data bases in the North American Southwest and documents all portions of the prehistoric period, beginning in the early Ceramic period (ca. A.D. 100–600).

Findings from the Roosevelt Community Development Study indicate the presence of an indigenous population that interacted and mixed with neighboring and migrant groups from the earliest periods of occupation onward (Elson et al. 1995). We focus here on archaeological evidence for the presence of different social groups during two periods of population movement: the Colonial period (ca. A.D.

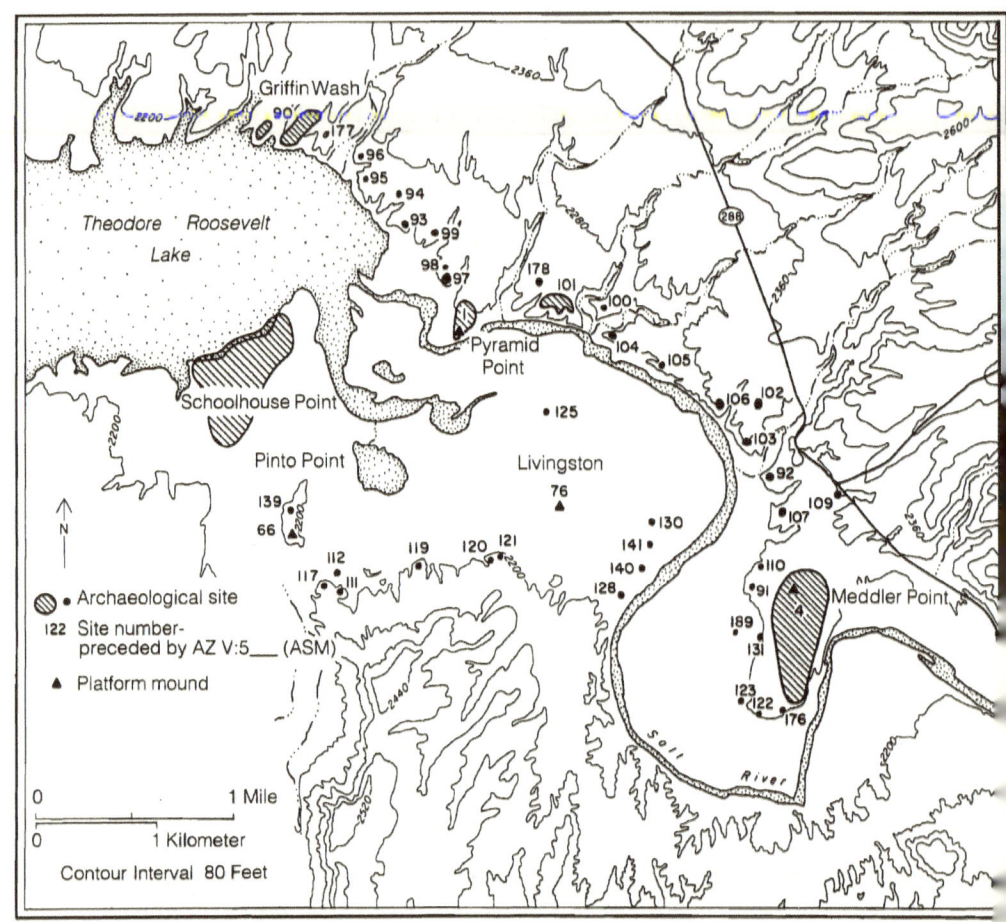

Figure 9.2. The Roosevelt Community Development Study (RCD) area and neighboring sites that date to the early Classic period (ca. A.D. 1150–1350).

750–950) and the Roosevelt Phase of the early Classic period (ca. A.D. 1250–1350). In each of these periods, patterning in technological styles of domestic architecture and of utilitarian ceramics provides evidence for the existence of multiple cultural groups within the study area.

Measuring Technological Style in Utilitarian Ceramics

Several production steps are present in the operational sequence for a manufactured good, such as an earthenware pot (Table 9.1; also see Mahias 1993, fig. 5.1).

TABLE 9.1
Steps in the Operational Sequence of Hand-Built Ceramic Manufacture

Operational Task	Types of Production Steps
Materials procurement	Collection of raw materials (clay, temper, slips, paints, glaze)
Materials preparation	Crushing (clay, temper, or both)
	Cleaning (clay) and/or size sorting (temper)
	Blending (clay and temper)
	Kneading (combination of clays or clay and temper mixture)
Primary forming techniques	Pinching and drawing
	Coiling
Secondary forming techniques	Beating/paddling
	Scraping
Decorative forming techniques	Smoothing → polishing
	Slipping
	Texturing (includes corrugation and incising)
	Painting or glazing
Drying and firing	Creation of fire clouds
	Use of reducing atmosphere
Postfiring treatments	Smudging

Modified from Rye 1981.

The amount of technological variation encoded in each step varies in predictable ways. The first task is materials procurement and preparation, which is guided as much by availability of local materials as it is by cultural preference. The second task involves the actual forming of the vessel, which establishes the general shape of the object and its basic proportions. Forming techniques are particularly resistant to change (see Gosselain, this volume; P. Rice 1984) and therefore a good indicator of social boundaries. The third task involves decorative techniques, which, as discussed above, may be a fickle indicator of group membership. The fourth and final task involves modifications that occur after the object has assumed its basic form and appearance.

Although the concept of "production steps" is not new, previous applications have generally concentrated on either descriptions of ceramic technology (Rye 1981) or on indexes of labor investment for particular types of ceramics (Feinman et al. 1981; Hagstrum 1988). Differences in production steps, however, can also be examined to understand technological styles. For example, several equivalent solutions that utilize different production steps may be employed to make vessels of a particular function. Ethnoarchaeological research in Guatemala (Reina and Hill 1978) illustrates a range of useable water jar shapes that vary geographically according to local terrain, water source, and cultural preference. Surface treatments on water jars are similarly variable: some traditional potters glaze their vessels' exterior surfaces, others impregnate their vessel interiors with organic materials, and still others use no surface treatment at all. This variation in technical choices at various points during the production process—rather than simply in one step within it (i.e., in decoration)—contains information regarding distinct technological styles.

Measuring Technological Style in Domestic Architecture

Domestic architecture in the Tonto Basin defines the domain of the household, the smallest corporate and economic unit in most prehistoric Southwestern communities. Architectural construction techniques reflect choices in technological style in much the same way as production steps reflect technical choices in ceramic manufacture. These choices are embedded within cultural preferences that are also resistant to change. Some aspects of the manufacturing sequence, such as construction materials, are affected by local resource availability. Other aspects, however, reflect culturally specific notions of construction and of the use of space.

Some methods described here for examining patterning in domestic architecture have been adopted from ceramic approaches. We can use the same production steps, from primary forming to finishing techniques, to analyze traditions of architectural construction. A second, and equally important, approach to domestic architecture relies on the use of physical space. Although technological constraints and environmental factors influence construction techniques and house form in residential architecture, conceptions of the proper organization of architecturally bounded space are highly conservative (Aldenderfer 1993; Collet 1987; Kus and Raharijaona 1990). For Hillier and Hanson (1984), this "ethnicity of space" differentiates between social groups across cultures (see also Kent 1990a). Although temporal and locational factors may also play a role in architectural design, differences between groups in the use of space largely reflect highly culture-specific behaviors (Kent 1990a:3; Rapoport 1990).

The following case studies briefly examine how cultural preferences are reflected in technological patterning in domestic architecture and in utilitarian ceramics. During the early Classic period, stylistic variation in decorated ceramics (Salado Polychromes, Cibola White Wares and White Mountain Redwares) and in monumental architecture (platform mounds) signaled broad interactional spheres that linked local systems into larger interactional networks (Stark 1995a; see Crown 1994). Technological variation in domestic goods, in turn, reflected local systems of cultural identity.

CASE STUDY 1: COLONIAL PERIOD MOVEMENT
INTO THE TONTO BASIN

Data from the RCD Study analysis suggest a migration of Hohokam-affiliated groups into the eastern Tonto Basin sometime during the Colonial period, between A.D. 700 and 800 (Clark 1995b; Elson et al. 1995; Stark, Clark, and Elson 1995). Evidence for migration is based on settlement morphology, architectural traits, and technological styles in utilitarian ceramics. These various lines of evidence suggest close resemblances to traditions in the Phoenix Basin, specifically those along the Gila River (Gregory and Huckleberry 1994; Haury 1976). This area is approximately 75–100 km southwest of the Tonto Basin.

Patterning in Colonial Period Domestic Architecture

Domestic residences in the eighth and ninth-century Tonto Basin consisted predominantly of large "houses-in-pits," illustrated at Locus B of the Hedge Apple site (Fig. 9.3). This architectural form contrasts with the much smaller pithouses found in the preceding early Ceramic period settlements (Fig. 9.4), which we believe represents indigenous populations derived from earlier Archaic period groups (Elson et al. 1995). Settlement morphology at the Meddler Point site (the largest Colonial period settlement in the lower Tonto Basin), consists of a central plaza and cremation cemetery, surrounded by trash mounds and pithouse courtyard groups (Craig and Clark 1994). The Meddler Point layout closely parallels settlement structure and the proper arrangement and use of domestic and ritual space described for contemporary Phoenix Basin settlements (see Wilcox 1991:259–262).

Patterning in Colonial Period Utilitarian Ceramics

Technological style in the utilitarian plain ware ceramic assemblage also closely parallels that found in traditions in the Phoenix Basin and particularly along the Gila River. Construction techniques for forming and finishing vessels are indistinguishable from those found in the Phoenix Basin utilitarian ware Gila Plain, Gila variety. Equally important is the similarity in raw material usage. Over 70 percent of the Colonial period plain ware ceramics are tempered with micaceous schist, a common temper type in Phoenix Basin ceramics. In all other periods in the developmental sequence, most Tonto Basin utilitarian ceramics were tempered with various types of nonmicaceous stream sands (Miksa and Heidke 1995).

Several factors suggest a nonlocal origin for the manufacture of these plain wares. For one, no raw material sources within a 30-km radius of the project area match the temper composition of the ceramics (Stark, Vint, and Heidke 1995).

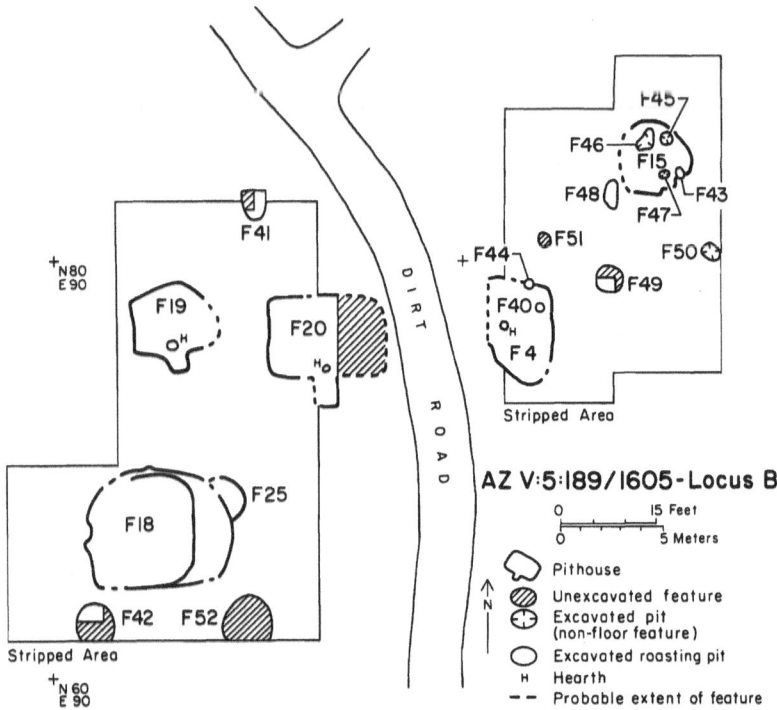

Figure 9.3. Example of "houses-in-pits" from the Colonial period (ca. A.D. 750–950). Locus B at the Hedge Apple site (AZ V:5:189/1605).

Thirty kilometers is an extremely conservative distance threshold that far exceeds previous estimates (Arnold 1985; Miksa and Heidke 1995) on how far traditional potters usually travel to procure temper resources. Moreover, no micaceous schist chunks (potential raw material for pottery-making) were recovered from any site in the project area. The nonlocal origin of the micaceous-tempered ceramics is also supported by paste compositional analyses, which differentiated sherds with micaceous schist temper from those made with locally available stream sands (Stark, Vint, and Heidke 1995).

One possible source zone for the temper lies along the Gila River in the Gila Butte and Santan Mountain areas, where micaceous-schist outcrops are found (Rafferty 1982). Temper composition in Tonto Basin Colonial period ceramics resembles (and may be identical to) materials from these areas, which lay two to three days' travel away from the project area.[1]

Both trade and migration were evaluated as mechanisms for transporting the plain wares into the Tonto Basin (Stark, Vint, and Heidke 1995). Plain ware ceramics circulate in utilitarian exchange spheres (Crown 1991b; Doyel 1991), as do

Social Boundaries in Tonto Basin Prehistory   219

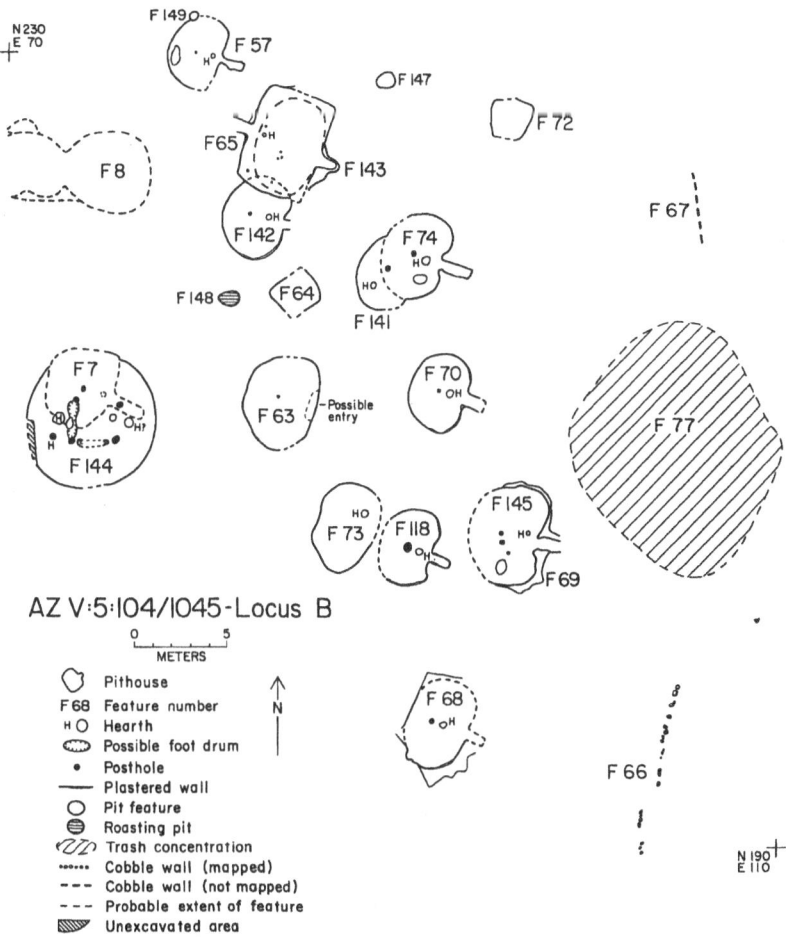

Figure 9.4. Example of pithouses from the early Ceramic period in the Tonto Basin (ca. 100 B.C.–A.D. 600). Excavated portion of Locus B at the Eagle Ridge site (AZ V:5:104/1045).

foodstuffs, baskets, and other common household goods. Ethnographic data suggest that travel time from one point to the next in such networks rarely exceeds a day's journey (Stark 1995a). Previous archaeological estimates for sizes of utilitarian exchange networks range from approximately 25 to 35 km, with a maximum of 50 km (Lightfoot 1979; Wilcox 1994). Application of these distances to the Colonial period sites suggests no areas where micaceous-tempered plain wares might be manufactured for exchange. Therefore, these pots seem more likely to have been moved by the migration of a different cultural group into the Tonto Basin than to have been procured through trade.

In the case of the Colonial period, studies of technological patterning provide a more satisfactory result than studies of stylistic patterning that previous researchers have used (Haury 1932) to posit migration into the Tonto Basin. Phoenix Basin Hohokam buff wares comprise over 20 percent of the total ceramic assemblage (Heidke 1995, fig. 6.4). These high buff ware frequencies contrast with frequencies found in later periods in the project area and at contemporary settlements in the upper Tonto Basin (about 50 km to the north). At these settlements (believed to be occupied by indigenous populations), buff ware ceramics constitute less than 5 percent of the ceramic assemblage (Elson et al. 1992). In contrast, decorated wares (predominantly buff wares) comprise just over 30 percent of the early Colonial period ceramic assemblage at the site of Snaketown on the Gila River (Haury 1965, fig. 107), a settlement that likely produced buff ware ceramics during this period.

CASE STUDY 2: CLASSIC PERIOD MIGRATIONS INTO THE TONTO BASIN

The era that began with the early Classic period Roosevelt phase (ca. A.D. 1250–1350) in the Tonto Basin marked a watershed throughout the greater Southwest. This 100-year interval was a period of accelerated cultural change that coincided with a period of climatic fluctuations and drought (Adams 1991; Crown 1994; Dean et al. 1985, 1994; Euler et al. 1979). Developments within the eastern Tonto Basin during this period reflect pan-regional demographic processes, notably regional abandonments, population dislocations, and shifting social boundaries (Stark, Clark, and Elson 1995).

Patterning in Classic Period Domestic Architecture

General architectural traditions, and conceptions of space, changed dramatically during this period. Surface masonry residential units, including both masonry compounds and room blocks (Fig. 9.5), were common in the Tonto Basin by the mid-thirteenth century, and possibly earlier (Clark 1995b). Distinctions between these two architectural types, which signify very different conceptions of the proper use of physical space (see Hillier and Hanson 1984), can be quantified and compared systematically. Multiple measures were used in the architectural analyses to differentiate between the two types of masonry architecture (see Clark 1995a; Stark, Clark, and Elson 1995). We focus here on the room contiguity index because it is the most robust method and the easiest to present in graphic form.

Social Boundaries in Tonto Basin Prehistory   221

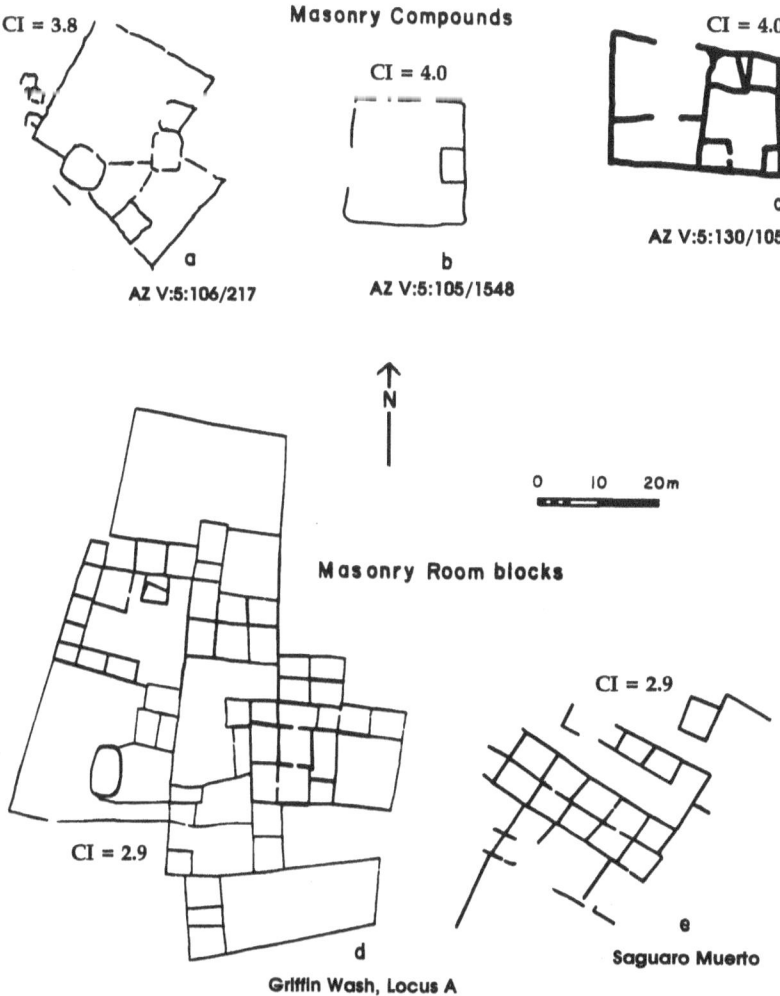

Figure 9.5. Examples of masonry compound and room block architectural forms from the RCD and Livingston Study areas (CI = contiguity index value).

Room Contiguity Measures

The room contiguity index measures the number of rooms that share walls with other rooms (Clark 1995a). The index is calculated by dividing the total number of room walls by the total number of rooms within an architectural unit, resulting in a value ranging between 2.0 and 4.0. A value of 4.0 represents a completely

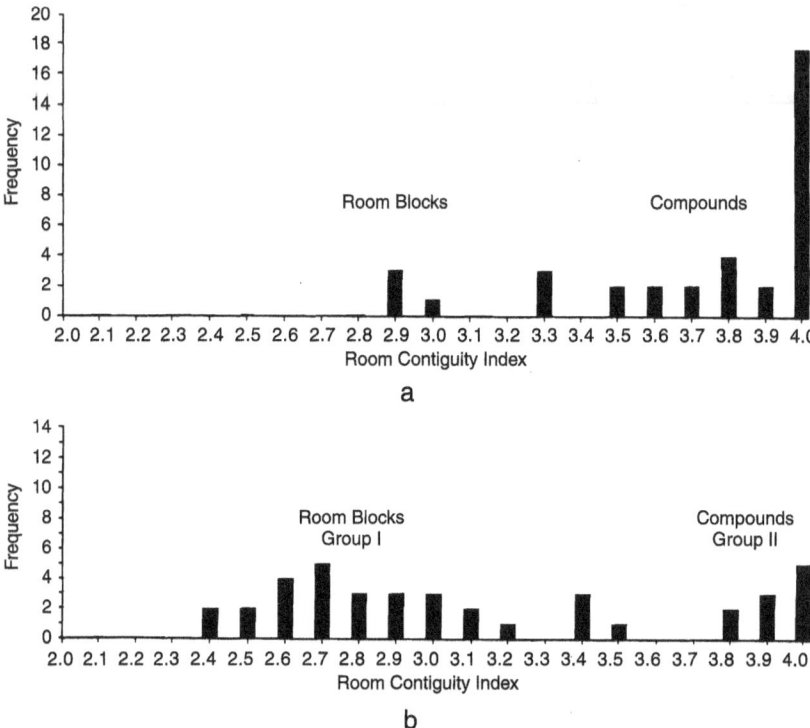

Figure 9.6. Room contiguity indexes for (a) the RCD and Livingston Study areas and (b) the general American Southwest.

noncontiguous arrangement (each room has four separate walls) while a value of 2.0 represents an infinitely large grid of contiguous rooms.

Room contiguity values from the eastern Tonto Basin local system (the RCD Study area and Arizona State University's Livingston Management Group and Schoolhouse Mesa) and the greater Southwest are presented graphically in Figure 9.6a–b. Contiguity indexes within the 3.5–4.0 interval indicate largely noncontiguous room arrangements and were defined as masonry compounds. Contiguity indexes of 3.3 or lower were defined as room blocks. These designations, derived from Tonto Basin sites, were then applied to a sample of 37 contemporary sites located throughout the greater Southwest (Fig. 9.6b). These values correspond very strongly with sites from southern Arizona with architecture previously defined as compounds (Group II), and sites from northern Arizona and parts of New Mexico and Colorado with architecture previously defined as pueblo room blocks (Group I). A boundary between sites with compounds and those with room blocks is illustrated in Figure 9.7. The Tonto Basin data are dispersed across the

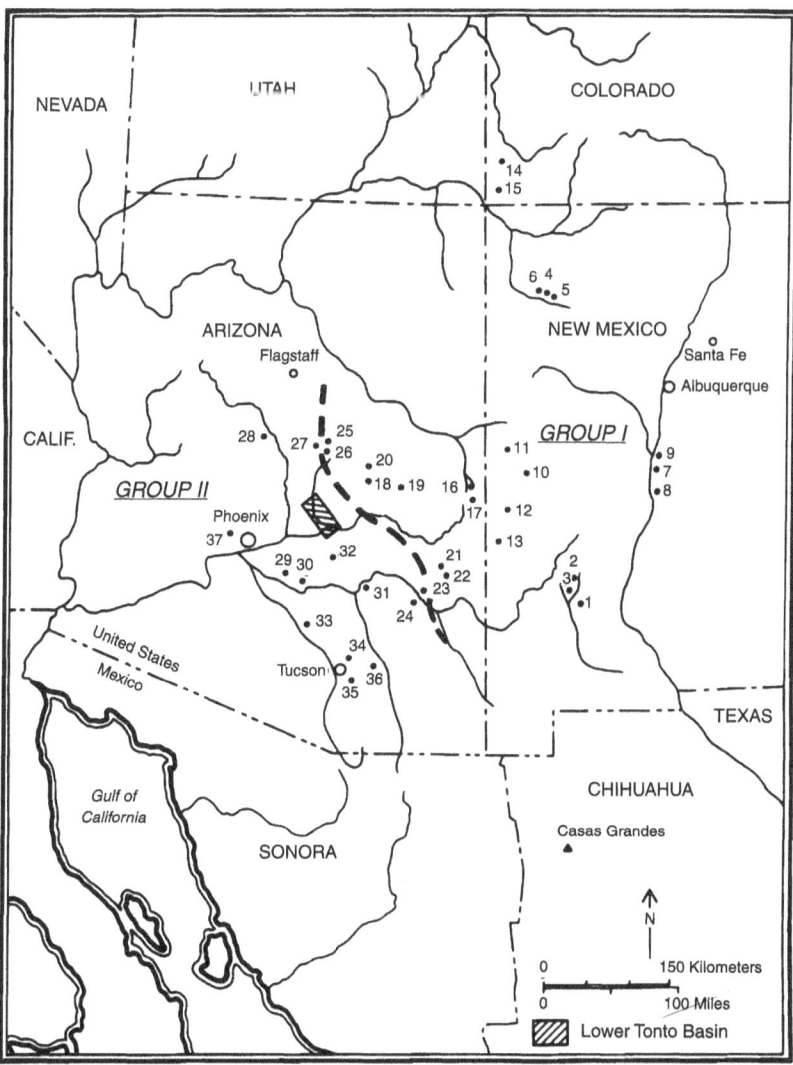

Figure 9.7. Spatial distribution of Group I (room block) and Group II (compound) residential units defined by the room contiguity index. Sites 1–27 contain Group I residential units, and Sites 28–37 contain Group II residential units.

entire spectrum of measurements, which suggests the presence of different cultural traditions. Furthermore, approximately 25 percent of the measured Tonto Basin sites or site components fall within an intermediate area between the two groups (Fig. 9.6b). This patterning may reflect the presence of two different architectural traditions within the same settlement (because the contiguity index is an averag-

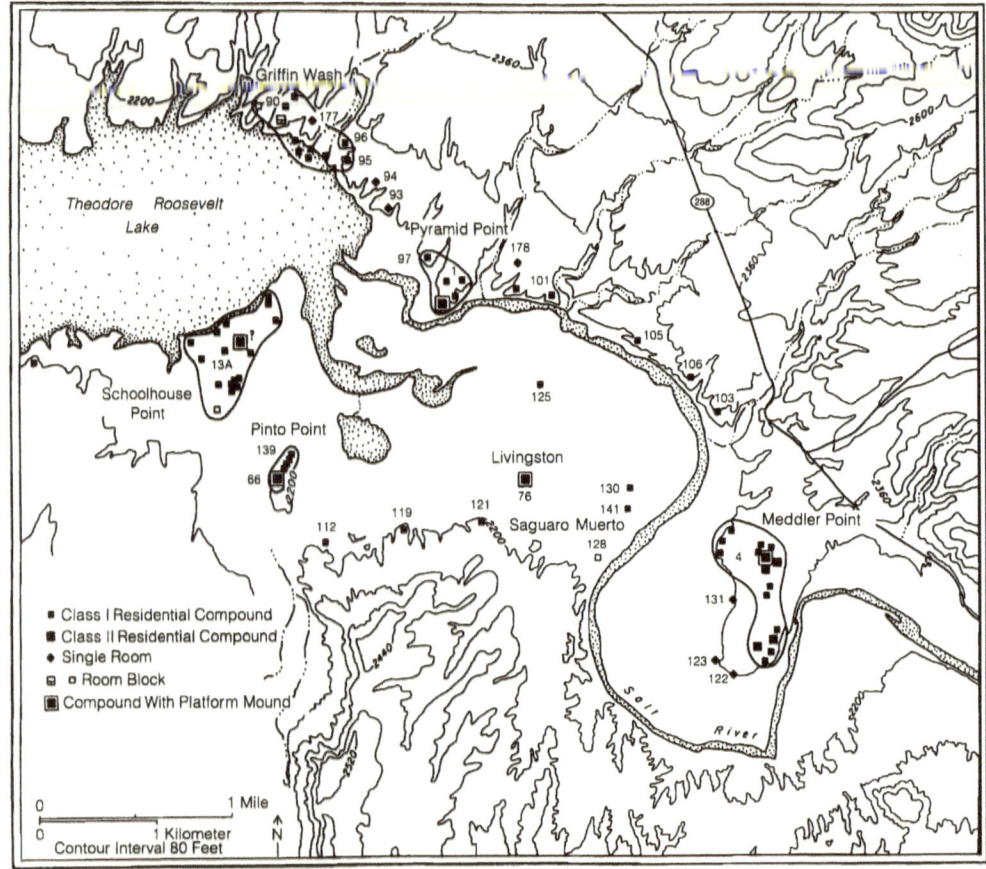

Figure 9.8. Distribution of Group I and Group II residential units in the eastern Tonto Basin during the early Classic period (ca. A.D. 1150–1350). Includes Roosevelt Community Development Study (RCD), Livingston, and Schoolhouse Point project areas and sites.

ing measure), as Redman (1993) has suggested for settlements within the nearby Payson Basin.

Differences in Settlement Morphology

Patterning of architectural technological styles and settlement morphology in the eastern Tonto Basin supports the hypothesis that the Classic period was characterized by immigration events, and that these immigrants used room block construction techniques. Large Classic period village settlements in the Tonto Basin (and southern Arizona in general) often included numerous, distinct compounds

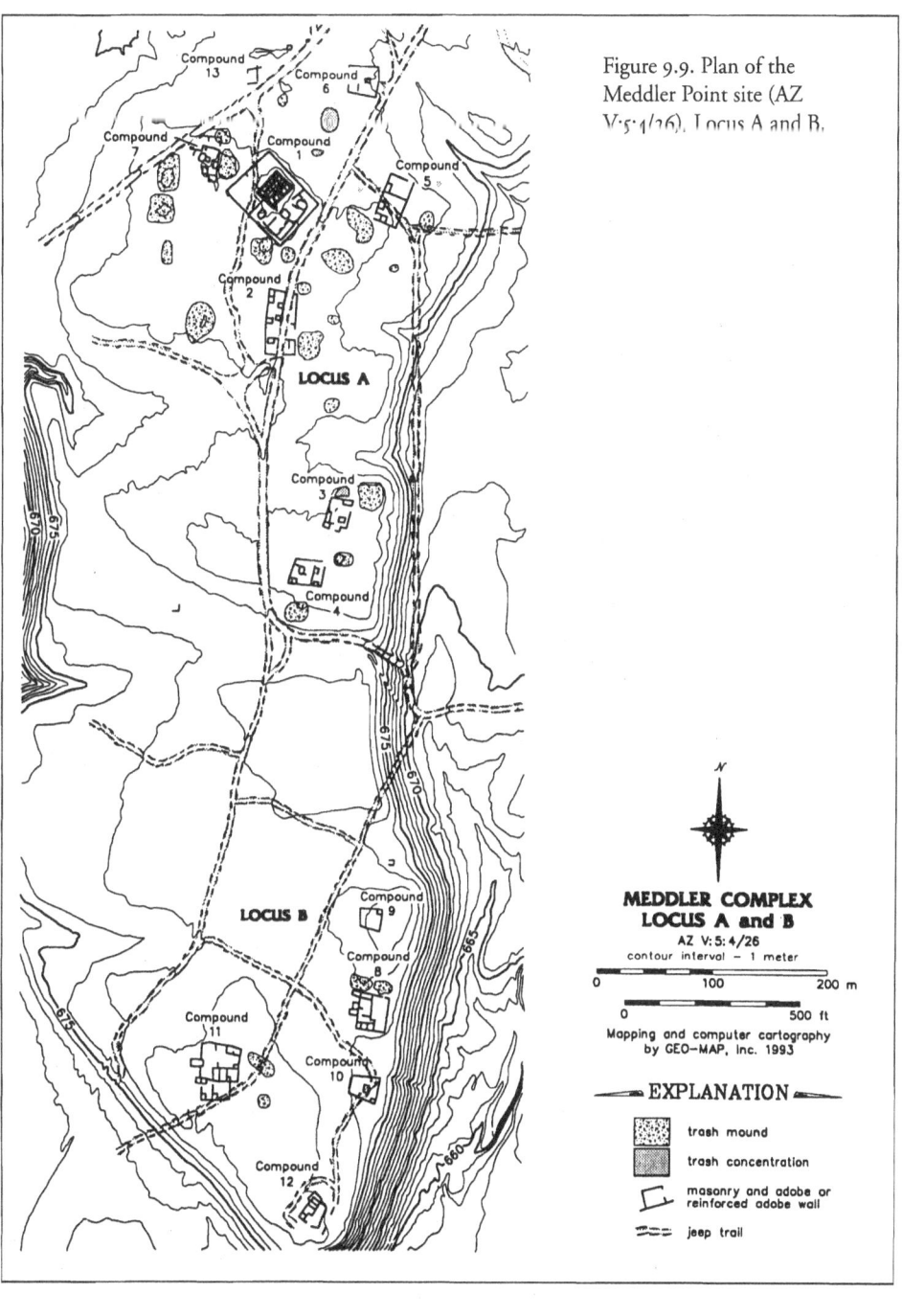

Figure 9.9. Plan of the Meddler Point site (AZ V:5:4/26), Locus A and B.

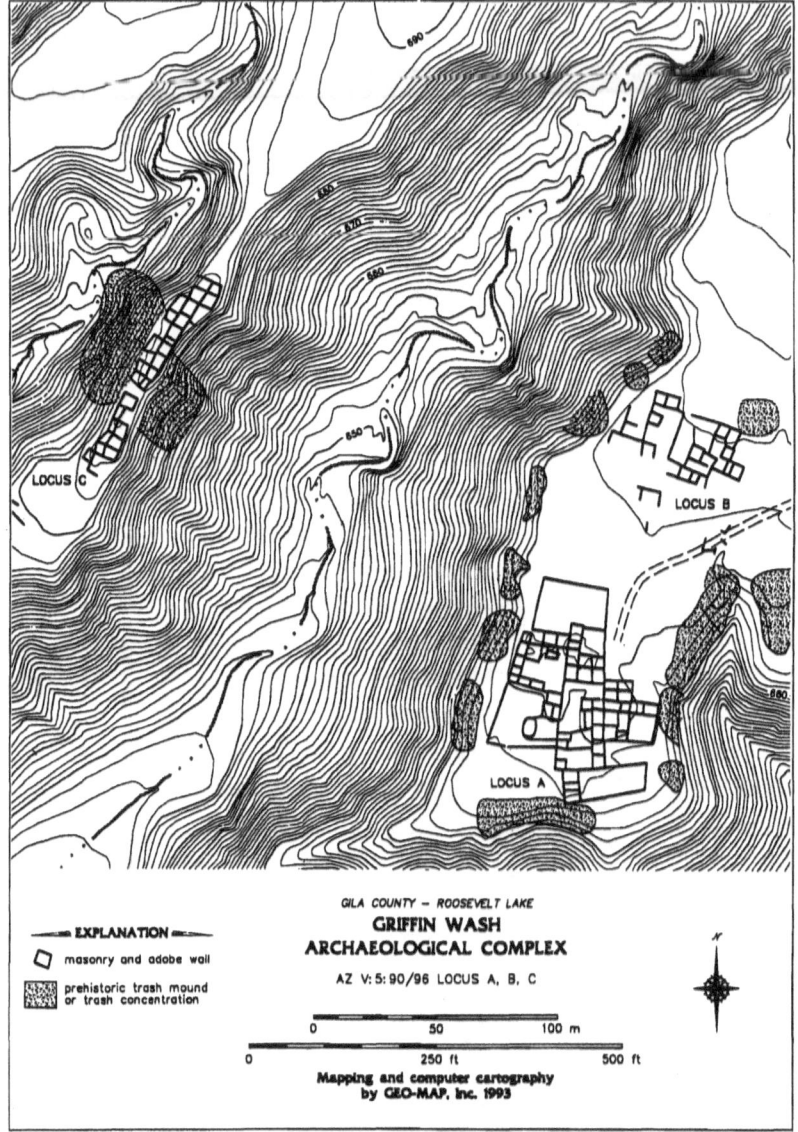

Figure 9.10. Plan of the Griffin Wash site (AZ V:5:90/96).

that were separated from one another by distances of up to 100 m. Some of the largest Classic period sites with compound architecture also have platform mounds, although some platform mound sites appear to have few associated residential units. This contrasts strongly with the spatial arrangement of room blocks that contain a similarly sized population within a much more restricted area.

Room block sites in the eastern Tonto Basin local system include Griffin Wash, Saguaro Muerto, and possibly several compounds within Locus B of the Meddler Point site (Fig. 9.8). Another room block, located within the Schoolhouse Point complex, may also have housed a migrant group (Lindauer 1995). In each case, portions of the settlement were constructed using architectural techniques that were anomalous for the Tonto Basin (Stark, Clark, and Elson 1995).

Differing conceptions of space among these settlements are most apparent when settlement morphology is compared between the early Classic period sites of Meddler Point and Griffin Wash. Meddler Point contains dispersed masonry compounds and an associated platform mound spread over a distance of approximately 1 km (Fig. 9.9). In morphology and appearance, Meddler Point represents the continuation of the earlier local tradition.[2] It shares certain similarities with contemporary sites in the Phoenix Basin as well. Griffin Wash, in contrast, represents a very different use of social and architectural space (Fig. 9.10). This site contains nucleated room blocks, lacks a platform mound, and has a pueblolike appearance. Griffin Wash shares similarities with pueblo settlements along the Mogollon Rim and in the Forestdale region to the northeast. Most important, Meddler Point and Griffin Wash were occupied at the same time, contained very similar population sizes, are situated less than 6 km apart, and were participants in the same local system.

If physical distance and barriers to circulation translate into social distance (see Hillier and Hanson 1984), then relations among room-block residents must have differed substantially from relations between compound residents. Noncontiguous rooms and other forms of architectural segregation in compound residential units emphasize either separation of activities in these areas or the separation of social units that resided within them. Contiguous room arrangements within room blocks, on the other hand, suggest greater integration and perhaps a more developed sense of community (see, for example, Hegmon 1989; Kent 1990a).

Patterning in Classic Period Utilitarian Ceramics

The most significant change in utilitarian ceramic production by the beginning of the Classic period involved the adoption of corrugated pottery. Temper compositional studies indicate that manufacture of plain ware and red-slipped pottery also continued during this period throughout the Tonto Basin (Stark and Heidke 1995). Corrugation technology, seen previously only in intrusive ceramics, became the predominant locally made culinary ware during the twelfth and thirteenth centuries. Most of it was manufactured using tempering materials from the Sierra Ancha (or eastern) side of the Tonto Basin (Miksa and Heidke 1995).

Previous studies (Doyel 1978:195; Tuggle 1982:27–28) reported limited quantities of corrugated ceramics from the late twelfth century in nearby regions: the Globe-Miami area (approximately 50 km to the south) and the Q Ranch region (approximately 50 km to the northeast). However, local production of utilitarian ceramics in the Tonto Basin surged after A.D. 1250 (Wood 1987). Corrugation technology appeared several centuries earlier (shortly after A.D. 900) in the Cibola and Mimbres regions east of the Tonto Basin, as well as in the Kayenta Anasazi area to the north (LeBlanc 1982; Mills 1987; Plog and Hantman 1986).

The construction of corrugated ceramics required substantial changes in the local technological style (Stark 1995c). Whereas the plain ware and red ware ceramics were made using paddle-and-anvil forming techniques, corrugated ceramics were manufactured using coil-and-scrape methods. Decorative forming techniques also varied, as corrugated ceramics required a textured surface finish not used previously to make culinary pottery. Corrugation technology was not widely practiced among Phoenix Basin Hohokam populations, suggesting that the technical knowledge required to corrugate ceramics was never successfully transferred into this area. At the same time, corrugation was the preferred decorative forming technique for cooking pots throughout the mountain and plateau regions during the late Prehistoric period.

Two styles characterize the Tonto corrugated tradition (see Wood 1987). One is an unslipped variety that consists primarily of jars, while the other is a slipped corrugated tradition that commonly includes interior smudging on both bowl and jar forms. Smudging was incorporated into the local technological style earlier in the sequence, and is seen on plain wares. Some corrugated ceramics recovered from Classic period contexts were also made in the Mogollon regions to the north (Mogollon Rim) and to the northeast (east-central Arizona). Examples of these three types—Tonto Corrugated (local), Linden Corrugated (Mogollon Rim), and McDonald Corrugated (east-central Arizona)—are illustrated in Figure 9.11.

Production steps used to manufacture Tonto Basin corrugated ceramics resembled, but did not duplicate, technologies of neighboring traditions. Corrugation coils in Tonto Basin ceramics are wide and exhibit a smeared or obliterated appearance, unlike the narrow indented clapboard coils found in contemporary traditions to the northeast. These technical traits (wide coils, uneven indentations) make Tonto Corrugated pottery distinctive from traditions of surrounding regions. Differences in the nature of available clays may have affected the technical expression of locally made corrugated ceramics. Local ceramic traditions, developed by emerging communities of potters, probably also shaped the technological style seen in this pottery. The obliteration technique became widespread in traditions to the north and east of the Tonto Basin by the middle to late thirteenth century.

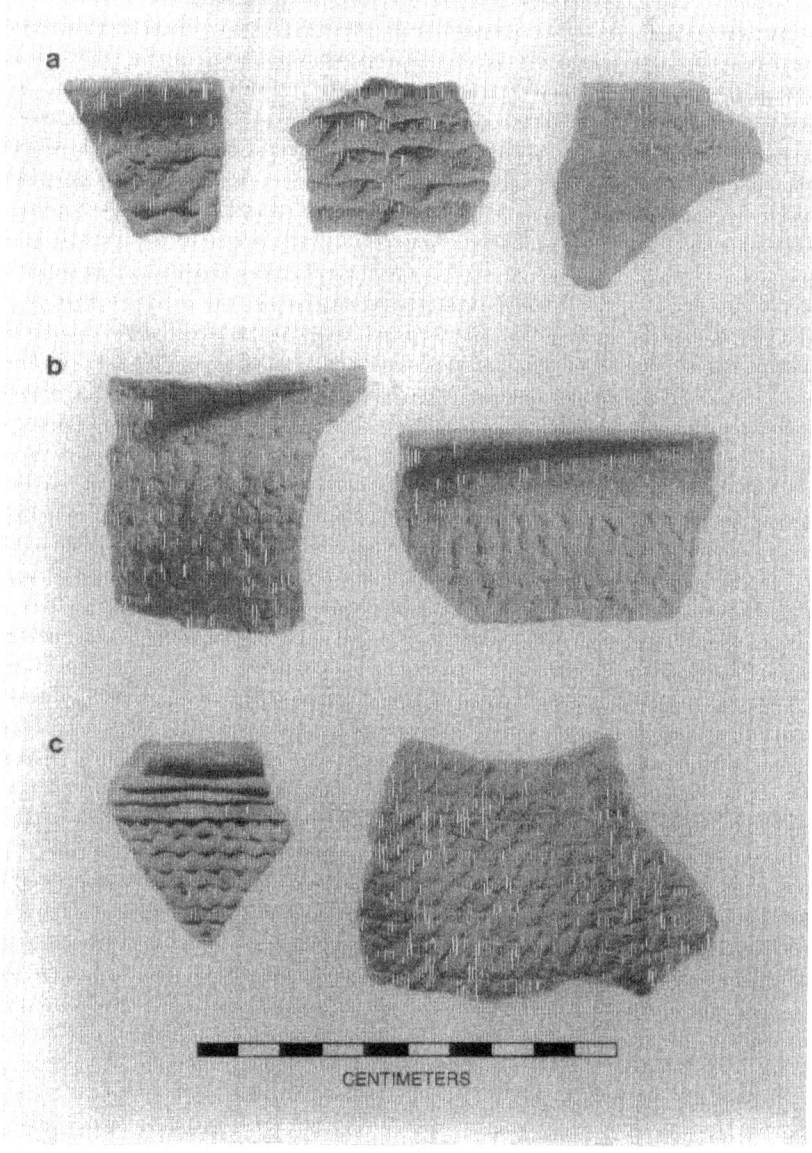

Figure 9.11. Varieties of thirteenth-century corrugated ceramics recovered from Classic period contexts: (a) Tonto Corrugated; (b) Linden Corrugated; (c) McDonald Corrugated.

SUMMARY AND CONCLUSIONS

Use of a technological approach provides new methods for exploring local social boundaries for the Colonial and Classic periods of the eastern Tonto Basin. Examination of architectural form, settlement morphology, and technological style of utilitarian ceramics suggest a movement of people from the Gila River area into the eastern Tonto Basin during the Colonial period. Although earlier studies (Doyel 1976; Haury 1932; Rice 1985) suggested a Colonial period migration into the Tonto Basin, they could not pinpoint more than a general geographic origin for these groups. Technological analyses of architecture and ceramics provide clues regarding source areas for these migrants. Application of a technological approach to early Classic period archaeological data point toward the existence of multiple cultural traditions in some settlements that functioned as parts of defined local systems.

The technological approach that we advocate provides an alternative to the information exchange model used widely in Southwestern archaeology. This is especially clear in the Classic period example, where comparisons of technical choices and ideas of space reveal a duality in architectural and ceramic patterning in villages within local systems. Although the possibility that such a duality had existed has been suggested previously (Gladwin and Gladwin 1935; Reid 1989; Whittlesey and Reid 1982), little empirical research had been undertaken to support this notion. Analytical methods are used in this method to evaluate previous ideas and provide substantive support for this interpretation.

Studies that rely on stylistic variation measured through decorative or other parameters, and that ignore technological styles, focus instead on broader interactional scales and overlook the local community. In the Classic period Tonto Basin, an exclusive focus on platform mound distribution at the expense of domestic architecture obscures fine-grained cultural variability that is evident in settlement morphology and contrasting uses of space. Likewise, the predominant emphasis on distributional patterning in decorated ceramics, particularly on Salado Polychromes, overlooks the existence of multiple variants in the local technological style.

Interest in understanding social boundaries continues unabated in Southwestern archaeology. One implication of this study is that patterning in stylistic variation (in monumental architecture or in decorated ceramics) sheds light on broad social systems whose boundaries may have been largely invisible to local populations. Contained within each of these larger regional or macroregional systems were numerous local systems whose participants were economically and socially interdependent. It is at this level—the local level—that issues of cultural affiliation probably mattered. Techniques offered in this study provide new methods for understanding local conceptions of group identity.

ACKNOWLEDGMENTS

Research for this paper was conducted through Desert Archaeology, Inc., for the Bureau of Reclamation in connection with modifications to the Roosevelt Dam. Fieldwork for the Roosevelt Community Development Study was conducted under terms of a permit from the Tonto National Forest. Special thanks are due to William Doelle, Catherine Cameron, and Stephen Lekson for their astute commentary on an earlier draft of this paper, and to James Bayman and David Gregory for thoughtful conversations on the topic. Thanks also go to Ronald Beckwith, Ellen Knight, and James Holmlund for generating figures used in this paper; we also thank Jo Lynn Gunness and Joe Singer for additional graphics assistance. We also thank Irina Hynes, Donna Breckenridge, and Elizabeth Black for help in the production process. Stark also thanks the Conservation Analytical Laboratory (Smithsonian Institution) for research support during the completion of this paper. We take full responsibility, however, for ideas expressed in this study.

NOTES

1. Times given for travel are based on the ethnographic work of Drennan (1984) who estimates approximately 36 km per day by foot.
2. Although a shift occurs from pithouse architecture to above-ground masonry architecture at the start of the Classic period, research strongly suggests that spatial patterns are maintained across this transition (Clark 1995a; Sires 1984, 1987). The pithouse courtyard group is transformed into the masonry compound room group, and spacing between the individual groups is similar. This continuity is supported by data from Meddler Point, where almost every Classic period compound sits atop an underlying pithouse cluster (Craig and Clark 1994).

10

Habitus, Techniques, Style: An Integrated Approach to the Social Understanding of Material Culture and Boundaries

MICHAEL DIETLER AND INGRID HERBICH

The detection and understanding of social boundaries of various types in the archaeological record is a problem with which archaeologists have grappled for many decades. Interest in this issue first developed during the period when archaeologists began to abandon the construction of artifact typologies linked to a unilineal evolutionary model as the primary goal of the discipline and to move instead toward the exploration of regional diversity (cf. Trigger 1989; Willey and Sabloff 1993). Initially, this involved the delineation of archaeological units called "cultures" (on the basis of the regional spatial distribution of associated stylistic similarities in selected aspects of material culture) and attempts to reconstruct a "history" of the interactions between these cultures, of their movements over the landscape, and of changes in their configurations (e.g., Childe 1929, 1956).

In Europe, especially, this research program can be traced to the fact that archaeology developed as a discipline largely as a backward extension of history, meaning essentially national history. Hence, the search for national origins in "ethnic" groups of the past was a strong stimulus in defining research goals (e.g., see Cleuziou et al. 1991; Dietler 1994; Härke 1991; Sklenár 1983). The interpretation of these archaeological constructs termed "cultures" in a fairly straightforward manner as the remains of ancient ethnic groups resulted from an absorption of ideas from nineteenth-century Romantic nationalist historiography and of the *Kulturkreis* concept developed by Kossina; and this was later bolstered by the organic model of culture derived from the functionalist tradition in social anthropology (cf. Herbich 1981; Shennan 1989; Trigger 1978; Veit 1989; Wotzka 1993).

During the 1960s, the overly simplistic aspects of this perspective were subjected to a withering attack by archaeologists such as Binford (1965) and Clarke (1968), and it appeared that the idea of a straightforward and predictable relationship between archaeological cultures and ethnic groups of the past had been dealt a fatal blow. However, while the conceptual framework under which this early research was launched has been largely abandoned, the reconstruction of the boundaries of ancient social groups has remained an important research goal for many archaeologists. What distinguishes a number of the more recent attempts to address this issue is a more explicit concern with the theoretical justification and methodology of the endeavor and, at least for some, a more nuanced conceptualization of the complex and fluid nature of social and cultural identities and of the contextual definition and negotiation of boundaries of various kinds. Nevertheless, it is our impression that progress in improving our understanding of this perennial issue has been elusive and often hampered both by the parochialism of different national and regional traditions of analysis and by the hasty adoption and perpetuation of simplistic, reductionist interpretive formulas.

More substantial progress in the pursuit of social groups and boundaries in the archaeological record, to the extent that this may be possible, requires that the problem be situated in a larger theoretical context that addresses the more general issue of the relationship between material culture and society. The imperative centrality of a social understanding of material culture stems from the twin facts that, as Trigger has noted, "prehistoric archaeology is the only social science that has no direct access to information about human behavior" (1989:357) and, as Appadurai has put it, things "constitute the first principles and the last resort of archaeologists" (Appadurai 1986a:5). All archaeological inference about past societies (including, potentially, the identification of social groups and boundaries) hinges critically upon an understanding of the relationship between material and nonmaterial aspects of culture and society: left with only remnants of the former, we seek to use them to perceive and comprehend the latter. That is the essence of the archaeological endeavor.

Clearly, archaeologists with an interest in the exploration of ancient social groups can approach them only through the delineation of material culture patterns and boundaries in the archaeological record. But, an interpretation of those patterns and boundaries requires a theoretical understanding of the full range of social processes that might have produced them. Moreover, this theoretical understanding should also guide the further improvement of appropriate strategies for the identification of patterns. In other words, the attempt to study social groups and boundaries of the past requires a coordinated self-conscious consideration of both: (1) the conceptual tools by which archaeologists define patterns, and (2) in what ways, and to what extent, the patterns they define may be related to social and cultural identity. However, coming to grips with these issues necessitates an exploration of two broader domains of anthropological inquiry. In the first place, it requires the pursuit of an understanding of the nature of material culture systems as social and historical phenomena. Secondly, it requires an ethnological understanding of the nature and reproduction of social groups, of the construction of identity, and of the nature and function of boundaries. For reasons of space, we focus here primarily on the former domain, and offer only a few comments on the latter.

THE SOCIAL UNDERSTANDING OF MATERIAL CULTURE

Over the course of the past couple of decades, archaeologists have become increasingly aware of the limitations of our rather rudimentary understanding of the crucial relationship between material and non-material aspects of culture and society. This has led a number of them to turn to the ethnographic study of material culture in living contexts, where both sides of this relationship can be observed, as a method of developing a set of theoretical tools by which to craft a more adequate window of entry for perceiving social relations and processes in ancient societies. The comments offered here stem from our experience of having undertaken several years of such "ethnoarchaeological" research among the Luo people of western Kenya.[1]

In approaching the study of material culture in an ethnographic context with the pragmatic desire to better perceive and comprehend ancient societies, two questions impose themselves as being fundamental. The first is "how does material culture originate in its social context?" That is, what are the social processes and structures that condition the production and reproduction of material culture? The second major question is "what social and technical roles does material culture serve and in what ways does material culture, in the performance of these roles, reciprocally affect social structures and processes?" An understanding of the interactive nature of the relationship is essential for comprehending its dynamics, the forces which direct the course of change. Moreover, it is only in the context of these larger questions that one can begin to engage in the more specific attempt to understand the role of material culture in the formation, expression, and reproduction of identity and to assess the feasibility of using remnants of material culture to identify social groups and boundaries of the past.

Of course, these are ambitious questions, and we are not proposing to be able to answer them definitively here. More modestly, we wish to explore a theoretical perspective that holds some promise in explicating some of the connections in this complex relationship and suggests further productive avenues of research. It is unlikely that a realistic general understanding of these issues will come easily or that such an understanding will produce some handy simple formula of ready utility to archaeologists (see also Lemonnier 1986, 1993b). Rather, we must be prepared to face squarely the complexity of the phenomenon and to commit ourselves to a rigorous long-term pursuit of the anthropological study of material culture.

Toward this end, we present here a critical comparative discussion of several recent archaeologically and ethnoarchaeologically derived approaches to the social understanding of material culture in order to identify both some key theoretical issues and various problems exhibited by these approaches in dealing with those issues. We then propose an alternative approach crafted from insights developed during our experience of ethnoarchaeological research in Africa and incorporating theoretically useful elements from (a) Bourdieu's (1977, 1980) theory of practice, (b) the study of material culture embodied in the French *technologie* or *techniques et culture* school, (c) the anthropology of consumption, and (d) an historically informed cultural economy perspective.[2] Finally, examples from the ethnoarchaeological research are used to demonstrate the utility of the proposed approach for understanding material culture as a social fact and to indicate the theoretical and analytical prerequisites to consideration of social groups and boundaries in the material record of the past.

MATERIAL CULTURE AND TECHNIQUES

In approaching the study of material culture one must begin by explicitly emphasizing a fundamental distinction between *things* and *techniques* which should be quite obvious, but which bears exhortation. The distinction is one between object and process. Things are physical entities that occupy space; they are what archaeologists recover as evidence. Techniques are those human actions that result in the production or utilization of things. From an archaeological perspective, they are one order of inference removed from things. Fortunately, things often preserve in their physical attributes and in their archaeological contexts clues that may inform, through a process of analogical interpretation, about the techniques employed in their creation and use. Moreover, things are made, exchanged, used, and discarded as part of human social activity. Hence, both things and techniques are embedded in and conditioned by social relations and cultural practice, and this fact holds out the promise that an understanding of this complex interrelationship may inform about society and culture in general.

However, this analytically and methodologically important distinction between things and techniques has been too often ignored by archaeologists (and even some ethnoarchaeologists) seeking to understand the social significance of material culture. Clearly, in order to adequately address the two fundamental questions posed earlier, one must look first at the ways that things are created and used in daily practice. The mediating process between things and society, and the key to understanding their reciprocal relationship, is techniques. Unfortunately, those archaeologists (at least within the Anglophone community) who have focused most intensely upon the interpretation of the social significance of things have tended to pay little serious attention to techniques, and (as will be explained below) have attempted to infer (or "read") social and cultural information directly from what is called the "style" of artifacts without a full investigation of and a realistic appreciation for the processes by which style is created (e.g., Hodder 1982; Wobst 1977). But style results most immediately from techniques; and it is only by studying techniques, with the full range of social and physico-technical constraints to which they respond, that we can arrive at an understanding of the social forces and relations that condition material culture.

The concept of style has played a major role in archaeological approaches to the social significance of material culture (see below), and it illustrates very well the importance of what might risk being viewed as an overly fastidious insistence on the distinction between things and techniques. The term style is often used to describe either of two phenomena. It may be used to designate characteristic ways of "doing things" (i.e., performing actions), what Mauss (1935) referred to as *techniques du corps*. Alternatively, it is more commonly used to designate characteristic patterns of material attributes in objects resulting from *some* of those ways of doing things (i.e., from techniques of production). We will refer to these two senses as *style of action* versus *material style*. Very frequently the two senses are conflated without a recognition of the difference or its significance. This issue will be more fully explored later, but the importance of the distinction may be briefly hinted at here by noting that not all explanations of why people perform actions in characteristic ways can be related directly to intended effects in a material product. Like history, although style is always the product of purposeful human action, it cannot be simply understood, or "read," as the consciously intended product of that action. Moreover, it is well to remember that the static, frozen pattern of traits which constitute material style is not the result of an instantaneous act of creation, but rather of a temporally extended process that is best conceptualized in the *chaîne opératoire* model developed in the French school of *technologie*.[3]

ARCHAEOLOGICAL APPROACHES TO THE SOCIAL SIGNIFICANCE OF MATERIAL CULTURE

The most common way that Anglophone archaeologists have attempted to deal with the social dimension of material culture has been to separate out three discrete aspects called "style," "technology," and "function" (cf. Braun 1983; Bronitsky 1986; Plog 1980a, 1983; Wright 1985). Frequently, analysts selectively specialize in the study of one of those aspects. By *technology*, in this narrow sense, is usually meant the techniques and materials used in the primary production of objects (most commonly, ceramics, stone tools, or metal goods). *Function* is usually taken to mean what might be called "utilitarian" or "instrumental" (as opposed to "social") function: it refers to those techniques that objects were designed to perform as "tools" acting upon matter. *Style* has a variety of meanings for archaeologists, although these are often somewhat ambiguously treated and are rarely very clearly or consistently defined. However, in the most general sense, whatever the differences in definition, it is usually considered to correspond to that aspect of material patterning which is thought to respond to primarily social and cultural demands or constraints (i.e., it serves a "social function" or is a residue of social action); hence it is the realm where most archaeological attention has been focused by those interested in the social significance of material culture. However, as the following discussion will show, exactly how one identifies style and how one interprets its social significance and role are subjects of considerable controversy.

Perhaps the most common way of identifying style has been to locate it negatively in relation to function and technology: it is thought to consist of those aspects of material patterning that remain after the latter two aspects have been accounted for. In other words, it is to be located in those attributes of objects that have no discernible role in affecting their utilitarian performance in the context of use (the domain of function) and that do not result from technical constraints in the context of their manufacture (the domain of technology). The presumption has been that these "residual" attributes were therefore included for reasons having to do with social processes. In the case of ceramics, since the possible effects of other attributes on performance are often difficult (or impossible) to evaluate, this has meant that studies of style have tended to focus almost exclusively upon what Sackett (1982) calls "adjunct form": traits that were presumably "added on" either to perform some social function or as a passive residue of social action. In practice, this has meant that the concept of ceramic style, by tacit definition, has become virtually synonymous with "decoration" (i.e., surface treatment patterns). In the case of stone tools, which have no obviously distinguishable decoration, style has generally been located in those patterned aspects of form that differ on implements assumed to be used for identical utilitarian functions.

This negative approach to identifying style, in both its manifestations, has some serious problems. If one is interested in understanding the full social significance and roles of objects, then a focus on decoration alone is unsatisfactorily narrow: it is now clear from ethnographic studies that decoration is of highly variable significance in relation to other physical attributes and its meaning cannot be comprehended in isolation (see Dietler and Herbich 1989, 1994b; Lemonnier 1986, 1990; Sackett 1982, 1990). If, as in the case of stone tools, one is attempting to compare implements of the same utilitarian function to detect their stylistic differences, it is, at a minimum, necessary to have some means of verifying the similarity of function which is independent of analogies based upon the formal characteristics one wishes to study. Moreover, as Sigaut (1991) has pointed out, from an ethnographic standpoint, this basic conceptualization of "function" and its relationship to form is naively oversimplified and severely limited.

These difficulties highlight the dangers of artificially separating style, function, and technology in this way and correlating these domains of material patterning with separate social and techno-utilitarian domains of action. While this analytical strategy may have limited heuristic utility for addressing some problems (e.g., see Wright 1985), it is not a productive approach for understanding the social significance of material culture. In so doing, one has unrealistically limited the possibilities for comprehending interrelationships between the domains and for perceiving techniques as "social facts." For example, such a conceptualization masks the role of social and cultural factors in conditioning technical choices and functional evaluations (Lechtman 1977). Moreover, ethnographic evidence has shown that whole "technical systems" are embedded in social processes and relations: both broad system strategies and choices made at all stages of *chaînes opératoires* of production and use are aspects of social action and cultural concepts that result in the production of material style (see Dietler and Herbich 1989; Lemonnier 1986, 1990, 1992). This is why, as an essential prerequisite to developing a social understanding of material culture, we have argued vigorously for a more integrated view of material style encompassing patterning in technological, formal, and decorative aspects (Dietler and Herbich 1989; Herbich 1987; Herbich and Dietler 1991) and for a corresponding approach to the production of material style based upon the concept of the *chaîne opératoire* (Dietler and Herbich 1989, 1994b). Among the principal advantages of this approach is that it allows one to view the production of material style as a temporally extended series of interrelated operational choices rather than as an instantaneous act of creation.

Most archaeologists share the assumption that material style is a key to understanding the social dimensions of material culture and would concur with Mauss (1930:470) in his statement that "Le domaine du social, c'est le domaine de la modalité."[4] However, granting that something like material style, by whatever definition, may be adequately identified within objects, there have been many conflicting interpretations of its social origin and roles. A primary division can be identified between those who see style as primarily a passive reflection of social behavior or of shared cultural concepts and those who see it as a more active "tool" in strategies of social action (cf. Sackett 1990; Wiessner 1990).

The former view of style as a passive residue represents a longstanding tradition in archaeological research. It was this implicitly accepted relationship that allowed Childe (1929) to identify archaeological "cultures" and correlate them with peoples. With the advent of the "New Archaeology" during the 1960s, craft-learning patterns were identified as a primary social mechanism underlying this view (albeit without much empirical investigation of the process in ethnographic contexts). Using an idealized version of this mechanism as a basis, several archaeologists (e.g., Deetz 1965; Longacre 1970) advanced interpretations of stylistic patterns as reflections of social organization (specifically, of post-marital residence patterns and kinship structures). These interpretations have now been shown to have suffered from a number of methodological and theoretical problems, including, most seriously, an overly static and stereotypic view of the process of craft learning and a rather limited understanding of its social context and relationship to material culture patterning (see Herbich 1981, 1987).[5]

Structuralist approaches to material culture (at least those which hold closest to the structuralism of Saussure and Levi-Strauss) may also be recognized as adopting an essentially passive, reflective view of style (cf. Deetz 1977; Glassie 1975; Hodder 1982:125–184). However, in this case, stylistic patterns are seen to be a surface manifestation of deep cognitive structures that also generate structures of social organization, myth, ritual, and other aspects of culture. Studies in this vein have tended to neglect intracultural variation as a significant phenomenon and to exclude the role in the production and reproduction of culture of socially situated subjects with different cultural competencies and different, often contradictory, interests. Where culture is viewed simply as a reflection, or an effect, of uniformly shared cognitive structure rather than as an historical social process, there is little scope within such an essentially static perspective for understanding change in either styles or society.

To a certain extent many of these problematic limitations are a result of the fact that little attention has been paid by structuralists to the actual processes by which material style is generated: the focus is on pattern rather than process. But a program that seeks to elucidate the relationship between different kinds of structures is severely limited to the extent that it fails to consider the activities that actually create the material manifestations of those structures. As Sahlins has observed more generally, "If structural/semiotic analysis is to be extended to general anthropology on the model of its pertinence to "language," then what is lost is not merely history and change, but practice—human action in the world" (Sahlins 1981:6). As will be discussed later, we hold this concept of practice as central to understanding the social significance of techniques.

Various other cognitive approaches to style have been developed without necessary reference to specific theoretical models purporting to explain the generative basis of the phenomenon as a social fact. Many of these, such as Hardin's (1984) studies of decorative design schemes and Washburn's (1977) approach to design symmetry, have at least contributed greatly to the advancement of our methodological sophistication in characterizing decorative aspects of material style for comparative analysis. Perhaps the most explicit explorer of the passive perspective on style has been Sackett (1982, 1990). Generally eschewing the term style, he has championed an alternative conception called "isochrestic variation" to describe the different ways people have of making and using things for similar purposes and the resultant characteristic combinations of traits that constitute the distinctive material culture patterns produced within what he calls "ethnic" groups.[6] A primary focus of his analysis has been the definition of where, in material patterning, style may be seen to "reside"; and he has made a valuable contribution in refuting the pervasive style/function/technology schism and corresponding focus on style-as-decoration in favor of a more realistic conception of the interrelated nature of material culture attributes. Sackett (1990), correctly, holds his definition of the locus of "isochrestic variation" to be independent of an explanation of its social origin. While recognizing the occasional active "iconological" use of material style for intentional signaling, he has argued persuasively that isochrestic variation should be viewed primarily as a result of the transmission within "ethnic" groups of largely unconscious perceptions of the way things should look and be used. Unfortunately, he has been reluctant to explore in satisfactory depth the ways in which social relations and processes condition the traditions guiding the production and reproduction of isochrestic variation. This hesitation most probably stems from a recognition of the limits of theorizing from an archaeological base: such understanding can only develop out of primary ethnoarchaeological research through which these social features can be observed. In any case, despite its insightful contributions, this perspective ultimately lacks explanatory power to provide a convincing social understanding of techniques and material culture.

Much of Sackett's argument was developed in critique of the more active approach to style. Again, there are several competing perspectives in this camp all of which agree in seeing style primarily and essentially as a medium of communication. Perhaps the most popular among American archaeologists has been a view of material style as a tool for "information exchange" (e.g., Hegmon 1992; Plog 1980a; Pollock 1983; Wobst 1977). Based upon a seminal paper by Wobst (1977), this approach hinges upon a narrow definition of style (as decoration) and a hidden premise that is an ethnocentric neoclassical economic argument (see Dietler and Herbich 1989 for a detailed critique). At the risk of schematizing to the point of caricature, the core of the argument runs as follows: style is seen as something "affixed" to objects at an extra "cost" in time and labor (Wobst 1977:326); it serves a social function of communicating information, but with greater "costs of emission" than other modes of communication (Wobst 1977:322); the target group of that information and its message content can therefore be inferred from a cost/benefit analysis of the energy expended in "stylistic behavior" (Wobst 1977:325). Because an investment in the use of style to communicate with people who are in daily personal contact would be a redundant "dysfunctional waste of energy and matter" (Wobst 1977:325), the most efficient use of such stylistic information has generally been located in the symbolic communication of group (particularly "ethnic") boundaries and identity to outsiders (Wobst 1977:328–330).

This perspective has some fatal flaws, and its application to archaeological cases has led to some rather curious conclusions, such as a direct (and necessary) correlation between increasing complexity of ceramic decoration and increasing complexity of political organization (e.g., Pollock 1983). Ethnographic studies have demonstrated that the exclusive identification of decoration as material style or as a communication medium is untenable (Dietler and Herbich 1989; Gosselain 1992b; Lemonnier 1986, 1990), and the economistic argument depends critically upon this correlation (one can only assess the energy "cost" invested in the production of style if it is viewed as something extra added on to an object). Moreover, this reductionist functional perspective pushed to its logical conclusion has the effect of explaining the creation of material style as an intentional strategy exclusively for communicating social boundaries. It tautologically confuses one potential eventual role of style with a primary constitutive function and hence the cause of its creation. As will be explained later, it shares with some of the "text"-analogy approaches to material style a basic confusion between the concepts of *communication* and *signification*. Again, a central defect with this program is that no attention was paid to understanding the social context of manufacture. Assumptions about the generation of material style were simply extrapolated from observations derived exclusively from the context of use. These problems stem ultimately from a failure to appreciate the distinction raised earlier between things and techniques.

A related defect in Wobst's (1977) argument is that the general principle of using a sliding scale of the relative visibility of material media as a gauge of their efficiency for broadcasting messages of identity was extrapolated largely from a narrow range of observations about clothing in Eastern European peasant communities (see Bogatyrev 1971, for a more nuanced discussion of the social categories expressed in Moravian folk costumes and the function of clothing in reiterating status and role distinctions). However, the reasons that clothing (and other bodily adornment) is so often a medium for the expression of identity has much less to do with its relative position on some abstract scale of visibility and efficiency than with its uniquely close association with the body and the social inscription of concepts of personhood (cf. Comaroff and Comaroff 1992:69–91; McCracken 1990:57–70;

Sahlins 1976:179; Turner 1969, 1980). Clothing (along with cosmetics, body painting, tattooing, scarification, jewelry, headdresses, etc.) is part of "the social skin"; the frontier between society and the self that "becomes the symbolic stage upon which the drama of socialization is enacted" (Turner 1980:112). Far from being a "dysfunctional waste of energy," the redundancy of bodily adornment in reiterating social status and role distinctions among closely interacting members of a group is an important mechanism for the naturalization of social categories and behavioral expectations in the formation of personal identity.[7]

An interesting alternative approach to material style as a communication medium proposed by Wiessner (1983, 1984, 1990) attempted to circumvent the severe limitations of the "information exchange" model (especially the exclusive focus on decoration and the ethnocentric economic reductionism) by identifying style in a more inclusive and positive way through its social function. She began with a restrictive definition of style as those aspects of material patterning that serve to communicate information about relative identity among individuals and groups. Style in this sense has two potential aspects called "emblemic" and "assertive." The former kind of style (emblemic) involves the existence of a distinct referent and conveys a clearly recognized message about the division of the social world into distinct groups with boundaries (cf. Davis 1985). The latter kind of style (assertive) has no distinct referent or clearly recognized meaning and is largely a matter of personal expression.

However, not all material culture patterning is necessarily related to these kinds of expression of identity. Hence, while this definition may be logically valid and operationally feasible in an ethnographic context such as the San case investigated by Wiessner, this is a problematical approach for archaeologists, which must lead them inevitably into tautology. This definition may be viable for ethnographers who can evaluate such communicative behavior directly (as Wiessner attempted to do in her ethnoarchaeological research). However, in the absence of a developed theory that explicitly demonstrates which aspects of material culture patterning will be consistently used for such communication and which will not, it is clearly untenable logically for archaeologists to identify material style within a set of artifacts on the basis of a social function that cannot be observed but which must instead be inferred from the data as an explanation of their patterning. Moreover, as Barth pointed out in his seminal article on ethnic boundary signaling, "one cannot predict from first principles which [cultural] features will be emphasized and made organizationally relevant by the actors" (Barth 1969b:14).

Other communication advocates operating outside Wobst's "information exchange" framework have also focused upon the concept of signaling ethnic boundaries. Hodder's (1979a) often cited ethnoarchaeological study in the Baringo area of Kenya, for example, posited a connection between the use of material culture to signal ethnic boundaries and the presence of "economic and social stress." If this were true and of more general validity, it would point the way toward developing the theoretical links that, by their absence, are a major problem for the archaeological application of Wiessner's approach. Unfortunately, the utility of Hodder's study is highly dubious. In the first place, there is an alarming disparity between the extremely brief period of fieldwork involved (and the methods employed) and the sweeping social interpretation offered. A few weeks of ethnoarchaeological work are simply not sufficient to give anything but a very superficial impression of the complex social forces underlying material patterns, particularly in the absence of investigation of the process and social context of production.[8] Moreover, Hodder neglected to define clearly what he meant by "economic and social stress" in the numerous articles in which he presented this material, and no adequate measure of its relative intensity (or, indeed, even an adequate demonstration that it actually existed) was ever offered in the cases where it was invoked. Furthermore, there was no attempt to demonstrate or explain the critical social link between this vague "stress" and the production and use of material culture. The study simply mapped the spatial distribution of objects of quite disparate ages and then fashioned a nebulous synchronic correlation which was asserted as causation without ever moving beyond the context of use (or indeed, even exploring that social context in any depth).

A number of recent studies focusing on the communication function of material culture have attempted to identify a broader range of roles than broadcasting identity, including particularly the use of material style in the representation of the social relations of power and strategies of ideological manipulation. These studies, although often designated as "post-processual," actually vary far too greatly in theoretical inspiration and methodology to be usefully lumped under this unfortunate polemical rubric. The insightful work of Miller (1985, 1987), for example, is quite compatible with the perspective developed in this paper, especially in its exploration of the relevance of Bourdieu's concept of practice to the social understanding of material culture and its grounding in solid ethnographic research. Others, however, despite programmatic statements to the contrary, may be seen to hold in common a fundamentally idealist central concept that views material culture as essentially a medium of symbolic expression (cf. Hodder 1982; Shanks and Tilley 1987; Tilley 1989). Consequently, the analysis of material culture has been directed toward a quest for its "meaning." Often this has involved the analogical perception of material culture as a form of text, conveniently enabling the use of analytical methods derived from semiotics and textual criticism to "read" the encoded meanings.

This latter approach is subject to some serious criticisms. In the first place, analysis of material style has once again generally been focused narrowly on decoration as the text to be read (cf. Hodder 1991; Shanks and Tilley 1987:146–171). There has been a failure to recognize the complexity of the ways that both objects and techniques are imbued with meaning, to situate technical activity within the scheme of analysis, and to examine the actual process of practice by which material style is created (see Lemonnier 1990, for a detailed critique; and Hodder 1991, for a particularly egregious example purporting to read the geometric decoration on calabashes as a form of silent discourse by women about their oppression). Moreover, the overly literal analogy between text and material culture and the consequent borrowing of methods of linguistic and textual analysis are highly suspect endeavors that stem from, among other problems, a confusion of signs and symbols (cf. Sperber 1975; Yengoyan 1985) and a confusion of the concepts of communication and signification.

One of the fundamental distinguishing features of symbols is precisely that they are not like language and are not subject to analysis by semiological methods: they don't "mean" but rather "evoke," and they are not articulated like language (Sperber 1975). Furthermore, material culture is not a text; it is not a coherent sequential string of connected signs with "referential meaning" (see Finegan and Besnier 1989:173–174) created expressly and exclusively as an instrument of communication. Material culture is embedded in systems of symbolic expression but also in systems of practical action on matter. Hence, although material culture participates in processes of *signification* (objects may provoke emotional and intellectual responses and be invested with significance of various kinds by users and makers), it is not primarily a system of *communication* like language. The relationship between the intentions of the maker of an object and the significance attached to that object in the context of consumption is far less direct and far more complex and ambiguous than in the reading of a text produced by a writer.

An understanding of the social origin and significance of material culture will not come from "reading" decorations as text (see Lemonnier 1990). It requires a dynamic, diachronic perspective founded upon an appreciation of differences in the contexts of both production and consumption (see Dietler and Herbich 1994a), upon an approach to material style centered on the *chaîne opératoire* concept, and upon a rigorous examination of the link between objects and techniques in the contexts where they are generated, reproduced, and transformed.

HABITUS AND TECHNIQUES

This quick survey was by no means intended to be exhaustive, but rather to schematically indicate a range of strategies archaeologists have developed for grappling with the crucial issue of the social dimensions of material culture and to identify some problems with these approaches. Most of the theoretical debate concerning this issue has, of course, developed in the context of ethnoarchaeological research where, unlike archaeological cases, it is possible to actually evaluate the plausibility of one's concepts through participant-observation of social activity (even if, alas, this potential is not always fully realized).

The most frequently invoked basic difference in approach, that between the passive and active conceptualizations of style, may be recognized as a manifestation of the persistent central paradigmatic dichotomy of the social sciences, that between structure and agency (cf. Bourdieu 1977, 1980; Giddens 1979; Ortner 1984; Sahlins 1976, 1985). Some scholars posit that stylistic patterns are predominantly an unconscious reflection of social or cultural phenomena. In the case of structuralist analyses, material culture patterns are thought to conform to deep cognitive structures underlying all social relations and cultural practices (e.g., Deetz 1977; Glassie 1975; Hodder 1982:125–184). Others see them as a largely unconscious behavioral reflection of social organization, interaction, or membership in "ethnic" or other social groups or categories (e.g., Longacre 1970; Sackett 1990; Washburn 1977). Those who emphasize a more action-centered view of material culture tend to view it largely as a medium of communication and to emphasize the manipulation of material symbols in strategies of group boundary maintenance, ideological representation of social relations, or cultural categorization (e.g., Hodder 1982; Plog 1980a; Wiessner 1983, 1984; Wobst 1977).

These views are not necessarily contradictory; they are merely partial. Some tend to perceive social action as determined directly from the level of structure, while others focus upon symbolic action without adequate reference to the manner in which broader socio-cultural forces and material conditions structure or constrain perceptions and decisions. However, a realistic theory of material culture as a social phenomenon that can address the two questions posed at the beginning of the paper must account for both structure and agency by showing how the two are mediated through practice: that is, both how practice is conditioned by structure and how it reshapes structure in the process of reproducing it.

For reasons which will be explained below, we believe that Bourdieu's (1977, 1980) theory of practice, and particularly his concept of the *habitus,* has a great deal of potential in pointing the direction toward such a bridging conceptualization for material culture.[9] However, in order to effectively bring his ideas to bear on the domain of material culture we must reestablish a more holistic conception of material style and techniques incorporating the concept of the *chaîne opératoire* and focus analysis concurrently on both the contexts of production and consumption. This requires a foundation strategy pioneered in the French tradition of *technologie* or *ethnologie des techniques* (cf. Cresswell 1976; Haudricourt 1987; Lemonnier 1976, 1986, 1990, 1992; Leroi-Gourhan 1943; Mauss 1936; Schlanger

1991; Sigaut 1987, 1991), which pays close attention to the process of making choices at all stages of the *chaîne opératoire* of production. Finally, in order to fully understand the nature of the demands to which techniques respond, we must reject a vision that seeks structure and meaning in homogeneously shared, bounded cultures in favor of a view of culture as an historical social process. Our perspective here is shaped particularly by the anthropological literature on consumption (e.g., Appadurai 1986a; Bourdieu 1984; Douglas and Isherwood 1979) and by the somewhat fractious theoretical positions that may be loosely gathered under the label historical anthropology and cultural economy (e.g., Comaroff and Comaroff 1992; Mintz 1985; Roseberry 1989; Sahlins 1985, 1994; Wolf 1982).

Material style can serve as a useful concept for archaeologists attempting to investigate the social role and meaning of material culture only if it is seen as the objectified result of techniques (rather than as straightforward objectified information); and more specifically it must be seen as the result of characteristic ranges of responses to interlinked technical, formal, and decorative choices made at all stages of a *chaîne opératoire* of production (Dietler and Herbich 1989). Understanding material culture as a social phenomenon, including the processes of stability and innovation within their historical trajectories, then becomes a matter of understanding the factors that condition these choices, their interrelations, and the reciprocal effects stemming from new choices made at various stages of the *chaîne opératoire*. This approach requires that we understand craftspeople as social actors (rather than simply as products/bearers of culture or as acultural adaptive engineers) and that we understand the production and use of objects as social activity.

The theoretical work of Bourdieu (1977, 1980) offers a means of situating both material culture and the *chaînes opératoires* and social actors responsible for its production and transformation within a framework that mediates structure and agency. Bourdieu has argued that people develop "dispositions" to act in certain ways through the influence of the structures of material conditions in which they live. These systems of durable dispositions, called *habitus,* can generate patterned actions that appear regulated as if resulting from rules, but which, in fact, operate without reference to, or symbolic mastery of, rules. Techniques, as with other patterns of social activity, are formed through the *habitus*. This involves the development through practice of "tendencies" and cultural perceptions of the limits of the possible in patterns of choice at all stages of *chaînes opératoires*. These dispositions of choice and perceptions of the possible in the technical domain are interwoven with similarly formed patterns of choice and perceptions in the domain of social relations and cultural categories in ways that evoke and reinforce each other such that they come to be perceived as "natural."

This is a particularly useful way of viewing the process by which material culture patterns are assimilated and reproduced, particularly in the case of prehistoric (and pre-industrial) societies of the type usually studied by archaeologists where craft learning generally takes place through observation and emulation without the use of a formally articulated set of rules (see Herbich 1987). The reproduction of material culture then becomes more realistically situated in social life. Techniques are not seen as, in some sense, a secondary "product" of social activity or of social strategies, but rather it is recognized that dispositions which generate action in all domains of social life are formed together in the course of practice. This perspective would, for example, shift the focus of analysis from seeing material style as something intentionally "added on" in order to signal group identity (as do advocates of the "information exchange" approach), to seeing the process by which a sense of group identity is formed and transformed as being coeval with and identical to the process by which a sense of techniques is formed and transformed. Both appear to be a part of the natural order. This avoids the problem of confusing function with intention by recognizing that, while all social action is purposeful, the larger patterns that we perceive are the often unintended consequences of many choices made by social actors following different strategies but linked by certain common structurally conditioned tendencies toward action.

It should be emphasized that the *habitus* is not a static concept: one of its most attractive features is that, as the "generative principle of regulated improvisations" (Bourdieu 1977:78), it allows the perception of how practice both reproduces and transforms structure as it adjusts to demands. Rather than seeing practice as predetermined by a static set of cultural concepts or structures (e.g., some sort of rigid mental template), the *habitus* is a dynamic relational phenomenon which is both an historical product and agent. This is because, as a set of learned dispositions that allow the solution of daily technical and social problems through a process of structured analogical reasoning, the solutions to these problems influence the development of the dispositions. A certain latitude in action is thus possible as people respond to practical demands. Practice may alter gradually without marked consequence as long as there continues to be a close fit between the objective conditions and the subjective organizational system of dispositions.

This relation of correspondence results in a state of unquestioning perception of the "naturalness" of the social and material world that Bourdieu calls "doxa." However, demands sometimes lead practice into responses that call this correspondence into question in certain areas, leading to the formation of a domain of self-conscious discussion between positions that Bourdieu calls "heterodoxy" and "orthodoxy." When the arbitrariness of some of what was accepted as implicit axiomatic knowledge is exposed in this way it results either in rationalization and systematization of what was formerly an unconscious set of dispositions, or in overt social conflict. This is one way in which material culture and techniques can have an unintended impact upon the social relations and structures which generated them, and this is particularly significant for social change when the arbitrariness of social institutions and practices embodying asymmetrical relations of power are exposed.

In order to understand the course of change in techniques and material culture, it is necessary to understand: (1) the apparatus that structures responses to technical and social problems, demands, or opportunities (i.e., the *habitus*), (2) the material conditions that influence the formation of the dispositions that constitute the *habitus*, and (3) the origin and nature of the problems or demands that provoke responses. It must be recognized that such demands originate simultaneously at several levels, from small-scale interpersonal relations to the supra-regional political economy. This means that, to reiterate once again, explanations of change that reside at the level of homogeneous, bounded cultural structures are inadequate. Concepts derived from anthropological work on consumption and historical cultural economy must be brought to play in exploring, at a range of scales in the contexts of both manufacture and use, the material conditions of the development of techniques and the transformation of strategies of choice in *chaînes opératoires*.

DEMONSTRATION OF THE APPROACH

In order to illustrate the utility of this perspective for understanding the social significance of techniques, it is necessary to very briefly examine a few empirical cases. The two examples used are drawn from ethnoarchaeological work among the Luo people of western Kenya.

Luo is a Nilotic language spoken by some two million people occupying a territory of about 10,000 sq km surrounding the Winam Gulf of Lake Victoria (Fig. 10.1). The regional settlement pattern is characterized by polygynous, patrilocal, three-generation extended families living in separate homesteads scattered over the countryside. Each homestead undergoes a life-cycle of foundation, growth, and abandonment, and the landscape is composed of interspersed homesteads representing all stages of the cycle. The economy is based upon horticultural production of a mixed assortment of grain, legume, and root crops. Fish from the gulf are traded widely over the area, and cattle also play an important role in the economy as a measure of wealth and as a source of milk and occasional meat. Cash cropping is poorly developed in the northwestern third of Luo territory where our study was focused (Siaya District), although it is more prevalent in the other two districts. In Siaya District the local economy is weakly linked to the national economy by a network of regular periodic rural markets and somewhat more strongly by a pattern of male wage-labor outside the region. The markets provide a source of cloth-

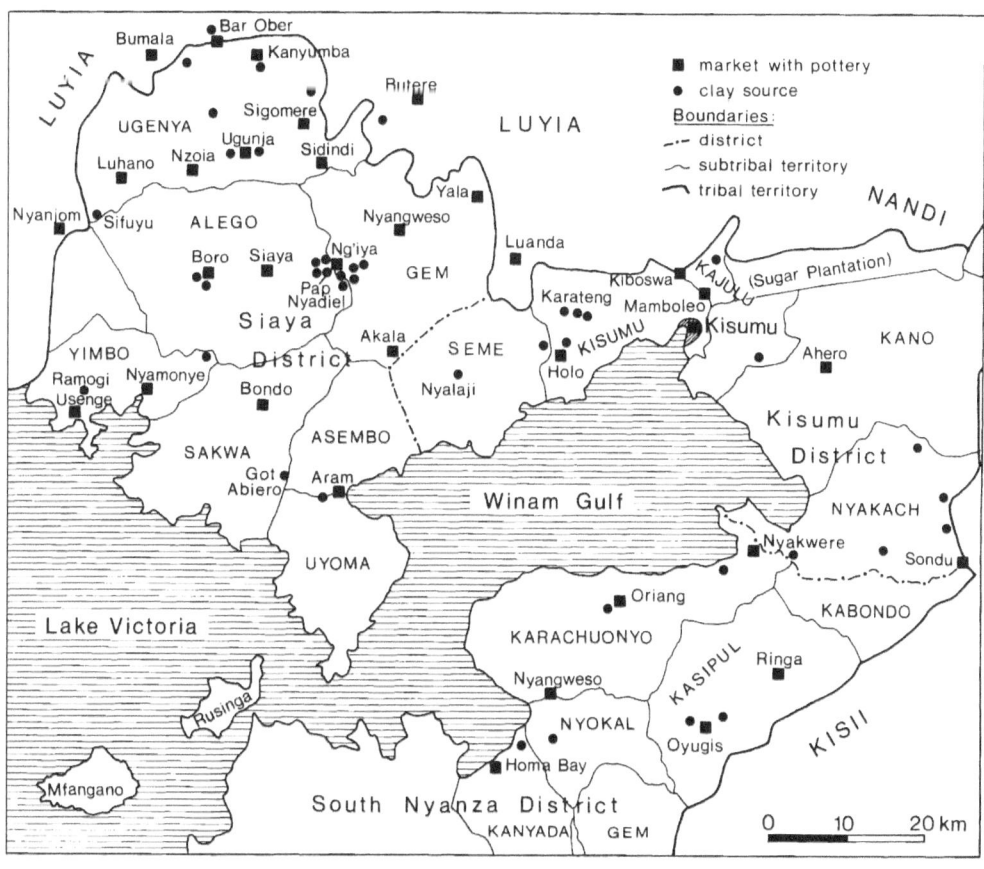

Figure 10.1. The Luo region (excluding some portions of south Nyanza) showing tribal and Luo subtribal boundaries, major markets with pottery, and major active clay sources associated with potter communities (from Dietler and Herbich 1989).

ing and other nonlocal commodities, but they do not result in a significant flow of crops or raw materials out of the region. They mostly serve to articulate local exchanges of crafts and foodstuffs.

The Luo traditionally had no central authority and, although there now exists a system of government-appointed administrative "chiefs," a strongly egalitarian political ethos has persisted. The current administrative divisions are based upon the territorial boundaries of the various Luo subgroups as they existed at the moment of the imposition of colonial administration. In the precolonial era, political organization was based upon fluid alliances among these shifting subgroups in an approximation of the segmentary lineage system (Evans-Pritchard 1949; Ogot 1967; Southall 1952).

The Luo produce a wide range of ceramic pots for household use, including cooking, storage, and serving vessels (see Dietler and Herbich 1989; Herbich 1981, 1987; Herbich and Dietler 1989, 1991). The pots are made exclusively by women, and moreover by a limited set of women who constitute a very small percentage of the total female population. In this sense, Luo pottery production is a specialized craft, although potters do not depend to any significant extent on the craft for their subsistence. Luo potters, like other Luo women, are full-time agriculturalists responsible for growing the food to feed their families.

Potters tend to live grouped in various network-clusters of homesteads that we call "potter communities." These women learn the craft after marriage from their mothers-in-law or other senior women in the husband's father's homestead, and this results in the production of distinctive local "micro-styles" (Herbich 1987). These micro-styles are defined not simply on the basis of decoration. Rather, they are characteristically patterned permutations of *technical, formal,* and *decorative* attributes. Figure 10.2 is an attempt to briefly and schematically convey a sense of the complex nature of these micro-style differences through selected examples of one pot category (a type of water storage/cooling vessel) from six different potter communities. Each drawing represents one typical example from a range of characteristic variants for this particular pot type produced within each community. (Figure 10.3 gives a corresponding sense of this internal variation by showing a typical range of variants for this pot type produced within the *same* potter community.) Each of the other pot types within the local repertoire for each community will also be distinctive from those produced by other communities, but not in the same way as the water pots: characteristic decorative motifs and structuring of the decorative field, for example, will differ considerably for different pot types within the same community (see Herbich and Dietler 1989, 1991 for a fuller explanation).

These patterns are the product of choices made at various stages of the *chaîne opératoire* of production (from clay procurement through to firing) by local sets of potters within a global population of Luo potters employing a very similar limited set of tools and basic techniques (Dietler and Herbich 1989). Each community producing a distinctive micro-style does so *not* through following a set of rules. In fact, although potters can distinguish their own local style from that of other communities, they are not usually able to self-consciously articulate the sets of attributes which constitute their style and they certainly do not teach or learn the craft this way. Moreover, the attributes that characterize pots falling within a style are not identical. Rather, potters share a set of learned dispositions that guide their perceptions of an acceptable range of variation in choices at the different stages of the *chaîne opératoire*.

This process can be partially situated within its social context by noting that these technical and aesthetic tendencies are learned as part of a general process of

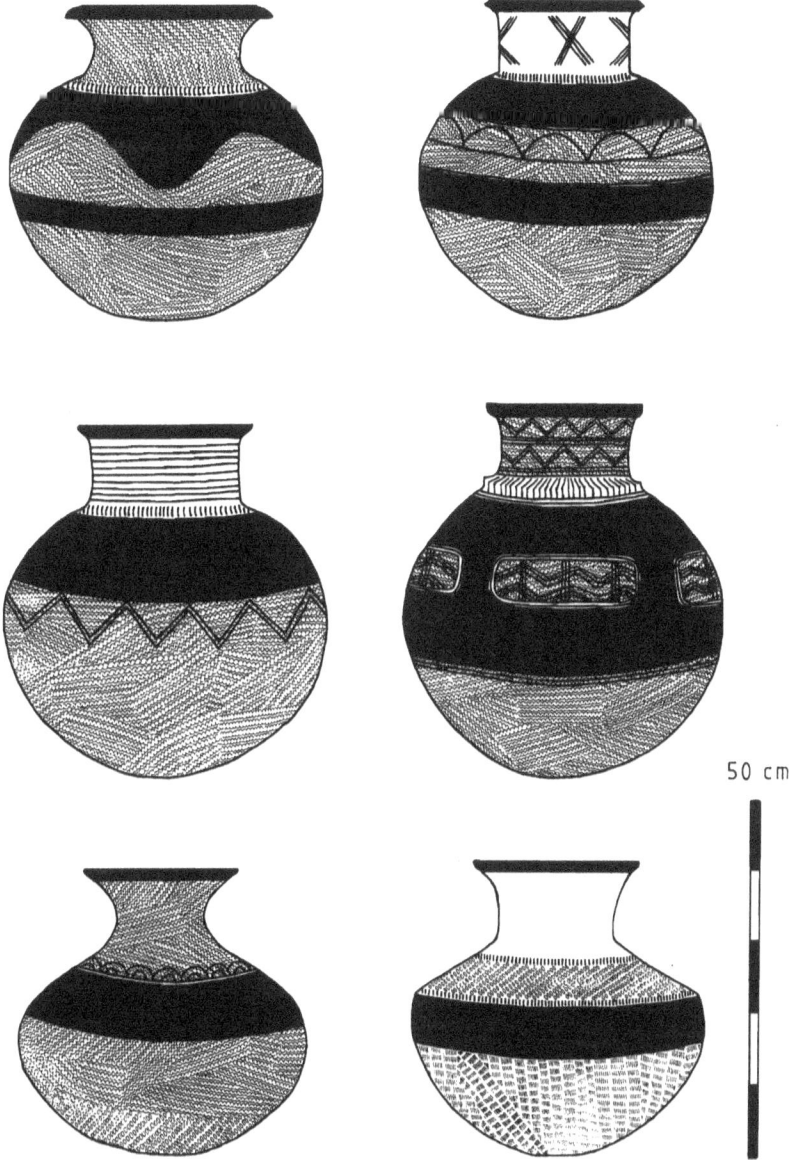

Figure 10.2. Examples of one pot category (water storage-cooling pot) from six different potter communities selected to schematically illustrate the nature of micro-style differences. Each drawing represents one example chosen from a range of characteristic variants produced in each community. Significant defining attributes include elements of form, decorative motifs, the structure of the decorative field, ceramic fabric, and technical aspects of production. Solid dark indicates burnished red-ochre paint (from Herbich 1987).

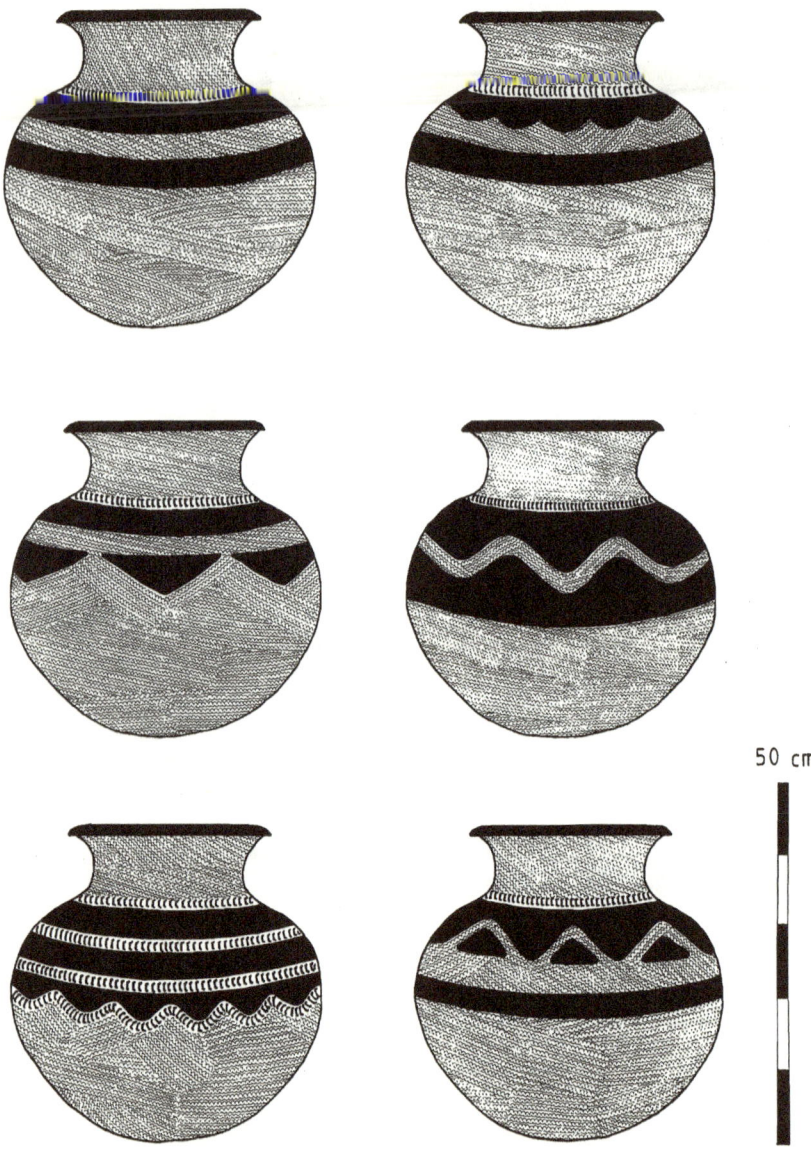

Figure 10.3. Selected examples of water storage-cooling pots produced within a single potter community near Ng'iya market, partially illustrating the range of variation. The upper left pot in Figure 10.2 is an additional example (from Herbich 1987).

post-marital resocialization that occurs for all new brides under the authority of their mothers-in-law and other senior women in the patrilocal homestead (Herbich 1987). The actual process of learning takes place in the context of normal domestic labor and is structured by networks of personal interaction and authority among friends, co-wives, and mothers-in-law. It would be misleading to view the material style produced in this way as a medium intended to communicate group identity (although individual choices at any stage of the *chaîne opératoire* may sometimes be directed by the expression of individual or group identity). Rather, it is a function of the personal transformation of *habitus* through practice responding to certain demands of social relations.

It is important to note that these potting styles are not static. The tendencies responsible for shared patterns of choice in the *chaîne opératoire* of production reproduce the local style, but not slavishly as if dictated by a rigid set of rules. Rather, they allow practice to continually respond to demands in ways that are conditioned by the dispositions. Changes may occur at any stage of the *chaîne opératoire* in response to a variety of demands, and this is the reason that styles of the different potter communities cannot be differentiated on the basis of a single trait (such as decorative motifs). The operation of the dispositions in the course of practice results in separate historical trajectories of ranges of acceptable production choices for different groups of potters and a corresponding polythetic overlapping distribution of traits over the region. Such choices are not random: all of the clusters of traits are compatible with the evolving structures which generated the *habitus*. However, none of them can be "read" like a text to yield the "meaning" of the style.

A few specific examples may help to clarify this perspective. Potters living in several potter communities around Ng'iya market (see Figure 10.1) all use ground sherd temper in the preparation of their potting paste. This practice is due not to any lack of suitable natural temper (or incompatibility of local natural temper with the local clays) but to a technical disposition which guides practice. The relatively high cost of obtaining pots to grind up for temper (one new pot is exchanged for two old ones), has stimulated a few potters to attempt experimenting with other tempering materials. However, they have not tried the abundant natural temper which is known by some to work for potters in other areas, but have instead tried grinding up other fired artificial products such as bricks and the burned soil from charcoal production mounds. Others use blanks of clay that they have fired alongside their pots to meet this need. This pattern of choice is clearly "cultural" in origin rather than physico-technically determined. However, in such a process of innovation, perception of the limits of the possible is constrained not by considerations of communication functions or identity concerns, but by the historically molded inclinations toward action of the *habitus*, which are tied in through their mutual origin to perceptions of "natural" social relations within the homestead and the society. While it might be tempting for some to read this rejection of natural temper in favor of artificially fired materials as a symbolic reflection of opposed concepts of "raw versus cooked" and "wild versus domesticated," etc., this is far too simplistic. Potters of the same Luo subtribe living a few kilometers away use natural temper instead of sherd temper. Such an interpretation is powerless to explain how such variations can arise within the same culture, while a practice approach yields more plausible insights. What is necessary is an historical understanding of the demands to which choices in the *chaîne opératoire* respond and how choices at other stages or times are subsequently constrained.

Many of these demands stem from the level of the political economy, while others reside at the level of personal relationships. As an example of the latter, certain aspects of decoration may be cited. Every potter has a slightly varied repertoire of decorative motifs which she shares variably with other potters in the community and applies in a relatively consistent spatial configuration with a common set of tools and techniques. However, potters often try subtle experiments with new motifs or arrangements. Whether these innovations become incorporated into the acceptable range of choices for decoration for a community depends upon several factors, such as market acceptance. However, one of the most influential of these factors appears to be personal relations among potters. The innovations of popular potters are more likely to be imitated by others than are those of unpopular potters. Antagonism between potters can also result in increasing differentiation of aspects of their decorative repertoires. In this way, personal relations in the context of manufacture among small groups of potters operating with common dispositions and tools can result over time in significant stylistic changes which have little or no resonance in the context of use.

In fact, careful analysis of the eventual spatial distributions of pots after they are purchased and carried to the contexts of consumption shows that most of the resulting micro-style zones cut across important social and cultural boundaries, including Luo subgroup boundaries and the border between the Luo and the neighboring Bantu-speaking Luyia (Figure 10.4 shows such a distribution for over 1,000 pots emanating from Ng'iya market and the Luo subgroup boundaries that are traversed). Moreover, not only are the borders of territories and groups, which are clearly important to people not reflected in the distribution of ceramic styles, but the boundaries of the style zones fall in areas that are of no cultural or social significance (see Dietler and Herbich 1994a). Even when the community of origin of a particular micro-style is clearly recognized, this is of little concern to the people who use the pots: ceramic style plays little role in the expression of group identity in the context of consumption.

In summary, the Luo ceramic data present us with two distinct and important phenomena. The initial production of ceramic style and the historical develop-

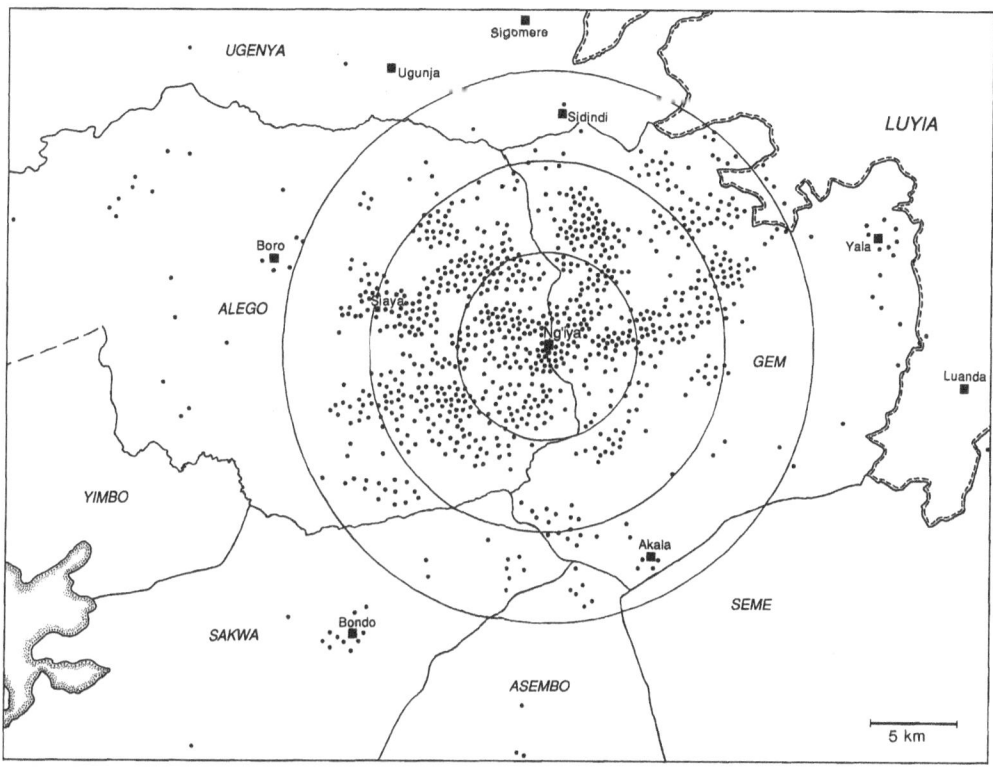

Figure 10.4. Distribution by primary consumers of 1,104 pots emanating from Ng'iya market (location precision is within 1 km) and the boundaries of major Luo subtribes traversed by the zone of distribution. The pattern represents aggregate data collected from multiple weekly market days over the course of several months. The rings represent radii of 5, 10, and 15 km distance from the market. Approximately 94 percent of the pots fall within a radius of 15 km (from Dietler and Herbich 1994a).

ment of changes in this domain are the result of traditions of production (shared dispositions guiding choices in the *chaîne opératoire* of production) characteristic of the different potter communities. These traditions are reproduced by women recruited from outside the potter community (through a system of patrilocal postmarital residence) by means of processes of craft learning in a domestic context and a more general resocialization after marriage (see Dietler and Herbich 1989; Herbich 1987; Herbich and Dietler 1991). Insofar as considerations of the expression of identity (group or individual) play a role in the creation of ceramic style, this is largely confined to the context of production and little understood outside these networks of personal interaction. Processes of distribution linking producers and consumers necessitate a change of context and of meaning; and the eventual spatial distributions of ceramic styles, which are so important to archaeologists, tend to override and obscure the meaning of style within the context of production.

Consequently, it must be admitted that for archaeologists neither the spatial distribution of ceramic styles nor regional resemblances in pot forms are necessarily very good indicators of group identity. Homogeneous style zones may pass across traditionally hostile borders and the boundaries of these style zones may bisect groups with a strong sense of mutual identity. And this is not only the case with obvious large-scale trade wares. The caveat applies even when, as with the Luo, such style zones are less than 30 km in diameter: a fairly typical pattern for pre- and protohistorians. A critical lesson for interpretation of ceramic data from this ethnoarchaeological case is the importance of understanding the distinction between the social contexts of production and consumption and the ways they are articulated (see Dietler and Herbich 1994a for a more detailed discussion).

Luo houses offer an interesting example of the reciprocal relationship between practice and structure. On a regional scale, they are in the process of undergoing a gradual transformation in form, technique of construction, and materials. The change is from houses of round plan to houses of rectangular plan (Fig. 10.5), and from wattle and daub construction on a post-built frame and a thatch roof to cement block construction with a corrugated iron roof. However, these aspects are not spreading as a package or at a uniform rate. While the traditional round house is found only in wattle and daub construction, the rectangular plan is found with wattle and daub construction and (far less commonly) in cement construction. Moreover, the wattle and daub version can be found with either a thatch roof or a corrugated iron roof. Some areas still have almost exclusively round houses, others are mixed (often in the same homestead), and in some areas it is rare to see a round house any more.

What is interesting about this situation is to see what demands are being responded to in these various changes, how inclinations of practice condition changes, and how some changes have unintended consequences for challenging "doxa" (the unquestioning perception of the "naturalness" of the social and material world) in other domains of social practice. All of these changes are ultimately a response to the impact of the world economy on the region, but in somewhat different ways. The rectangular form is an adaptation to the adoption of European furniture, particularly beds and tables, which do not fit very well in a round house. While it is also felt by many people to be "modern," it does not carry particularly heavy symbolic weight as a sign of unusual prestige or wealth (although, to a certain extent, the furniture "implied" by a rectangular house may have this effect). Because of their cost, corrugated iron roofs and, especially, cement construction do clearly and directly carry such implications.

The spatial and temporal organization of the Luo homestead (called *dala*) is an

Habitus, *Techniques, Style* 257

Figure 10.5. A Luo homestead *(dala)* with both circular and rectangular versions of houses. This homestead is in the early phase of its life-cycle (with a still short hedge-fence and few houses). The smallest structures are granaries, and the houses near the front gate are those occupied by unmarried sons of the founder.

extremely complex symbolic representation of the genealogical structure and the relations of authority in both the homestead and society that can be only briefly alluded to here (for a fuller account see Dietler and Herbich 1993; Herbich and Dietler 1993). In a very schematic way, one can say that seniority and authority, both generational and structural, is represented literally by elevation, with lower seniority individuals building downslope from more senior individuals. Lines of structural opposition and alliance between co-wives, and in the broader kinship and political system, are correlated with house placement on alternating sides of the homestead (see Fig. 10.6). Moreover, relations of seniority and authority are also represented and naturalized through temporal sequences of house construction, repair, and a host of daily activities and rituals that take place in the homestead. In view of the importance of spatio-temporal relationships, it may appear somewhat surprising that a change in house *form* could be accomplished with so little apparent concern or turmoil. Yet this seems to be the case. Experiments with changes in the position of the houses due to land shortage have been a cause of

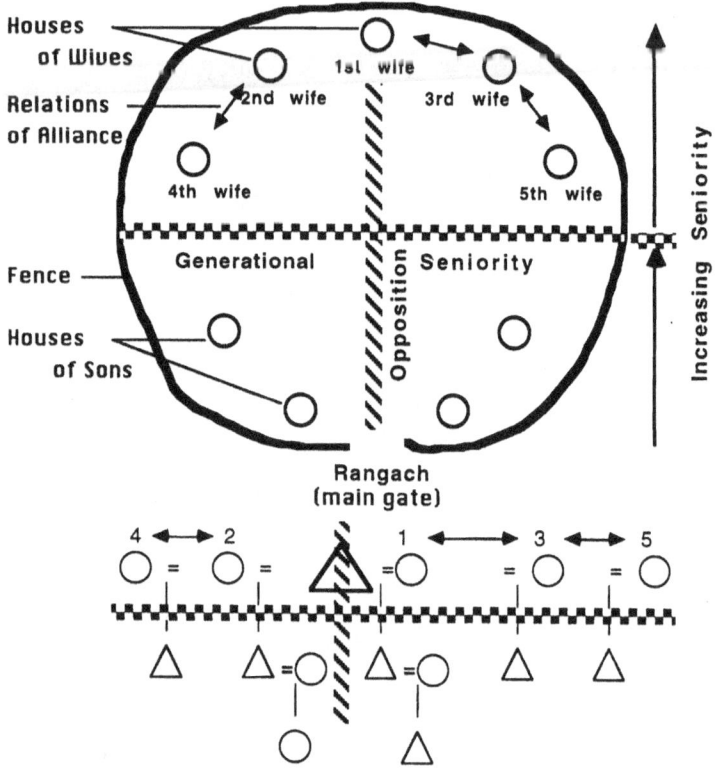

Figure 10.6. An idealized Luo homestead coupled with a kinship diagram showing the relationship of occupants and a schematic representation of lines of structural opposition and authority expressed in the spatial organization of houses (adapted from Herbich and Dietler 1993). The upper half is occupied by the male founder and his wives, each with her own house. Immediately succeeding wives are considered to be in a relationship of structural tension, while relations of cooperation are assumed to exist within the group of even-numbered wives and within the group of odd-numbered wives. Houses in the lower half of the homestead are occupied by unmarried sons and by married sons and their wives. A daughter-in-law is expected to cooperate with her mother-in-law and allied co-wives.

considerable anxiety, of concerns about supernatural consequences, and of discussions attempting to rationally establish an orthodoxy. But the form of the house appears to be a feature open to substantial variation without much comment or concern.

It should not be imagined that the spatial arrangement of the homestead is inflexible: in fact, there are possibilities for all kinds of contingencies in practice that even make the underlying regularity of structure somewhat difficult to perceive for an outside observer. However, the range of choices is decidedly limited by the *habitus* and reinforced by ritual and the threat of supernatural sanctions. There are certain innovative responses of practice (such as, for reasons of land shortage, a man building a house behind that of his grandfather, i.e., upslope from it within the homestead) which have called the axiomatic nature of "doxa" into question and provoked a discussion of the logic of practice, particularly among senior men. Space and time are so important in the context of settlement organization because dispositions governing the relationship of houses are closely linked in their reproduction with dispositions governing the structure of seniority, kinship, and authority. The spatial structure of the homestead (as well as ritual and temporal sequences involved with acts of founding, building, etc.) constitutes a powerful symbolic representation of the structure of social relations because it forms the physical environment in which the *habitus* producing the perceptions of the "natural" order of social relations is formed in the course of daily social life. Changes in the form of the house have been less critical in this case because they have had less of an impact on daily relations of interaction. Moreover, the internal structural relations of domestic space remain unchanged from round to rectangular houses. There is still a female side of the house with the hearth and an opposite side for visitors. This structural principle even determines the proper function of rooms as the innovation of internal divisions appear in houses.

Curiously, changes in the materials from which houses are made have had more profound social consequences than the form of the house. This is, again, because it affects practices that underlie dispositions governing social relations. Traditionally, thatched roofs can be built and repaired only by men, while walls must be regularly smeared with clay by groups of women at least once a year. The need for periodic roof repair reinforces relations of dependence between women (the "owners" of houses) and men, and the smearing parties reinforce relations of mutual support and dependence among women. As a further illustration, upon the death of a male head of the homestead, none of the other co-wives is allowed to have her house repaired until the first wife's house has been repaired; and this is allowed only after she has undergone a ritual to mark the end of mourning. It is clear that the more permanent construction materials, by eliminating such practical needs for repair, may have a profound impact on relations of authority and dependence. Moreover, when the owner of a house dies, others are not allowed to occupy it: the house must simply be left to deteriorate and fall down or be pulled down. As noted earlier, homesteads themselves undergo a regular cycle of occupation by a three-generation family (a man, his wives, his sons and their wives and children). They are abandoned after the death of the founding generation and converted to farmland by the sons of the original male head of the homestead. These sons are obligated by custom to move out of their houses in the father's homestead and found their own homesteads when their own sons are ready for marriage. The building of expensive permanent houses is creating strains in this system, which is again intimately tied in with kinship and political structures.

CONCLUSION

The intent of this paper was to suggest and explore a way of circumventing some of the problems that have troubled archaeologists attempting to grapple with the crucial relationship between material culture and society. Penetrating this relationship and understanding some of the social forces that produce material culture patterning is a fundamental requisite to engaging in the more specific attempt to understand the role of material culture in processes of identity formation, expression, and reproduction and to assess the feasibility of using remnants of material culture to identify social groups and boundaries of the past. This must, of course, be coupled with an improved understanding of the complex nature of social group identity and the shifting contextual definition of boundaries that takes full account of the political-economic dimensions of the process of interaction between groups. The nature of group definition, the strategies for signaling exclusion and belonging, and the shifting salience of different cultural elements in these strategies vary greatly according to the relative asymmetry of power relations among interacting groups (see Comaroff and Comaroff 1992:49–67).

It was suggested that the artificial division between style, technology, and function has become excessively reified and is not heuristically useful for the purpose of understanding the social dimension of material culture. These concepts and the approaches they entail cannot, for example, produce plausibly cohesive answers to the essential questions posed at the beginning of the paper: how does material culture originate in its social context and how does material culture reciprocally condition social structures and processes? A more integrated approach incorporating elements of the French tradition of *technologie,* with its emphasis upon techniques as the mediating factor between objects and society and upon understanding choices and demands at various stages of *chaînes opératoires,* holds much greater promise in producing a realistic perspective of the complexities of material culture patterning. Moreover, a theory of practice modeled on the work of Bourdieu may allow us to situate techniques more readily within their social context as products and producers of *habitus.* It provides a "temporal" approach to techniques that links structure and agency as mutually generative forces. Both the objective material conditions that generate dispositions and the demands to which practice responds are best approached from a perspective that views culture as an historical social process within a larger world of social and economic relationships.

Such a synthetic approach, it is hoped, may propel us to a new realization of the complex nature of the problem facing us. A social understanding of techniques is a crucial issue for archaeologists; we must address it realistically through both empirical ethnographic research and the development of theory. Our ability to propose and evaluate plausible interpretations of the past, including the delineation and understanding of social groups and boundaries, depends upon the progress of this endeavor.

ACKNOWLEDGMENTS

This chapter is a revised version of a paper that was originally presented in 1992 at a multidisciplinary symposium on *Genèse sociale des techniques—Genèse technique des hommes* at the estate of the Fondation des Treilles, near Tourtour, France, and subsequently published in French as "Habitus et reproduction sociale des techniques: L'intelligence du style en archéologie et en ethnoarchéologie" (Dietler and Herbich 1994b). We would like to express our gratitude to Bruno Latour and Pierre Lemonnier for the invitation to participate in that symposium, to all the participants for their helpful comments and camaraderie during the week we were assembled together, and to the Fondation des Treilles for its exceptional hospitality. We also wish to thank Miriam Stark for her invitation to include this paper in the current volume and for editorial comments.

NOTES

1. The research was carried out between April 1980 and January 1983. Funding was provided by the National Science Foundation, the L. S. B. Leakey Foundation, The Wenner-Gren Foundation, and the Boise Fund of Oxford University, for which we are extremely grateful. Our thanks also to the National Museums of Kenya, the Office of the President of Kenya, and the British Institute in Eastern Africa. *Erokamano maduong'* to the Luo people and to our field assistants Monica Oyier, the late Elijah Oduor Ogutu, and most especially Rhoda Onyango.
2. It should be emphasized for non-Francophone readers that our approach is not an orthodox recapitulation of the French *technologie* program, and it should not be used as a handy substitute for reading that rich and important body of literature. Rather, ours is a distinctive perspective that incorporates concepts derived from that "school" into a framework rooted in our ethnoarchaeological experience and informed by anthropological theory in the domains of consumption, historical cultural economy, and practice. One important difference between our approach and the *technologie* school is that our introduction of Bourdieu's theory of practice into the discussion of material culture is an issue of clear disagreement with most of its practitioners (we have had the pleasure of amiably wrangling about this and various

points of accord and disagreement over the past decade at various symposia and seminars at the University of Paris and the École des Hautes Études en Sciences Sociales).
3. The *chaîne opératoire* (or "operational sequence") is an analytical concept first developed by Leroi-Gourhan (1964), who was himself inspired by the work of Mauss (e.g., 1935). The concept has been further refined and elaborated in slightly different ways by various practitioners of the French *technologie* school, for whom it is a fundamental tool for approaching the anthropological study of techniques (cf. Cresswell 1976; Lemonnier 1986, 1992). We prefer to employ the original French term rather than a translated equivalent in order to indicate this historically specific usage. Basically, a *chaîne opératoire* is a technical process composed of a series of operations that result in the production of an object.

   The description of particular *chaînes opératoires* is a very effective way of illuminating the series of choices involved at all stages of the process of production, of revealing the cultural and physico-technical context of those choices, and of characterizing differences in technical systems. Such description involves the sequential specification of the materials and tools used, the actions performed and their names (if any), as well as the identity of the maker and the place, time, and context of production. See Dietler and Herbich (1989) for an example of the application of the *chaîne opératoire* to the analysis of pottery production among the Luo of Kenya (cf. Gosselain 1992b; van der Leeuw 1993).

   The term *technologie* is a potential source of confusion for a mixed French and English audience. Sigaut (1985, 1987, 1991) defines it as a social science of technical facts, or what might be called the "anthropology of techniques" (cf. Haudricourt 1987; Lemonnier 1983, 1986, 1989, 1990, 1992, 1993b; Schlanger 1991). This use of the term does not correspond to its common English use (or rather, given the root meanings of the parts of the word, its misuse) and certainly not to its accepted meanings within the Anglophone anthropological and archaeological community (cf. Basalla 1988; Fagan 1981; Pacey 1990; Renfrew and Bahn 1991; Rice 1987; Wright 1985). Unfortunately, nor is there, as yet, a developed or even recognized domain of Anglophone anthropological research corresponding to the French tradition denoted by this term.
4. The sense of this phrase is difficult to convey in translation, but it may be rendered approximately as "The domain of the social is the domain of repetitive similarity."
5. The same pattern of localized ceramic microstyles can result from several different systems of post-marital residence because design concepts are not immutably fixed in childhood. Potters can learn new patterns of production choices, and in patrilocal societies where women marry at a young age, they may learn potting only after marriage. Hence a patrilocal system with women recruited exogamously from outside the area of the potter community and a mother-in-law/daughter-in-law learning process will result in the same pattern of localized microstyles as a matrilocal system with women remaining in place and a mother-daughter learning process (Herbich 1981, 1987).
6. Sackett uses the term "ethnic" group to indicate, very generally, social groups at a wide variety of possible scales. This is a somewhat idiosyncratically broad usage even among archaeologists who tend to use the term in a much less specific and rather different sense than cultural anthropologists (e.g., Barth 1969b; Bentley 1987; Comaroff and Comaroff 1992:49–67; Nash 1989).

7. A further major difficulty with the application of Wobst's (1977) information exchange approach to ceramics is that the principle of relative visibility as a key to the investment of energy in "stylistic behavior" runs counter to common practice. Among the Luo, as elsewhere, one finds highly decorated pots that never leave the kitchen and are so covered with soot that one can barely see the original decoration. Clearly, the desire to communicate with outsiders cannot be invoked as an explanation of aesthetic elaboration of this kind, and considerations of efficient exchange of information are irrelevant.
8. In terms of its inappropriateness for achieving social and cultural understanding and, critically, for situating objects in their sociocultural context, work of this kind is really the ethnographic equivalent of Flannery's famous "telephone booth" excavation. As archaeologists, we must learn to resist quick simplistic formulas and become as sophisticated and critically demanding in our assessment of ethnoarchaeological studies as we are in evaluating excavations; and, as ethnoarchaeologists, we need to standardize a more rigorous set of field methods that includes long-term participant-observation and ethnohistorical research.
9. Although the term "habitus" has been employed in related ways by earlier anthropologists (e.g., Mauss 1935), we use it in the very specific sense defined by Bourdieu (1977, 1980) as part of his theory of practice. While recognizing the merits of other similar approaches mediating structure and agency, such as the structuration theory of Giddens (1979), we prefer to use Bourdieu's approach both because its formulation predated these other theories and because Bourdieu has been intensely concerned to develop, emend, and explicate the theory through repeated applications to a broad range of empirical domains.

11

Technology, Style, and Social Practices: Archaeological Approaches

MICHELLE HEGMON

Archaeological research in the past few decades has challenged traditional ways of thinking about material culture, and the exploration of technology and social boundaries put forth in this volume continues that challenge. Some of the major developments in recent Anglo-American archaeological thought about material culture can be summed up in two phrases: "style has function" (Wobst 1977) and "technology has style" (Lechtman 1977). That these declarations, once called somewhat radical, now appear to be patently obvious says much about the expansion of archaeological concepts and thinking since the mid 1970s. In this essay, in order to put the papers in this volume in perspective, I review where the study of technology and material culture has been and where I believe it is going. I begin with archaeological approaches to style and technology and then consider the concepts of technological choices and social boundaries and their interrelationship.

STYLE HAS FUNCTION

Archaeological literature on style has been developed and reviewed extensively in recent years (e.g., Carr and Neitzel 1995; Conkey and Hastorf 1990; Gebauer 1987; Hegmon 1992; Plog 1983), and this is not the place for another review. Instead, I will briefly discuss several issues regarding style that are relevant to this volume's focus on technology and social boundaries. In part, I hope that researchers now focusing on the social components of technology can gain some insights, perhaps even learn some lessons, from archaeological approaches to style.

First, as are many concepts close to anthropologists' hearts, style is difficult to define, although most archaeologists probably think they know what they mean by the term. If we define style as that part of variation not determined by functional constraints, we end up with style as a by-product, and we exclude understandings of style as a choice between functionally equivalent alternatives, an understanding important to considerations of technology (see discussion by Dietler and Herbich, this volume). If we define style, as I am most comfortable doing, as "a way of doing things" (Hegmon 1992:518) we create a concept that is perhaps too broad to be useful or meaningful (though Mauss's [1936] similar concept of a way of "doing things" has been revived in recent research on technological choices and systems [Dietler and Herbich, this volume; Lemonnier 1992]). Finally, if we define style as that part of formal variability in material culture that is culturally significant (following Wobst 1977:321), we end up with a concept that is good to think about but difficult to apply archaeologically. Most importantly, although flexibility about definitions is usually wise, we must be careful not to combine our definitions in a circular way. That is, we must not *identify* style archaeologically as variation we cannot explain according to functional criteria (definition #1), and then *assume* that the style we have identified is culturally significant (definition #3) (for a related discussion, see Carr 1995a).

Second, recent literature has emphasized that style is not just a passive by-product of cultural norms or mental templates. Style *does* something. According to earlier, more strictly functional accounts, style functioned (efficiently) by conveying information (Wobst 1977). Although these accounts can be criticized for their extreme functionalism and for assuming that people passively played out their roles in their cultural systems, at least style was understood as being a part of those systems. More recently, a number of accounts examine the ways in which style—or variation in material culture generally—can be understood as part of the strategies people pursue in their social and cultural context (e.g., Hegmon 1995; Miller 1985; Shanks and Tilley 1987; Wiessner 1983, 1984).

Third, most authors who have wrestled with the problem of style seem to have come to the conclusion that style is best understood at many levels, or that there are many kinds of style (see summary in Hegmon 1992:522-524; Carr 1995b). Some kinds of style do play a role in the definition of groups and group boundaries (what Wiessner [1983, 1984] calls emblemic style or Macdonald [1990] calls protocol). However, other kinds of style (including Wiessner's assertive style or Macdonald's panache) are better understood as expressions of individual identity. Style can express a person's status, either as an individual or as a member of a group. And sometimes style seems to have little to do with social distinctions and instead is best understood as an expression of cultural understandings of the universe (see David et al. 1988; also Franklin's [1986] concept of stochastic style).

Finally, the study of style often forces archaeologists to wrestle with complex social concepts. At least since Childe's (1951) *Social Evolution,* we have been warned that archaeological "cultures" (often defined in terms of styles) are not necessarily equivalent to ethnic groups or any kind of human grouping (see review in Shennan 1989). As MacEachern (this volume) notes, archaeology may not be well-suited to do ethnography of the past. We can work towards understanding the societies of the past, but perhaps not in the same way that societies of the present can be understood. Specifically, many ethnographic concepts—such as society or kinship—refer to kinds of interactions and beliefs, not entities or things. If we use those concepts in our investigations of the past, we must guard against reifying the concepts and looking for "things" that do not really exist (see discussion in Eriksen 1991:127–128).

For example, the "ceramic sociology" studies of the 1960s and early 1970s (e.g., Hill 1970; Longacre 1970) were innovative, optimistic, and (to me) inspirational studies that attempted to develop methods by which archaeologists could learn about the details of past social organization, particularly kinship systems. Because of their importance, these studies have been scrutinized from every angle and often criticized (for reviews of this literature, see Hegmon 1992; Plog 1978). Most of the criticisms have to do with methodology (temporal control, design classification, etc.), but the theoretical problems are more relevant here. That is, at the same time that archaeologists were attempting to *find* prehistoric kinship systems, social anthropologists were questioning the existence of kinship as a social *fact,* and asking how professed kinship principles actually relate to kinship practice (see, for example, Barnes 1962; Goodenough 1956; Kelly 1976; Stanislawski and Stanislawski 1978). The problem is that the kinship systems had been reified and turned into things, rather than being understood as potentially problematic concepts in the dynamics of social relations.

TECHNOLOGY HAS STYLE

Archaeologists' broadening understanding of style was paralleled by a broadening understanding of technology, ushered in by the series of papers in Lechtman and Merrill (1977). No longer was technology viewed only as a means to an end, a form of material adaptation. Rather, Lechtman (1977) made it clear that technology had a style of its own, and that technology could only be understood within its social and cultural context. Here I review several approaches that relate technology and society and that are fairly closely tied to the issue at hand, that is technology and social boundaries. My focus necessarily, but unfortunately, excludes work on other aspects of technology and society including the growing literature on the sociality of tool use (Ingold 1993b); Marxist perspectives on the mode of production (e.g., Bender 1978; Hindess and Hirst 1975; Ingold 1986); a long history of research on the organization of production and specialization (Costin 1991; Rice 1981); perspectives on how the value and meaning of goods are social products (Appadurai 1986b; Douglas and Isherwood 1979); and recent work on the multiple factors involved in technological organization (Nelson 1991).

In a series of papers, James Sackett put forth a perspective on material culture variation critical to the understanding of technology and boundaries (this perspective is reviewed and summarized in Sackett [1990] and it is most extensively developed in Sackett [1982]). Sackett described what he called isochrestic variation, isochrestic meaning "equivalent in use." When there are several functionally equivalent ways to make or do something, people choose between those ways based in large part on their cultural traditions. Those choices are isochrestic choices, and the resulting patterning is what Sackett calls isochrestic variation. Although isochrestic variation is sometimes described as a kind of style, this is not Sackett's meaning. Instead, he argues that style is a subset of isochrestic variation, specifically, the part of isochrestic variation that has ethnic significance, whether or not it is actively used to signal ethnicity. In a sense, Sackett is arguing that technological choices, as well as decoration, can have style and can provide archaeologists with information about social boundaries.

In 1977 Heather Lechtman articulated the idea that technology has style, drawing from her work on Andean technologies, particularly metallurgy. She argued that Andean metallurgy cannot be understood in terms of Old World utilitarian models, which emphasize properties such as strength and hardness. Instead, Andean metallurgy, which focuses on soft metals such as gold and silver and seems to put great emphasis on the color of final products, can only be understood in terms of the Andean symbolic system and conception of the universe. That is, Andean smiths achieved gold and silver colors not by gilding (i.e., applying a thin coat of the precious metals to the surface of objects) but by complex techniques of enriching the metals on the surface of alloys (gold and copper or silver and copper). Thus, gold and silver were intrinsic and essential parts of the alloy mixtures and the object themselves, they were not just surface coverings. Similarly, Andean textile production also emphasized the incorporation of designs into the structure of the cloth. Lechtman argues "that key technologies associated with cloth and metal production shared stylistic modes, perhaps because those modes are expressions of cultural ideals, incorporating ideological concerns of the society at large" (Lechtman 1993:273).

Lechtman's concept of technological style involves normative cultural behavior and the rules behind that behavior (1977:12). However, Lechtman's concept of technological style is quite different from Sackett's concept of sometimes passive isochrestic variation. Technological style actively involves symbolic structures as well as attitudes of the community and artisans. Furthermore, technological style does more than simply express ideology; the technology (and artisans who make it) play a role in perpetuating and possibly changing status relations and basic ideological concepts (Dobres and Hoffman [1994:217–219] develop this perspective on Lechtman's work). In general, Lechtman opened avenues of exploration for archaeologists by demonstrating that at least some aspects of prehistoric technology could best be understood from an ideational perspective, and she laid the groundwork for exploring the social role of technology.

Pierre Lemonnier (1986, 1989, 1992) has also developed an understanding of the relationship between technology and society, drawing from his ethnographic studies of the technical systems of the Anga in Papua New Guinea. He conceives of technological activity as the interplay between matter, energy, objects, gestures in sequence, and knowledge. Lemonnier places much emphasis on the operational sequences of technology, not just the final product. He argues that those operational sequences involve technological choices. Furthermore, those choices—including some, such as sequences of gestures, that have no obvious effect on the final material product—are the product of social learning processes and may be social actions, sometimes used to mark group distinctions.

Like Lechtman, Lemonnier is interested in technological styles and in relating those styles to a society's world view (see Dobres and Hoffman 1994:217–221 for a comparison of the two). However, the two researchers emphasize different aspects of the relationship of technology and society. Specifically, although Lechtman considers how technology is used to mark status distinctions, much of her emphasis is on how ideology and world view shaped Andean technologies. Furthermore, although much of her research focused on understanding how Andean metals were produced, her explanation of metallurgical practices emphasizes the properties of the final product, that is, the fact that gold and silver are part of the essence of alloy metals. In contrast, although Lemonnier considers the relationship between technology and Anga conceptions of the universe, his emphasis is on the dynamic interplay between technology—particularly operational sequences—and social groups. Some of the differences between Lechtman's and Lemonnier's approaches may be theoretical, and other differences may result from archaeological vs. ethnographic research. In addition, the researchers' understandings may be a factor of intrinsic differences in the kinds of technologies they studied. That is, the production of precious metals by specialists in a stratified society may be much more stable and less open to social manipulation than the production of everyday items in a tribal society. These kinds of differences may provide a useful avenue of exploration for anthropologists interested in the relationship between society and technology.

In a recent statement, Dobres and Hoffman (1994) attempt to advance anthropologists' understanding of social agency in prehistoric technology. They argue that "the final goal of technological studies is not to describe microscale prehistoric *activities*, but to understand microscale social *processes*" (Dobres and Hoffman 1994:213, emphasis in original). They argue that practice theory (see Ortner 1984), including Giddens's (1984) concept of structuration and Bourdieu's (1977) concept of *habitus*, can be used by archaeologists to understand the dynamic interplay between technology and society. Specifically, both the operational sequences and the products of technology are part of a society's structure and/or *habitus*. At the same time, people are part of those operational sequences, and people produce things. Thus people—active social agents—participate in the reproduction and change of their society (see also Dietler and Herbich, this volume; Moore 1986).

Ingold (1993a, 1993b) makes an important distinction between technique and technology, a distinction relevant to an understanding of agency and technology. To Ingold, technique involves "the embodied skills of human agents" while technology comprises "the operational principles embodied in the external apparatus of production" (Ingold 1993a:342). In a sense, technique is to technology as agency is to structure; an understanding of either depends on a consideration of the relationship. If we focus too strongly on technology, as he defines it, we risk objectifying or reifying it, turning it into an object separate from human agency. Furthermore, Ingold suggests that a focus on objective, explicit technology in contrast to more subjective experiential technique is a product of Western society's increasing dependence on mechanical systems. If he is correct, then archaeologists' emphasis on systems of technology may involve an imposition of our own social constructs on the past and may cause us to underestimate the importance of human agency in the form of individual skills and techniques, at least this would be the interpretation from the perspective of critical theory. The result of this emphasis on technology and systems "is tantamount to a *disembedding* of technical relations from their matrix in human sociality, leading to the modern opposition between technology and society" (Ingold 1993b:436).[1]

I find the concepts of agency and structuration, or practice and *habitus* to be helpful in my own thinking about what society is and how material culture can be understood as part of complex social dynamics (Hegmon 1995). However, I believe that we must guard against simply adopting these new (stylish) words and applying them to old concepts. That is, *habitus* is not the same thing as culture, and agency cannot be equated with the behavior of an individual in an optimality model. Practice theory demands that we consider what Giddens (1991) calls the duality (the dynamic interplay) of structure and agency or *habitus* and practice. Furthermore, we may benefit by drawing what is best or most useful from different approaches. For example, Bourdieu's (especially 1977, 1990) is probably most useful for understanding practice or human agency and how it is directed by *habitus*; in contrast, Giddens (especially 1984) is probably more useful in helping us understand how agency affects the structure and social change at a larger scale, the process he calls structuration. Other important perspectives on the interplay—or dialectic—of structure and agency are developed in historical Marxist accounts (e.g., Callinicos 1988; Thompson 1991), and in Archer's (1988) work on culture and agency.

While the link between such complex social theory and archaeological data is not obvious or direct, neither is the link between social boundaries and material culture. If we, as archaeologists, wish to understand the relationship between a social process and material culture, we must be prepared to understand the complexity of that social process, and in this endeavor we will be well served by current social theory including practice theory. Archaeologists have begun to apply this theory to gain new insights about past social dynamics (e.g., Clark and Blake 1994). Bourdieu's concept of *habitus*, which was developed as part of his study of Berber architecture, holds particular promise for research on material culture (see Hodder 1992:73–77). Unfortunately, many archaeologists are still waiting to see practice theory concepts fully used in the interpretation of prehistoric material culture. Although Dobres and Hoffman (1994) cite a number of excellent studies that provide information regarding technology and society (see also Dobres 1995), they provide only one example that explicitly uses the concept of human agency. Unfortunately, that study (M. Johnson 1993) concerns *historic* English housing. Ironically, the author of that study has previously criticized archaeologists for failing to put the concept of agency into practice (M. Johnson 1989). Dietler and Herbich (this volume) also provide a thorough discussion of *habitus* and technological choices, though their examples are ethnoarchaeological, not prehistoric.

A growing literature on sociotechnical systems (mostly outside of anthropology) may be extremely useful to archaeologists interested in the relationship between technology and society. Much of this work is summarized by Pfaffenberger (1992). A sociotechnical system is more than the procedures used to achieve some technical end; the concept draws in the ideology and the social relations involved and it grants agency to the people involved. One example of such a sociotechnical system is the water control system in Bali, which involves water temples and priests as well as irrigation procedures (Pfaffenberger 1992:509–510). Sociotechnical systems are not necessarily perfectly efficient in a utilitarian sense; in fact, they cannot be understood without accounting for the material symbols that they involve. Sociotechnical systems "produce power and meaning as well as goods" (Pfaffenberger 1992:502). The concept of sociotechnical systems may be useful to researchers interested in the concept of practice. Specifically, the creation of a sociotechnical system involves structuration in that social actors work within the system but also can work to change the system (Pfaffenberger 1992:500).

SOCIAL BOUNDARIES: THE CONCEPT

At least two general approaches to the archaeology of social boundaries are represented by the authors of this collection of papers. First, the overall thrust of the volume, as expressed by the editor, is on discovering how technological choices *reflect* social boundaries (Stark, this volume). This approach takes the existence of social boundaries as unproblematic. It is assumed that boundaries (like the kinship systems studied by the ceramic sociologists) were social facts of the past. Emphasis here is on developing detailed methods that will allow us to identify those social facts. The chapter by Dietler and Herbich generally fits with this first approach but expands it in important ways. That is, rather than assuming a relationship between social boundaries and technology, Dietler and Herbich explore that relationship theoretically and help us understand why and how it exists in certain cases.

A second approach represented in this volume attempts more directly to wrestle with the very concept of social boundaries. This perspective is probably best articulated by Goodby (this volume, p. 161) when he says "social boundaries are abstractions and ideological constructs, recognized differently and for different reasons by people on the basis of their perceived identity, interests, and social context" (see also MacEachern's introduction, and Hitchcock and Bartram's musing about fences). The point here is that if archaeologists mean to study social boundaries in the past, we need to wrestle with the concept, with what social boundaries are and might have been. Here social theory, in all its complexity, must play a role.

Some perspective on these contrasting approaches can be gained from the literature on ethnicity, which is closely related to the concept of social boundaries. As MacEachern (this volume) notes, archaeologists are fond of citing Barth's (1969a) *Ethnic Groups and Boundaries*. At least two major arguments directly relevant to archaeology are developed in that volume. First, Barth, and others in the volume, emphasize structure over content. That is, they are more concerned with a boundary as an expression of some sort of difference than with the specifics of what lay on either side of that boundary. This strikes an optimistic note for archaeologists, who are more likely to be able to discover the existence of boundaries than to understand the specific content of bounded units.

Second, the volume solidified anthropologists' and other social theorists' understanding of ethnic groups and boundaries as social constructs, not social facts. "Ethnic groups are categories of *ascription* and identification by . . . actors themselves, and thus have the characteristic of organizing interaction between people" (Barth 1969a:10, emphasis in original). Barth emphasized that there is no one-to-one correspondence between ethnic units and cultural similarities. "The features that are taken into account (in marking ethnic boundaries) are not the sum of 'objective' differences, but only those which the actors themselves regard as significant" (Barth 1969a:14). And while the literature on ethnicity is vast and somewhat contentious, there does seem to be a consensus regarding this conclusion; that is, ethnicity involves *self-conscious* identification (e.g., Bentley 1987; Eriksen 1991; Kumar and Kadirgamar 1989:vii; Moerman 1965). In other words, ethnicity does not simply exist; it is something that people do. Eriksen (1991:127–128) provides a summary of the problem of reifying ethnic groups as fixed entities, and Schortman (1989) provides a discussion of the way in which named concepts influence research. Perhaps somewhat ironically, Barth's later work, rarely cited by archaeologists, has emphasized this aspect of how people create their own cultural and social realities (e.g., *Cosmologies in the Making* 1987).

This understanding of ethnicity as "something that people do" is potentially problematic for archaeologists interested in the study of boundaries. Clearly such an understanding adds to the complexity of our study. It also supports the questioning of what boundaries are, part of the second group of studies (those by Goodby, MacEachern, and Hitchcock and Bartram), and is relevant to approaches concerned with human agency (e.g., Dietler and Herbich, this volume; Dobres and Hoffman 1994). On the one hand, we archaeologists cannot simply do ethnography in the past. At the same time, a perspective that questions what boundaries are also opens new possibilities for research. If we can identify under what conditions certain kinds of material—including technological—differences exist and how those differences change over time and space, we might be able to add a new perspective to the study of social boundaries (Buchignani 1987). Furthermore, archaeology should be able to provide a historical perspective, something Barth's work has been criticized for lacking (e.g., Wolf 1982).

Research on ethnicity has long involved debates between contrasting sets of ideas. The content vs. structure dichotomy, discussed above with regard to Barth's work, has also taken form as a debate between primordial and instrumental theories. Primordialists generally argue that ethnic differences arise as a result of differences in language and appearance. In contrast, instrumentalists argue that groups create their ethnic identities for instrumental (i.e., political and economic) reasons. For summaries of some of these debates see Barth (1969a), Bentley (1987), Eriksen (1991), and A. S. Brooks et al. (1993).

Recent work on ethnicity has attempted to resolve some of these debates by accepting at least some parts of "all of the above." That is, according to Eriksen (1991), in an article entitled "The Cultural Contexts of Ethnic Differences," we must take into account both the formal structures and the content within those structures; furthermore, we must understand the interrelationship between the content and structure. This kind of solution has also been expressed in terms of practice theory (Bentley 1987; see also Geary 1983; Worsley 1984:242–243). Specifically, ethnicity can be understood in terms of the interaction (or dialectic) between, on the one hand, existing (but not immutable) structures or ethnic groups, and, on the other hand, the strategies of actors in a given social context. To put it in other terms, *habitus,* which molds—often at a subconscious level—what people are, contributes to the creation of apparent ethnic differences (the primordialist view). Ethnic symbols are open to manipulation (the instrumentalist view), though they did not come into being simply for manipulative purposes. Finally, the *habitus* is reproduced and potentially changed by the actions of those people, whether or not ethnicity is purposely altered. This kind of dynamic approach to ethnicity may be useful for archaeologists interested in practice theory.

A final point I wish to raise regarding ethnicity is that many scholars have argued that it is essentially a product of state-level societies (Smith 1986) or even of industrialism, market forces, and colonialism (Gellner 1983; Pryor 1989; see summary in Shennan 1989:11–17). This argument is not unlike Fried's (1968) questioning of the concept of the tribe. If these authors are correct, the concept of ethnicity may be irrelevant for most archaeologists studying pre-state societies (see Suttles 1987). At the same time, archaeologists studying very complex societies and/or colonialism may be positioned to provide a unique and important perspective on the origin of ethnicity (Brumfiel 1994; Goodby this volume; McLaughlin 1987).

The point is not that most archaeologists should simply avoid the word "ethnicity," but rather, we should be wary of the concept it invokes, especially in research on pre-state societies. That is, ethnicity connotes all-encompassing marked and bounded groups, and it may be that such clear-cut groups did not exist in much of the past. At the very least, if we wish to assume that such bounded groups did exist, we need to justify our assumption. A prehistoric material culture boundary may be indicative of some kind of social boundary, but such a material boundary should not automatically be equated with an *ethnic* boundary without further information on social processes. For example, the concept of ethnicity involves intergenerational self-identity that is sustained, and even emphasized, in contact with other ethnic groups. Thus one possible archaeological correlate of ethnicity might be that it persists intergenerationally, as it clearly does in protohistoric-historic New England (Goodby, this volume). In contrast, in the American Southwest, there are several well-documented examples of groups migrating into new areas and bringing a distinct material culture tradition with them (e.g., Adams and Hays 1991; Haury 1958; Stark, Clark, and Elson 1995). However, it is not possible to evaluate whether the material culture differences persisted intergenerationally, because the migrants apparently left the area after a decade or so or the occupation as a whole ended.[2]

In arguing that not all archaeologists can study ethnicity, I do not intend to limit the archaeological enterprise. Instead, I suggest that archaeologists may be particularly well positioned to consider social forms no longer prevalent in today's world. That is, social groups certainly did exist in the past, but they may have had structures that were different from and more flexible than those of ethnic groups (see also MacEachern, Welsh and Terrell, and Stark et al., this volume). It is likely that people have always defined themselves in terms of material culture differences (including technology) and that part of that definition involved group identity (Wiessner 1984) . However, rather than being a member of a bounded group, people may also have had networks of overlapping identities. Again, I urge archaeologists to question the concept of boundaries and to consider other kinds of social relationships.

THE RECOGNITION OF BOUNDARIES

Despite what I have just said, social boundaries appear to exist in the world today, even if many of these boundaries were created or reified by state societies and colonial administrations. What can we learn from these? Do the social boundaries have any relationship to technological choices or to other aspects of material culture? Unfortunately, the answer, almost universally, is "sometimes." The point is not to present a cautionary tale, but to explore at *which* times certain kinds of boundaries have expression in certain kinds of material or technology. In the following discussion, although I emphasize insights gained by studies of technology, I also draw from studies of style. This is because I do not wish to assume, a priori, that either style or technology is better suited for the study of social boundaries.

First, a number of studies have found that various technological differences do seem to correspond to social differences or ethnic boundaries. For example, Lemonnier (1986) argues that various aspects of technical systems mark differences between Anga groups in New Guinea. Some components of these technical systems—particularly sequences of gestures—may be culturally significant but do not result in any material differences that archaeologists could hope to recover. Furthermore, some aspects of technical systems and their products mark social distinctions other than group boundaries; for example, the length and material of a skirt may relate to the wearer's age and initiatory state (Lemonnier 1986:160). Magne and Matson (1987) found that differences in lithic assemblages paralleled Athabascan Salish ethnic differences in interior British Columbia. Gosselain (1992b and this volume) concludes that some, though by no means all, aspects of pottery technology do correspond to ethnic differences in central Cameroon. His exploration of *why* those particular aspects of technology seem to be most significant socially is particularly valuable, and is explored further below. Similarly, in northern Cameroon, Sterner (1989) found that pottery morphology and use patterns with ethnicity, while pottery decoration is better understood as an expression of world view (see also David et al. 1988).

Chilton (this volume) also finds a correlation between social and technological differences in the Late Woodland of eastern North America, although I suspect that those technological differences are better explained in terms of differences in subsistence strategies. In studies in east Africa, Larick (1986, 1991) found that there was an association between ethnicity and spear technology and morphology, but he argues that the ethnic distinctions are passively constituted. Culturally, the importance of spears has to do with age and gender distinctions. Similarly, Childs (1991) found that differences in traditional African iron-smelting furnaces are not unrelated to group differences, but that the symbolism expressed by the furnaces varies greatly and often has more to do with cosmologies and world view than with the expression of social distinctions.

Other studies have found that technologies tend to cross-cut ethnic or other social boundaries. For example, in her study of pottery from third millennium B.C. Southwest Asia, Wright found that a common technology cross-cuts groups defined by a series of material culture differences (including differences in pottery designs). She concludes that while style may be a medium for social expression "technologies do not function in this way. Rather they transfer easily . . . without the encumbrances of cultural barriers" (Wright 1985:22). Similarly, in research on the American Northwest Coast, Croes (1989) found that stone and bone technologies appear to cross-cut ethnic boundaries that are indicated by styles of basketry and cordage. In both cases, it may be that shared technologies relate to a larger social entity than shared styles; unfortunately, we have only limited information on what kinds of social units (ethnic groups?) are associated with those shared styles.

Goodby (this volume) found that in the contact period in New England technological styles do not correspond to social boundaries, though he suggests that the technological differences may relate to older social differences that were transformed by colonialism. Also, among the Luo of Kenya, Dietler and Herbich (1989, 1994a, this volume) found that microstyles of pottery (the microstyle distinction in part includes technological style) result from networks of interaction among the potters but bear little or no relationship to ethnic identity. Finally, MacEachern (this volume) found little correspondence between social groups and ceramic traditions in the Mandara Mountains region of Cameroon and Nigeria.

Studies of style (other than technological style) confound the picture even more. Clearly we cannot assume a relationship between stylistic similarity and degree of social interaction (Graves 1985; Hardin 1984; Plog 1978). However, there may be a link between learning context and stylistic similarity, especially when the learning context is somewhat formalized (Longacre 1991a). Numerous studies have documented how style can be used to mark social, ethnic, or language group distinctions, for example, San arrows (Wiessner 1983), Kalinga vessel shapes (Longacre 1991b), and Baringo District pots, stools, and dress (Hodder 1982). At the same time, Kalinga vessel decoration (Graves 1985) and Baringo District spears and gourds (Hodder 1982) cross-cut those group distinctions. Other kinds of style in the same contexts seem to be more important as means of expressing the self rather than the group (e.g., San beadwork [Wiessner 1983, 1984]). Finally, in some cases style may be best understood as means of expressing and reinforcing world view, unrelated to specific social distinctions (e.g., Cameroonian pottery decorations [David et al. 1988]).

In summary, it is clear that all kinds of meanings are conveyed by various aspects of technology, by technological styles, and by style in general, including decoration. Sometimes the decoration is imbued with meaning and expressions of world view while less heavily loaded technological differences result from different ways of doing things and thus do mark social boundaries (e.g., the Cameroonian pottery studied by David et al. [1988] and Sterner [1989]). This kind of technology may actually *reflect* social boundaries and is probably comprehensible in terms of Sackett's (1982) concept of isochrestic variation. In other cases, technologies—such as Andean and African metallurgy (Lechtman 1977; Childs 1991), and African spears—are imbued with profound symbolic meanings. When these meanings do not cross-cut cultural boundaries there results a correlation of technology and boundary, but that correlation is an epiphenomenon of other kinds of expression. Sometimes aspects of technology seem to be a form of emblemic style (Wiessner 1983) in that they specifically mark group distinctions, as among the Anga studied by Lemonnier (1986). Technologies are sometimes shared across social groups and thus cross-cut styles (Wright 1985). Conversely, particular aspects of technologies may be shared among only a small group of producers who are only a small sub-set of any recognized social group (e.g., the Luo pottery studied by Dietler and Herbich 1989, this volume). In these cases, shared technologies may be byproducts of interaction at different levels, but do not necessarily correlate with the boundaries of any recognized social group.

Given more time and space, the list of possibilities could probably grow significantly. The point, however, is not to obfuscate, but to push ahead. And here I am optimistic. Although the relationship between social and cultural processes and material culture is complex, there do seem to be some regularities that can perhaps be understood if we ask—as Gosselain (this volume) does—*why* certain aspects of material culture seem to relate to certain aspects of society and culture. Carr's work on a "middle range theory of artifact design" (1995a, 1995b) is particularly valuable in this regard, because he attempts to understand why and how different aspects of material culture (both technological and stylistic) vary. This kind of investigation holds a great deal of promise for archaeologists and others concerned with technology and other aspects of material culture.

For this reason, I list some of the regularities that I perceive, particularly with regard to the chapters in this volume, and that others have noted (see also Hegmon 1992; Jones and Hegmon 1990). I do not suggest that these regularities are universal, merely that they bear more investigation. (1) Complex technologies (e.g., metallurgy) seem to be heavily loaded symbolically; (2) Very simple decorations may not be particularly significant at a conscious level, and as a result may be useful indicators of interaction or production contexts; (3) Aspects of production (or technological sequences) that are taught in a relatively formal context may be useful indicators of learning contexts (see Longacre 1991a; Gosselain 1992b, this volume); (4) Decorations that cross-cut a number of media (what DeBoer [1991] calls pervasive style) are likely to be symbolically meaningful; (5) Elaboration of nonvisible objects (such as pots used only in inner rooms or medallions worn under clothing) are likely to communicate not with distant others but with oneself, with one's domestic unit, and sometimes with the spirits (David et al. 1988); (6) Highly visible elaboration is likely to involve communication with others (*sensu* Wobst 1977), although that communication need not always involve group distinctions; (7) Material (including domestic ceramics and architecture) that is used in and that structures everyday domestic life may be particularly relevant to the concept of *habitus*; that is, such material plays an important role in defining who people are socially (see Cameron, Dietler and Herbich, this volume).

CONCLUSIONS: SO WHERE DO WE GO FROM HERE?

At some points, especially with regard to the discussion of boundaries, I fear my discussion may have been critical and somewhat pessimistic. At other points, particularly regarding the search for various kinds of regularities, I hope I have sounded a more optimistic note. The study of material culture, in all of its manifestations, is complex, but that complexity can lead to a richness in our understanding. To this end, I offer three bits of advice regarding the study of technological choices and social boundaries.

First, I am optimistic that if we ask enough "why?" questions we will be able to develop a more comprehensive understanding of the relationship between material culture and social practices. Gosselain's work (1992b, this volume) is an excellent example. He notes that of the many aspects of pottery technology, only some (forming techniques) seem to relate to ethnic differences. Furthermore, he explains this relationship by noting that only forming techniques are taught by hands-on instruction. Another example is suggested by Dietler's and Herbich's (this volume) discussion of Luo architecture, as well as the growing literature on *habitus* and architecture in general (e.g., Bourdieu 1977; Kent 1990b; Moore 1986). Dietler and Herbich note that the Luo change from round to rectangular houses carries little symbolic weight because it has little effect on the spatial organization of the Luo homestead, and it is the spatial organization rather than the geometric form that is symbolically loaded. The archaeological implication is that spatial organization, because it is so much a part of *habitus,* might be a better indication of cultural differences and changes than simply house form or exterior. I have noted other possible regularities at the end of the last section, and I am convinced that exploration of these sorts of regularities will be fruitful for archaeologists interested in the very complex relationship between material culture and society (see Carr and Neitzel 1995).

Second, if we archaeologists intend to focus on certain kinds of social concepts—such as boundaries or ethnicity—in the past, we must be sure we have a good understanding of what those concepts involve. Specifically, we must ask if those concepts really represent things or entities that we can discover in the past. In order to do this, we must involve ourselves in the literature—outside of archaeology—that deals with these concepts, including the literature of sociocultural anthropology and social theory.

Finally, technology clearly has been ignored in considerations of social processes, and we need to make up for this deficit, as this group of papers does. However, we must guard against assuming that technology (whether technological style or technological choices) is *necessarily* a better avenue of exploration than style or other aspects of material culture. Researchers focusing on style have (mostly) stopped arguing about which perspective on style is best and have usefully combined different perspectives. I hope new arguments that pit style against technology do not take their place. Whether style or technology is the best subject will depend on the research question and on the particular study. A combined approach (possibly including style, technology, and function [see Dietler and Herbich, this volume]) will probably be preferable in many cases.

ACKNOWLEDGMENTS

I am grateful to Miriam Stark, the volume editor, for inviting me to participate in the symposium and the volume. In the spirit of intellectual progress, she has opened her work to possibly critical perspectives; may we all have the courage to do the same. I also thank Bob Bolin, Jan Downy, George Cowgill, and Peggy Nelson for discussing some of these issues with me and providing me with useful references and comments.

NOTES

1. Lechtman (1993:245-246) also emphasizes the distinction between technique and technology, though from a different perspective. She argues that the sophistication of Andean technologies (agricultural production, accounting, metallurgy, weaving) has been underestimated because of contemporary emphasis on the relative simplicity of the techniques—that is tools—and lack of consideration of nonmaterial aspects of human intelligence and energy. Although Ingold and Lechtman use different definitions of techniques and technology, both appear to be arguing for more consideration of the human role in technology.
2. The absence of material culture differences should not be interpreted as convincing evidence of an absence of social differences. In fact, ethnic co-residence in the Southwest during the historic period appears to have few material correlates (Cordell and Yannie 1991; Dozier 1966).

References Cited

Adams, E. C.
1983 The Architectural Analogue to Hopi Social Organization and Room Use and Implications for Prehistoric Southwestern Culture. *American Antiquity* 43:44–61.
1991 *The Origin and Development of the Pueblo Katsina Cult.* University of Arizona Press, Tucson.

Adams, E. C., and K. A. Hays
1991 *Homol'ovi II: Archaeology of an Ancestral Hopi Village, Arizona.* Anthropological Papers No. 55. University of Arizona Press, Tucson.

Adams, R. McC.
1996 *Paths of Fire: An Anthropologist's Inquiry into Western Technology.* Princeton University Press, Princeton, New Jersey.

Adler, M. A.
1992 Fathoming the Scale of Anasazi Communities. Paper presented at the session, Issues of Geographic and Demographic Scale in the Southwest, 3rd Southwest Symposium, Tucson, Arizona.

Aldenderfer, M. S. (editor)
1993 *Domestic Architecture, Ethnicity, and Complementarity in the South-Central Andes.* University of Iowa Press, Iowa City.

Allen, B.
1977 Kombio and Yambes. In *Pottery of Papua New Guinea: The National Collection,* edited by B. Egloff, pp. 68–71. Trustees of the Papua New Guinea National Museum and Art Gallery, Port Moresby.

Allen, K. M. S.
1992 Iroquois Ceramic Production: A Case Study of Household-Level Organization. In *Ceramic Production and Distribution: An Integrated Approach,* edited by G. J. Bey III and C. A. Pool, pp. 133–154. Westview, Boulder, Colorado.

Allen, K. M. S., and E. B. W. Zubrow
1989   Environmental Factors in Ceramic Production: The Iroquois. In *Ceramic Ecology, 1988: Current Research on Ceramic Materials*, edited by C. C. Kolb, pp. 61–95. BAR International Series 513, Oxford.

Ambler, C.
1988   *Kenyan Communities in the Age of Imperialism*. Yale University Press, New Haven, Connecticut.

Appadurai, A.
1986a   Introduction: Commodities and the Politics of Value. In *The Social Life of Things: Commodities in Cultural Perspective*, edited by A. Appadurai, pp. 3–63. Cambridge University Press, Cambridge.
1986b   *The Social Life of Things*. Cambridge University Press, Cambridge.

Archer, M. S.
1988   *Culture and Agency: The Place of Culture in Social Theory*. Cambridge University Press, Cambridge.

Arnold, D. E.
1971   Ethnomineralogy of Ticul, Yucatan Potters: Etics and Emics. *American Antiquity* 36(1):20–40.
1975   Ceramic Ecology of the Ayacucho Basin, Peru: Implications for Prehistory. *Current Anthropology* 16(2):183–204.
1981   A Model for the Identification of Non-Local Ceramic Distribution: View from the Present. In *Production and Distribution: A Ceramic Viewpoint*, edited by H. Howard and E. Morris, pp. 31–44. BAR International Series 120, Oxford.
1985   *Ceramic Theory and Cultural Process*. Cambridge University Press, Cambridge.
1989   Patterns of Learning, Residence and Descent among Potters in Ticul, Yucatan, Mexico. In *Archaeological Approaches to Cultural Identity*, edited by S. J. Shennan, pp. 174–184. Unwin Hyman, London.
1991   Can We Go Beyond Cautionary Tales? In *The Ceramic Legacy of Anna O. Shepard*, edited by R. L. Bishop and F. W. Lange, pp. 321–345. University Press of Colorado, Niwot, Colorado.
1993   *Ecology and Ceramic Production in an Andean Community*. Cambridge University Press, Cambridge.

Arnold, P. J.
1991   *Domestic Ceramic Production and Spatial Organization: A Mexican Case Study in Archaeology*. Cambridge University Press, Cambridge.

Aronson, M., J. M. Skibo, and M. T. Stark
1994   Production and Use Technologies in Kalinga Pottery. In *Kalinga Ethnoarchaeology: Expanding Archaeological Method and Theory*, edited by W. A. Longacre and J. Skibo, pp. 83–111. Smithsonian Institution Press, Washington, D.C.

Austen, L.
1947   Rattan Cuirasses and Gourd Penis-Cases. *Man* 47:91–92.

Baker, V. G.
1980   Archaeological Visibility of Afro-American Culture: An Example from Black Lucy's Garden, Andover. In *Archaeological Perspectives on Ethnicity in America: Afro-American*

and Asian-American Culture History, edited by R. L. Schuyler, pp. 29-37. Baywood Publishing, New York.

Baldwin, S. J.
1987   Roomsize Patterns: A Quantitative Method for Approaching Ethnic Identification in Architecture. In *Ethnicity and Culture,* edited by R. Auger, M. F. Glass, S. MacEachern, and P. H. McCartney, pp. 163-174. The University of Calgary Archaeological Association, Calgary.

Barley, N.
1994   *Smashing Pots: Feats of Clay from Africa.* The British Museum Press, London.

Barlow, K.
1985   The Role of Women in Intertribal Trade among the Murik of Papua New Guinea. *Research in Economic Anthropology* 7:95-122.

Barlow, K., L. Bolton, and D. Lipset
1986   *Trade and Society in Transition Along the Sepik Coast.* Report on Anthropological Research in the East Sepik and Sundaun Province, P.N.G. July-August 1986, Sepik Documentation Project. The Australian Museum, Sydney.

Barnard, A.
1979   Kalahari Bushman Settlement Patterns. In *Social and Ecological Systems,* edited by P. C. Burnham and R. F. Ellen, pp. 131-144. Academic Press, London.
1992   *Hunters and Herders of Southern Africa: A Comparative Ethnography of the Khoisan Peoples.* Cambridge University Press, Cambridge and New York.

Barnard, A., and A. Good
1984   *Research Methods in the Study of Kinship.* Academic Press, New York.

Barnes, J. A.
1962   African Models in the New Guinea Highlands. *Man* 62:5-9.

Barreteau, D.
1987   Un essai de classification lexico-statistique des langues de la famille tchadique Parlées au Cameroun. In *Langues et cultures dans le bassin du Lac Tchad,* edited by D. Barreteau, pp. 43-78. Colloques et seminaires de l'Orstom. Éditions de l'Orstom, Paris.

Barreteau, D., R. Breton, and M. Dieu
1984   Les langues. In *Le nord du Cameroun: Des hommes, une région,* edited by J. Boutrais, J. Boulet, A. Beauvilain, P. Gubry, D. Barreteau, M. Dieu, R. Breton, C. Seignobos, G. Pontie, Y. Marguerat, A. Hallaire, and H. Frechou, pp. 159-180. Éditions de l'Orstom, Paris.

Barreteau, D., and L. Sorin-Barreteau
1988   La poterie chez les Mofu-Gudur: Des gestes, des formes et des mots. In *Le milieu et les hommes: recherches comparatives et historiques dans le bassin du Lac Tchad,* edited by D. Barreteau and H. Tourneux, pp. 287-339. Éditions de l'Orstom, Paris.

Barth, F.
1969a   *Ethnic Groups and Boundaries: The Social Organization of Culture Difference.* Little, Brown and Company, Boston.
1969b   Introduction. In *Ethnic Groups and Boundaries,* edited by F. Barth, pp. 9-38. Little, Brown and Company, Boston.
1987   *Cosmologies in the Making.* Cambridge University Press, Cambridge.

Barton, F. R.
1910 The Annual Trading Expedition to the Papuan Gulf. In *The Melanesians of British New Guinea*, edited by C. G. Seligman, pp. 96–120. Cambridge University Press, Cambridge.

Bartram, L. E.
1993 An Ethnoarchaeological Analysis of Kua San (Botswana) Bone Food Refuse. Ph.D. dissertation, Department of Anthropology, University of Wisconsin, Madison. University Microfilms, Ann Arbor, Michigan.

Basalla, G.
1988 *The Evolution of Technology*. Cambridge University Press, Cambridge.

Bastin, Y., A. Coupez, and B. de Halleux
1983 Classification lexicostatistique des langues bantoues (214 relevés). *Bulletin des séances de l'académie royale des sciences d'outre-mer* 27(2):173–199.

Bedaux, R. M. A.
1986 Recherches ethno-archéologiques sur la poterie des Dogon (Mali). In *Op Zoek naar mens en materiële cultuur*, edited by H. Fokken, P. Banga, and M. Bierma, pp. 117–148. Universiteitsdrukkerij R.U.G., Gröningen.

Bender, B.
1978 Gatherer-Hunter to Farmer: A Social Perspective. *World Archaeology* 10:204–222.

Bentley, G. C.
1987 Ethnicity and Practice. *Comparative Studies in Society and History* 29:24–55.

Binford, L. R.
1965 Archaeological Systematics and the Study of Culture Process. *American Antiquity* 31(2):203–210.
1973 Interassemblage Variability—the Mousterian and the "Functional" Argument. In *The Explanation of Culture Change*, edited by C. Renfrew, pp. 227–254. Duckworth, London.
1978a Dimensional Analysis of Behavior and Site Structure: Learning from an Eskimo Hunting Stand. *American Antiquity* 43(3):330–361.
1978b *Nunamiut Ethnoachaeology*. Academic Press, New York.
1980 Willow Smoke and Dogs' Tails: Hunter-Gatherer Settlement Systems and Archaeological Site Formation. *American Antiquity* 45(1):4–20.
1982 The Archaeology of Place. *Journal of Anthropological Archaeology* 1(1):5–31.
1983 *Working at Archaeology*. Academic Press, New York.

Bleed, P.
1986 The Optimal Design of Hunting Weapons: Maintainability or Reliability. *American Antiquity* 51(4):737–747.

Blumenschein, H.
1958 Further Excavations and Surveys in the Taos Area. *El Palacio* 65:107–111.

Bogatyrev, P.
1971 *The Functions of Folk Costume in Moravian Slovakia*. Translated by R. G. Crum. Mouton, The Hague.

Bordes, F., and D. de Sonneville-Bordes
1970   The Significance of the Variability in Paleolithic Assemblages. *World Archaeology* 2:61–73.
Bourdieu, P.
1977   *Outline of a Theory of Practice.* Cambridge University Press, Cambridge.
1980   *Le sens pratique.* Les Éditions de Minuit, Paris
1984   *Distinction: A Social Critique of the Judgement of Taste.* Harvard University Press, Cambridge, Massachusetts.
1990   *The Logic of Practice,* translated by R. Nice. Stanford University Press, Stanford, California.
1993   Race and Ethnicity in America. *Anthro Notes* (National Museum of Natural History Bulletin for Teachers) 15(3):1–15.
Bradford, W.
1968   *History of Plymouth Plantation 1620–1647,* vol.1. Russell and Russell, New York.
Bradley, B. A.
1992   Excavations at Sand Canyon Pueblo. In *The Sand Canyon Archaeological Project: A Progress Report,* edited by W. D. Lipe, pp. 79–97. Occasional Paper no. 2. Crow Canyon Archaeological Center, Cortez, Colorado.
Braithwaite, M.
1982   Decoration as Ritual Symbol: A Theoretical Proposal and an Ethnographic Study in Southern Sudan. In *Symbolic and Structural Archaeology,* edited by I. Hodder, pp. 80–88. Cambridge University Press, Cambridge.
Braun, D. P.
1983   Pots as Tools. In *Archaeological Hammers and Theories,* edited by J. A. Moore and A. S. Keene, pp. 107–134. Academic Press, New York.
1987   Coevolution of Sedentism, Pottery Technology, and Horticulture in the Central Midwest, 200 B.C.–A.D. 600. In *Emergent Horticultural Economies of the Eastern Woodlands,* edited by W. F. Keegan, pp. 153–216. Center for Archaeological Investigations, Occasional Paper no. 7. Southern Illinois University, Carbondale.
Breternitz, C. D., D. E. Doyel, and M. P. Marshall
1982   *Bis sa ani: A Late Bonito Phase Community on the Escavada Wash, Northwest New Mexico.* Navajo Nation Papers in Anthropology no. 14. Window Rock, Arizona.
Breternitz, C. D., and M. P. Marshall
1982   Summary of Analytical Results and Review of Miscellaneous Artifacts from *Bis sa ani* Pueblo. In *Bis sa ani: A Late Bonito Phase Community on the Escavada Wash, Northwest New Mexico,* vol. 2, pt 1, edited by C. D. Breternitz, D. E. Doyel, and M. P. Marshall, pp. 433–448. Navajo Nation Papers in Anthropology no. 14. Window Rock, Arizona.
Broadhurst, H.
1975   Oma, Potter of Vanimo. *Pottery in Australia* 14:10–15.
Bronitsky, G.
1986   The Use of Materials Science Techniques in the Study of Pottery Construction and Use. *Advances in Archaeological Method and Theory,* vol. 9, edited by M. B. Schiffer, pp. 209–276. Academic Press, New York.

Brooks, A. S., F. L. Collier Jackson, and R. R. Grinker
1993 Race and Ethnicity in America. *Anthro Notes* (National Museum of Natural History Bulletin for Teachers) 15(3):1–15.

Brooks, E.
1946 Pottery Types from Hampden County, Massachusetts. *Bulletin of the Massachusetts Archaeological Society* 7(4):78–79.

Brooks, G.
1993 *Landlords and Strangers: Ecology, Society and Trade in Western Africa, 1000–1630.* Westview Press, Boulder, Colorado.

Brown, D. E.
1982 Chihuahuan Desertscrub. In *Desert Plants: Biotic Communities of the American Southwest—United States and Mexico,* edited by D. E. Brown, pp. 169–179. Published by the University of Arizona for the Boyce Thompson Southwestern Arboretum, Tucson.

Brown, D. E., and C. H. Lowe
1980 *Biotic Communities of the Southwest.* USDA Forest Service General Technical Report RM-78, 1 p. map. Rocky Mountain Forest Experiment Station, Fort Collins, Colorado.

Brown, J.
1989 Central Bantu: Agikuyu, Akamba. In *Kenyan Pots and Potters,* edited by J. Barbour and S. Wandibba, pp. 86–96. Oxford University Press, Nairobi.

Brumbach, H. J.
1975 "Iroquoian" Ceramics in "Algonkian" Territory. *Man in the Northeast* 10:17–28.

Brumfiel, E. M.
1992 Distinguished Lecture in Archaeology: Breaking and Entering the Ecosystem-Gender, Class and Faction Steal the Show. *American Anthropologist* 94(3):551–567.
1994 Ethnic Groups and Political Development in Ancient Mexico. In *Factional Competition and Political Development in the New World,* edited by E. M. Brumfiel and J. W. Fox, pp. 89–102. Cambridge University Press, Cambridge.

Buchignani, N.
1987 Ethnic Phenomena and Contemporary Social Theory: Their Implications for Archaeology. In *Ethnicity and Culture,* edited by R. Auger, M. F. Glass, S. MacEachern, and P. H. McCartney, pp. 15–24. Proceedings of the Eighteenth Annual Chacmool Conference, Archaeological Association of the University of Calgary, Calgary, Alberta.

Burnham, P., E. Copet-Rougier, and P. Noss
1986 Gbaya et Mkako: Contribution ethnolinguistique à l'histoire de l'Est-Cameroun. *Paideuma* 32:87–128.

Byers, D. S., and I. Rouse
1960 *A Re-examination of the Guida Farm.* Bulletin of the Archaeological Society of Connecticut no. 30.

Callinicos, A.
1988 *Making History: Agency, Structure and Change in Social Theory.* Cornell University Press, Ithaca, New York.

Cameron, C. M.
1991 Architectural Change at a Southwestern Pueblo. Ph.D. dissertation, Department of Anthropology, University of Arizona, Tucson. University Microfilms, Ann Arbor.

1995 Migration and the Movement of Southwestern People. *Journal of Anthropological Archaeology* 14:104–124.
Campbell, A., and M. Main
1991 *Western Sandveld Remote Area Dwellers: Socio-Economic Survey, Remote Area Development*. Report to the Central District Administration, Serowe and the Remote Area Development Program, Ministry of Local Government and Lands, Gaborone, Botswana.
Camps, G.
1974 *Les civilizations préhistoriques de l'Afrique du Nord et du Sahara*. Doin, Paris.
Carr, C.
1993 Identifying Individual Vessels with X-Radiography. *American Antiquity* 58(1): 96–117.
1995a Building a Unified Middle-Range Theory of Artifact Design. In *Style, Society, and Person: Archaeological and Ethnological Perspectives*, edited by C. Carr and J. Neitzel, pp. 151–170. Plenum Press, New York and London.
1995b A Unified Middle-Range Theory of Artifact Design. In *Style, Society, and Person: Archaeological and Ethnological Perspectives*, edited by C. Carr and J. Neitzel, pp. 171–258. Plenum Press, New York and London.
Carr, C., and J. E. Neitzel (editors)
1995 *Style, Society, and Person: Archaeological and Ethnological Perspectives*. Plenum Press, New York and London.
Cashdan, E.
1979 *Trade and Reciprocity among the River Bushmen of Northern Botswana*. Ph.D. dissertation, Department of Anthropology, University of New Mexico, Albuquerque. University Microfilms, Ann Arbor, Michigan.
1983 Territoriality among Human Foragers: Ecological Models and an Application to Four Bushman Groups. *Current Anthropology* 24(l):47–66.
1984 G//ana Territorial Organization. *Human Ecology* 12(4):443–463.
Casimir, M. J., and A. Rao (editors)
1992 *Mobility and Territoriality: Social and Spatial Boundaries among Foragers, Fishers, Pastoralists, and Peripatetics*. Berg Publishers, New York and London.
Cassedy, D., P. Webb, T. Mills, H. Mills, and Garrow and Associates
1993 New Data on Maize Horticulture and Subsistence in Southwestern Connecticut. Paper presented at the 33rd meeting of the Northeastern Anthropological Association, Danbury, Connecticut.
Ceci, L.
1979 Maize Cultivation in Coastal New York: The Archaeological, Agronomical and Documentary Evidence. *North American Archaeologist* 3(1):45–74.
1982 Method and Theory in Coastal New York Archaeology: Paradigms of Settlement Pattern. *North American Archaeologist* 3(1):5–36.
1990 Radiocarbon Dating "Village" Sites in Coastal New York: Settlement Pattern Change in the Middle to Late Woodland. *Man in the Northeast* 39:1–28.
Chapin, H. M.
1931 *Sachems of the Narragansett*. Rhode Island Historical Society, Providence.

Charlton, T.
1969   Texcoco Fabric-Marked Pottery, Tlateles, and Salt-Making. *American Antiquity* 34(1):73–76.

Chazan, M.
1995   Conceptualization of Time and the Development of Paleolithic Chronology. *American Anthropologist* 97(3):457–467.

Childe, V. G.
1929   *The Danube in Prehistory.* Oxford University Press, Oxford.
1951   *Social Evolution.* Watts and Company, London.
1956   *Piecing Together the Past: The Interpretation of Archaeological Data.* Routledge and Kegan Paul, London.

Childers, G. W.
1976   *Report on the Survey/Investigation of the Ghanzi Farm Basarwa Situation.* Government Printer, Gaborone, Botswana.
1981   *Western Ngwaketse Remote Area Dwellers: A Land Use and Development Plan for Remote Settlements in Southern District.* Government Printer, Gaborone, Botswana.

Childs, S. T.
1991   Style, Technology, and Iron Smelting Furnaces in Bantu-Speaking Africa. *Journal of Anthropological Archaeology* 10:332–359.

Childs, S. T., and D. Killick
1993   Indigenous African Metallurgy: Nature and Culture. *Annual Reviews of Anthropology* 22:317–337.

Chilton, E. S.
1991   The Goat Island Rockshelter: New Light from Old Legacies. M.A. thesis. Department of Anthropology, University of Massachusetts, Amherst.
1994   Confronting Complexity: Material Culture in Historical Context. Paper presented at the annual meeting of the Society for American Archaeology, Anaheim, California.
1996   *Embodiments of Choice: Native American Ceramic Diversity in the New England Interior.* Ph.D. dissertation, Department of Anthropology, University of Massachusetts, Amherst. University Microfilms, Ann Arbor, Michigan.

Clark, G. A.
1994   Migration as an Explanatory Concept in Paleolithic Archaeology. *Journal of Archaeological Method and Theory* 1(4):305–343.

Clark, J. E., and M. Blake
1994   The Power of Prestige: Competitive Generosity and the Emergence of Rank Societies in Lowland Mesoamerica. In *Factional Competition and Political Development in the New World,* edited by E. M. Brumfiel and J. W. Fox, pp. 17–30. Cambridge University Press, Cambridge.

Clark, J. J.
1995a   Domestic Architecture in the Early Classic Period. In *The Roosevelt Community Development Study: New Perspectives on Tonto Basin Prehistory,* edited by M. D. Elson, M. T. Stark, and D. A. Gregory, pp. 251–305. Anthropological Papers no. 15. Center for Desert Archaeology, Tucson, Arizona.

1995b   The Role of Migration in Social Change during the Early Classic Period. In *The Roosevelt Community Development Study: New Perspectives on Tonto Basin Prehistory*, edited by M. D. Elson, M. T. Stark, and D. A. Gregory, pp. 369–384. Anthropological Papers no. 15. Center for Desert Archaeology, Tucson, Arizona.

Clark, V. H.

1992   Decorated Ceramic Analysis. In *The Rye Creek Project: Archaeology in the Upper Tonto Basin*, vol. 2, *Artifact and Specific Analyses*, by M. D. Elson and D. B. Craig, pp. 17–88. Anthropological Papers no. 11. Center for Desert Archaeology, Tucson, Arizona.

Clarke, D. L.

1968   *Analytical Archaeology*. Methuen, London.

Cleuziou, S., A. Coudart, J.-P. Demoule, and A. Schnapp

1991   The Use of Theory in French Archaeology. In *Archaeological Theory in Europe: The Last Three Decades*, edited by I. Hodder, pp. 91–128. Routledge, London and New York.

Coles, J.

1973   *Archaeology by Experiment*. Charles Scribner's Sons, New York.

Collett, D.

1987   A Contribution to the Study of Migrations in the Archaeological Record: The Ngoni and Kololo as a Case Study. In *Archaeology as Long-Term History*, edited by I. Hodder, pp. 105–116. Cambridge University Press, Cambridge.

Collins, S. M.

1975   *Prehistoric Rio Grande Settlement Patterns and the Inference of Demographic Change*. Ph.D. dissertation, Department of Anthropology, University of Colorado, Boulder. University Microfilms, Ann Arbor, Michigan.

Colton, H. S.

1939   *Prehistoric Culture Units and Their Relationships in Northern Arizona*. Museum of Northern Arizona Bulletin 17. Flagstaff, Arizona.

1953   *Potsherds: An Introduction to the Study of Prehistoric Southwestern Ceramics and Their Use in Historic Reconstruction*. Museum of Northern Arizona Bulletin 25. Flagstaff, Arizona.

Comaroff, J.

1987   Of Totemism and Ethnicity: Consciousness, Practice and the Signs of Inequality. *Ethnos* 52:301–323.

Comaroff, J., and J. Comaroff

1992   *Ethnography and the Historical Imagination*. Westview Press, Boulder, Colorado.

Conkey, M. W.

1989   The Place of Material Culture Studies in Contemporary Anthropology. In *Perspectives on Anthropological Collections from the American Southwest: Proceedings of a Symposium*, edited by A. L. Hedlund, pp. 13–32. Anthropological Research Papers no. 40. Arizona State University, Tempe.

1990   Experimenting with Style in Archaeology: Some Historical and Theoretical Issues. In *The Uses of Style in Archaeology*, edited by M. W. Conkey and C. A. Hastorf, pp. 5–17. Cambridge University Press, Cambridge and New York.

Conkey, M. W., and C. A. Hastorf (editors)
1990   *The Uses of Style in Archaeology.* Cambridge University Press, Cambridge and New York.

Cooke, C. K.
1979   The Stone Age in Botswana: A Preliminary Survey. *Arnoldia Rhodesia* 27(81):1–32.

Cooke, C. K., and M. L. Patterson
1960   Stone Age Sites, Lake Dow Area, Bechuanaland. *South African Archaeological Bulletin* 15(59):119–122.

Copet-Rougier, E.
1987   Du clan à la Chefferie dans l'est du Cameroun. *Africa* 57(3):345–363.

Cordell, L. S.
1979a   *Cultural Resources Overview of the Middle Rio Grande Valley, New Mexico.* U.S. Government Printing Office, Washington D.C.
1979b   Prehistory, Eastern Anasazi. In *Handbook of North American Indians*, vol. 9, edited by A. Ortiz, pp. 131–151. Smithsonian Institution Press, Washington, D.C.
1989   Northern and Central Rio Grande. In *Dynamics of Southwest Prehistory*, edited by L. S. Cordell and G. J. Gumerman, pp. 293–336. Smithsonian Institution Press, Washington, D.C.
1995   Tracing Migration Pathways from the Receiving End. *Journal of Anthropological Archaeology* 14:203–211.

Cordell, L. S., and V. J. Yannie
1991   Ethnicity, Ethnogenesis, and the Individual: A Processual Approach toward Dialogue. In *Processual and Postprocessual Archaeologies: Multiple Ways of Knowing the Past*, edited by R. W. Preucel, pp. 96–107. Center for Archaeological Investigations Occasional Paper no. 10. Southern Illinois University, Carbondale.

Costin, C. L.
1991   Craft Specialization: Issues in Defining, Documenting, and Explaining the Organization of Production. *Archaeological Method and Theory* 3:1–56.

Courty, M. A., and V. Roux
1995   Identification of Wheel Throwing on the Basis of Ceramic Surface Features and Micro-Fabrics. *Journal of Archaeological Science* 22(1):17–50.

Cowgill, G. L.
1982   Clusters of Objects and Associations between Variables: Two Approaches to Archaeological Classification. In *Essays on Archaeological Typology*, edited by R. Whallon and J. A. Brown, pp. 30–56. Center for American Archaeology Press, Evanston, Illinois.

Craig, D. B., and J. J. Clark
1994   The Meddler Point Site. In *The Roosevelt Community Development Study*, vol. 2, *Meddler Point, Pyramid Point, and Griffin Wash Sites*, edited by M. D. Elson, D. L. Swartz, D. B. Craig, and J. J. Clark, pp. 1–198. Anthropological Papers no. 13. Center for Desert Archaeology, Tucson, Arizona.

Creamer, W., with C. M. Cameron, and J. D. Beal
1993   *The Architecture of Arroyo Hondo Pueblo, New Mexico.* School of American Research, Santa Fe, New Mexico.

Cresswell, R.
1976   Avant propos. *Techniques et Culture* 1:5–6.
Croes, D. R.
1989   Prehistoric Ethnicity on the Northwest Coast of North America: An Evaluation of Style in Basketry and Lithics. *Journal of Anthropological Archaeology* 8:101–130.
Cronon, W.
1983   *Changes in the Land: Indians, Colonists, and the Ecology of New England*. Hill and Wang, New York.
Crown, P. L.
1991a   The Hohokam: Current Views of Prehistory and the Regional System. In *Chaco and Hohokam, Prehistoric Regional Systems in the American Southwest*, edited by P. L. Crown and W. J. Judge, pp. 135–157. School of American Research Press, Santa Fe, New Mexico.
1991b   The Role of Exchange and Interaction in Salt-Gila Basin Hohokam Prehistory. In *Exploring the Hohokam*, edited by G. J. Gumerman, pp. 383–415. University of New Mexico Press, Albuquerque.
1994   *Ceramics and Ideology: Salado Polychrome Pottery*. University of New Mexico Press, Albuquerque.
Crown, P. L., J. D. Orcutt, and T. A. Kohler
1996   Pueblo Cultures in Transition: The Northern Rio Grande. In *The Prehistoric Pueblo World, A.D. 1150–1350*, edited by M. A. Adler, pp. 188–204. University of Arizona Press, Tucson.
Cushing, F. H.
1889   Report to the Secretary of the Interior re: Casa Grande, February 13, 1889. Ms. envelope 418, Hedge-Cushing Collection. Southwest Museum, Los Angeles.
Damon, F., and R. Wagner (editors)
1989   *Death Rituals and Life in the Societies of the Kula Ring*. Northern Illinois University Press, DeKalb.
David, N.
1992a   The Archaeology of Ideology: Mortuary Practises in the Central Mandara Highlands, Northern Cameroon. In *An African Commitment: Papers in Honour of Peter Lewis Shinnie*, edited by J. Sterner and N. David, pp. 181–210. University of Calgary Press, Calgary.
1992b   Integrating Ethnoarchaeology: A Subtle Realist Perspective. *Journal of Anthropological Archaeology* 11(4):330–359.
David, N., K. Gavua, A. S. MacEachern, and J. Sterner
1991   Ethnicity and Material Culture in North Cameroon. *Canadian Journal of Archaeology* 15:171–177.
David, N., and H. Hennig
1972   The Ethnography of Pottery: A Fulani Case Seen in Archaeological Perspective. *McCaleb Module in Anthropology* 21:1–29.
David, N., and A. S. MacEachern
1988   The Mandara Archaeological Project: Preliminary Results of the 1984 Season. In *Le*

milieu et les hommes: recherches comparitives et historiques dans le bassin du Lac Tchad, edited by D. Barreteau and H. Tourneux, pp. 51–80. Éditions de l'Orstom, Paris.

David, N., and J. Sterner
1993 Water and Iron: Phases in the History of Sukur. Paper presented at the 1993 Mega-Chad Conference, Frankfurt, May 1993.

David, N., J. Sterner, and K. Gavua
1988 Why Pots Are Decorated. *Current Anthropology* 29:365–389.

Davis, D.
1985 Hereditary Emblems: Material Culture in the Context of Social Change. *Journal of Anthropological Archaeology* 4:149–176.

De Crits, E.
1994 Style et technique: Comparaison interethnique de la poterie Sub-Saharienne. In *Terre cuite et société: La céramique, document technique, économique, culturel*, edited by F. Audouze and D. Binder, pp. 343–350. Éditions APDCA, Juan-les-Pins.

de Vos, G., and L. Romanucci-Ross
1975 Ethnicity: Vessel of Meaning and Emblem of Contrast. In *Ethnic Identity: Cultural Continuities and Change*, edited by G. de Vos and L. Romanucci-Ross, pp. 363–390. Mayfield, Palo Alto, California.

Dean, J. S.
1969 *Chronological Analysis of Tsegi Phase Sites in Northeastern Arizona*. Papers of the Laboratory of Tree-Ring Research, no. 3. University of Arizona, Tucson.
1988 The View from the North: An Anasazi Perspective on the Mogollon. *The Kiva* 53:197–199.

Dean, J. S., W. H. Doelle, and J. D. Orcutt
1994 Adaptive Stress, Environment, and Demography. In *Themes in Southwest Prehistory*, edited by G. Gumerman, pp. 53–86. School of American Research Press, Santa Fe, New Mexico.

Dean, J. S., R. C. Euler, G. J. Gumerman, F. Plog, R. H. Hevly, and T. N. V. Karlstrom
1985 Human Behavior, Demography, and Paleoenvironment on the Colorado Plateaus. *American Antiquity* 50(3):537–554.

DeBoer, W. R.
1986 Pillage and Production in the Amazon: A View through the Conibo of the Ucayali Basin, Eastern Peru. *World Archaeology* 18:231–246.
1990 Interaction, Imitation, and Communication as Expressed in Style: The Ucayali Experience. In *The Uses of Style in Archaeology*, edited by M. W. Conkey and C. A. Hastorf, pp. 82–104. Cambridge University Press, Cambridge.
1991 The Decorative Burden: Design, Medium, and Change. In *Ceramic Ethnoarchaeology*, edited by W. A. Longacre, pp.144–161. University of Arizona Press, Tucson.

DeCorse, C. R.
1989 Material Aspects of Limba, Yalunka and Kuranko Ethnicity: Archaeological Research in Northeastern Sierra Leone. In *Archaeological Approaches to Cultural Identity*, edited by S. J. Shennan, pp. 125–140. Unwin Hyman, London.

Deetz, J. F.
1965   *The Dynamics of Stylistic Change in Arikara Ceramics.* Illinois Studies in Anthropology no. 4. University of Illinois Press, Urbana.
1977   *In Small Things Forgotten: The Archaeology of Early American Life.* Anchor, Garden City, New Jersey.

Delneuf, M.
1991   Un champ particulier de l'expérimentation en céramique: Les ateliers de poterie traditionnelle du Nord-Cameroun. In *25 ans d'etudes technologiques en préhistoire,* pp. 65–82. Éditions APDCA, Juan-les-Pins.

Demeritt, D.
1991   Agriculture, Climate, and Cultural Adaptation in the Prehistoric Northeast. *Archaeology of Eastern North America* 19:183–202.

Denbow, J. R.
1983   *Iron Age Economics: Herding, Wealth, and Politics along the Fringes of the Kalahari Desert during the Early Iron Age.* Ph.D. dissertation, Department of Anthropology, Indiana University, Bloomington.

Denny, P.
1994   Archaeological Correlates for Iroquoian and Siouan. Paper presented at the Iroquois Conference, Rensselaerville, New York.

Dialo, B., M. Vanhaelen, and O. P. Gosselain
1995   Plant Constituents Involved in Coating Practices among Traditional African Potters. *Experientia* 51:95–97.

Diehl, R. A.
1983   *Tula, the Toltec Capital of Ancient Mexico.* Thames and Hudson, Ltd., London.

Dietler, M.
1994   "Our Ancestors the Gauls:" Archaeology, Ethnic Nationalism, and the Manipulation of Celtic Identity in Modern Europe. *American Anthropologist* 96:584–605.

Dietler, M., and I. Herbich
1989   *Tich Matek:* The Technology of Luo Pottery Production and the Definition of Ceramic Style. *World Archaeology* 21:148–164.
1993   Living on Luo Time: Reckoning Sequence, Duration, History, and Biography in a Rural African Society. *World Archaeology* 25:248–260.
1994a Ceramics and Ethnic Identity: Ethnoarchaeological Observations on the Distribution of Pottery Styles and the Relationship between the Social Contexts of Production and Consumption. In *Terre cuite et société: La céramique, document technique, économique, culturel.* XIVe Rencontre internationale d'archéologie et d'histoire d'Antibes, edited by D. Binder and F. Audouze, pp. 459–472. Éditions APDCA, Juan-les-Pins.
1994b Habitus et reproduction sociale des techniques: L'intelligence du style en archéologie et en ethnoarchéologie. In *De la préhistoire aux missiles balistiques: L'intelligence sociale des techniques,* edited by B. Latour and P. Lemonnier, pp. 202–227. Éditions la Découverte, Paris.

Dieu, M., and P. Renaud (editors)
1983   *Situation linguistique en Afrique Centrale, inventaire préliminaire: Le Cameroun.*

Atlas Linguistique de l'Afrique Centrale, ACCT-CERDOTOLA-DGRST, Yaounde.

Dilliplane, T.
1980 European Trade Kettles. In *Burr's Hill: A 17th Century Wampanoag Burial Ground in Warren, Rhode Island*, edited by S. Gibson, pp. 79–84. Haffenreffer Museum of Anthropology, Brown University, Providence, Rhode Island.

Dincauze, D. F.
1975 Ceramic Sherds from the Charles River Basin. *Bulletin of the Archaeological Society of Connecticut* 39:5–17.
1976 *The Neville Site: 8,000 Years at Amoskeag, Manchester, New Hampshire*. Peabody Museum Monographs 4. Harvard University, Cambridge, Massachusetts.
1991 A Capsule Prehistory of Southern New England. In *The Pequot: The Fall and Rise of an American Indian Nation*, edited by L. Hauptman and J. Wherry, pp. 19–32. University of Oklahoma Press, Norman.

Dincauze, D. F., and R. M. Gramly
1973 Powisett Rockshelter: Alternative Behavior Patterns in a Simple Situation. *Pennsylvania Archaeologist* 43(1):41–61.

Di Peso, C. C.
1974 *Casas Grandes: A Fallen Trading Center of the Gran Chichimeca*. The Amerind Foundation, Dragoon, Arizona.

Dobres, M.-A.
1995 Gender and Prehistoric Technology: On the Social Agency of Technological Strategies. *World Archaeology* 27:25–49.

Dobres, M.-A., and C. R. Hoffman
1994 Social Agency and the Dynamics of Prehistoric Technology. *Journal of Archaeological Method and Theory* 1(3):211–258.

Doelle, W. H., and D. B. Craig
1992 Prehistoric Demography in the Tonto Basin. In *Research Design for the Roosevelt Community Development Study*, edited by W. H. Doelle, H. D. Wallace, M. D. Elson, and D. B. Craig, pp. 81–88. Anthropological Papers no. 12. Center for Desert Archaeology, Tucson, Arizona.

Doolittle, W. E.
1988 *Pre-Hispanic Occupance in the Valley of Sonora, Mexico: Archaeological Confirmation of Early Spanish Reports*. Anthropological Papers of the University of Arizona no. 48. The University of Arizona Press, Tucson.

Douglas, M. T., and B. Isherwood
1979 *The World of Goods*. Basic Books, New York.

Doyel, D. E.
1976 Salado Cultural Development in the Tonto Basin and Globe-Miami Areas, Central Arizona. *The Kiva* 42:5–16.
1978 *The Miami Wash Project: Hohokam and Salado in the Globe-Miami Area, Central Arizona*. Contribution to Highway Salvage Archaeology in Arizona no. 52. Arizona State Museum, University of Arizona, Tucson.

1980 Hohokam Social Organization and the Sedentary to Classic Transition. In *Current Issues in Hohokam Prehistory*, edited by D. Doyel and F. Plog, pp. 23–40. Arizona State University Anthropological Research Papers no. 23. Tempe, Arizona.

1991 Hohokam Cultural Evolution in the Phoenix Basin. In *Exploring the Hohokam. Prehistoric Desert Peoples of the American Southwest*, edited by G. J. Gumerman, pp. 231–278. University of New Mexico Press, Albuquerque.

Dozier, E. P.
1966 *Hano: A Tewa Indian Community in Arizona.* Holt, Rinehart and Winston, New York.

Draper, P.
1975 !Kung Women: Contrasts in Sexual Egalitarianism in Foraging and Sedentary Contexts. In *Toward an Anthropology of Women*, edited by R. Reiter, pp. 77–109. Monthly Review Press, New York.

Drennan, R. D.
1984 Long-Distance Transport Costs in Pre-Hispanic Mesoamerica. *American Anthropologist* 86:105–112.

Dunnell, R. C.
1978 Style and Function: A Fundamental Dichotomy. *American Antiquity* 43(2):192–202.

Dutton, T. (editor)
1982 *The Hiri in History: Further Aspects of Long Distance Motu Trade in Central Papua.* Australian National University Pacific Research Monograph no. 8. Australian National University, Canberra.

Dykeman, D. D.
1982 Architecture of the *Bis sa ani* Community. In *Bis sa ani: A Late Bonito Phase Community on the Escavada Wash, Northwest New Mexico*, vol. 2, pt. 2, edited by C. D. Breternitz, D. E. Doyel, and M. P. Marshall, pp. 835–869. Navajo Nation Papers in Anthropology no. 14. Window Rock, Arizona.

Dyson-Hudson, R., and E. A. Smith
1978 Human Territoriality: An Ecological Reassessment. *American Anthropologist* 80(1):21–41.

Egloff, B. (senior editor)
1977 *Pottery of Papua New Guinea: The National Collection.* Trustees of the Papua New Guinea National Museum and Art Gallery, Port Moresby.

Elmberg, J.-E.
1966 The Popot Feast Cycle: Acculturated Exchange among the Mejprat Papuans. *Ethnos* (supplement to vol. 30).

Elson, M. D., M. T. Stark, and D. A. Gregory (editors)
1995 *The Roosevelt Community Development Study: New Perspectives on Tonto Basin Prehistory.* Anthropological Papers no. 15. Center for Desert Archaeology, Tucson, Arizona.

Elson, M. D., M. T. Stark, and J. M. Heidke
1992 Prelude to Salado: Preclassic Period Settlement in the Upper Tonto Basin. In *Proceedings of the Second Salado Conference, Globe, Arizona, 1992*, edited by R. C. Lange and S. Germick, pp. 274–285. Occasional Paper of the Arizona Archaeological Society, Phoenix.

Engelbrecht, W.
1972   The Reflection of Patterned Behavior in Iroquois Pottery. *Pennsylvania Archaeologist* 42(3):1–15.
1978   Ceramic Patterning between New York Iroquois Sites. In *The Spatial Organization of Culture*, edited by I. Hodder, pp. 141–152. University of Pittsburgh Press, Pittsburgh, Pennsylvania.

Erdweg, M. J.
1902   Die Bewohner der Insel Tumleo, Berlinhafen, Deutsch-Neuguinea. *Anthropologische Gesellschaft in Wien, Mitteilungen* 32(3 folge 2 band):274–310, 317–399.

Ericson, J. E., D. Read, and C. Burke
1972   Research Design: The Relationship between the Primary Functions and the Physical Properties of Ceramic Vessels and Their Implications for Ceramic Distribution on an Archaeological Site. *Anthropology UCLA* 3:84–95.

Eriksen, T. H.
1991   The Cultural Contexts of Ethnic Differences. *Man* 26:127–144.

Euler, R. T., G. J. Gumerman, T. N. V. Karlstrom, J. S. Dean, and R. H. Hevly
1979   The Colorado Plateaus: Cultural Dynamics and Paleoenvironment. *Science* 205:1089–1101.

Evans-Pritchard, E. E.
1949   Luo Tribes and Clans. *Rhodes-Livingstone Journal* 7:24–40.

Fagan, B.
1981   *In the Beginning: An Introduction to Archaeology.* Little, Brown and Company, Boston.
1995   *People of the Earth: An Introduction to World Prehistory.* 8th edition. HarperCollins, New York.

Feinman, G. M.
1989   Tinkering with Technology: Pitfalls and Prospects for Anthropological Archaeology. In *Pottery Technology: Ideas and Approaches,* edited by G. Bronitsky, pp. 217–220. Westview Press, Boulder, Colorado.

Feinman, G. M., S. Upham, and K. Lightfoot
1981   The Production Step Measure: An Ordinal Index of Labor Input in Ceramic Manufacture. *American Antiquity* 46(4):871–874.

Fenton, W.
1940   Problems Arising from the Northeastern Position of the Iroquois. *Smithsonian Miscellaneous Collections* 100:159–251.
1978   Northern Iroquoian Culture Patterns. In *Handbook of the North American Indians,* vol. 15, *The Northeast,* edited by B. G. Trigger, pp. 296–321. Smithsonian Institution Press, Washington, D.C.

Ferguson, L.
1992   *Uncommon Ground: Archaeology and Early African America, 1650–1800.* Smithsonian Institution Press, Washington, D.C.

Fernstrum, K. W.
1990   Exchange and Information Processing, Problems of Archaeological Measurement: A Case

Study from New Guinea. Ph.D. dissertation, Department of Anthropology, Southern Illinois University, Carbondale. University Microfilms, Ann Arbor, Michigan.

Fewkes, J. W.
1906 The Sun's Influence on the Form of Hopi Pueblos. *American Anthropologist* (n.s.) 8(1):88–100.
1910 Notes on the Occurrence of Adobes in Cliff-Dwellings. *American Anthropologist* 12(3):434–436.
1911 *Antiquities of the Mesa Verde National Park, Cliff Palace.* Bureau of American Ethnology Bulletin 51. Smithsonian Institution, Washington, D.C.

Finegan, E., and N. Besnier
1989 *Language: Its Structure and Use.* Harcourt Brace Jovanovich, New York.

Finlayson, W. D.
1977 *The Saugeen Culture: A Middle Woodland Manifestation in Southwestern Ontario.* Archaeological Society of Canada Paper 61. National Museum of Man, Ottawa.

Fish, P. R., and S. K. Fish
1991 Hohokam Political and Social Organization. In *Exploring the Hohokam, Prehistoric Desert People of the American Southwest,* edited by G. J. Gumerman, pp. 151–175. University of New Mexico Press, Albuquerque.

Fishman, J. A.
1989 *Language and Ethnicity in Minority Sociolinguistic Perspective.* Multilingual Matters 45. Multilingual Matters Ltd., Clevedon, England.

Fitch, J. M., and D. P. Branch
1960 Primitive Architecture and Climate. *Scientific American* 203:134–144.

Foster, G. M.
1965 The Sociology of Pottery: Questions and Hypotheses Arising from Contemporary Mexican Work. In *Ceramics and Man,* edited by F. R. Matson, pp. 43–61. Aldine Publishing Company, Chicago.

Fowler, A. P., and J. R. Stein
1992 *Anasazi Regional Organization and the Chaco System.* Maxwell Museum of Anthropology Anthropological Papers no. 5. University of New Mexico, Albuquerque.

Fowler, W. S.
1945 Motifs of Ceramic Design in Massachusetts: A Proposed Plan of Research. *Bulletin of the Massachusetts Archaeological Society* 6:64.
1966 Ceremonial and Domestic Products of Aboriginal New England. *Bulletin of the Massachusetts Archaeological Society* 27(3–4):33–68.
1974 Two Indian Burials in North Middleboro. *Bulletin of the Massachusetts Archaeological Society* 35(3–4):14–18.

Foy, W.
1902 Ethnographische Beziehungen zwischen Britisch- und Deutsch-Neu-Guinea. *Globus* 82:379–383.

Frank, B. E.
1993 Reconstructing the History of an African Ceramic Tradition: Technology, Slavery

and Agency in the Region of Kadiolo (Mali). *Cahiers d'Etudes Africaines* 33(3): 381–401.

Frankel, H.
1978 *Canoes of Walomo.* Institute of Papua New Guinea Studies, Boroko.

Franklin, N. R.
1986 Stochastic vs. Emblemic: An Archaeologically Useful Method for the Analysis of Style in Australian Rock Art. *Rock Art Research* 3:121–124.

Fried, M. H.
1968 On the Concepts of 'Tribe' and 'Tribal Society.' In *Essays on the Problem of Tribe*, edited by J. Helm, pp. 3–22. University of Washington Press, Seattle.
1975 *The Notion of Tribe.* Cummings Publishing, Menlo Park.

Funk, R. E.
1967 Garoga: A Late Prehistoric Iroquois Village in the Mohawk Valley. In *Iroquois Culture, History, and Prehistory: Proceedings of the 1965 Conference on Iroquois Research*, edited by E. Tooker, pp. 81–84. New York State Museum and Science Service, Albany.
1976 *Some Recent Contributions to Hudson Valley Prehistory.* New York State Museum Memoir 22, Albany.

Galis, K. W.
1955 *Papua's van de Humboldt-Baai: Bijdrage tot een ethnographie.* J. N. Voorhoeve, The Hague.

Gallay, A.
1994 Sociétés englobées et traditions céramiques: Le cas du pays Dogon (Mali) depuis le XIIIe siècle. In *Terre cuite et société: La céramique, document technique, economique, culturel*, edited by F. Audouze and D. Binder, pp. 435–457. Éditions APDCA, Juan-les-Pins.

Gallay, A., E. Huysecom, and A. Mayor
1994 *Peuples et céramiques du delta intérieur du Niger.* Département d'Anthropologie et d'Ecologie de l'Université de Genève, Genève.

Gann, D. W.
1992 The Adobe Pueblo Site: Preliminary Testing at AZ J:14:316 (ASM). Ms. on file, Homol'ovi Research Program. Arizona State Museum, Tucson.

Gavua, K.
1990 Style in Mafa Material Culture. Unpublished Ph.D. dissertation, Department of Archaeology, University of Calgary, Calgary.

Geary, P. J.
1983 Ethnic Identity as a Situational Construct in the Early Middle Ages. *Mitteilungen der Anthropologischen Gesellschaft in Wien* 133:15–26.

Gebauer, A. B.
1987 Stylistic Analysis: A Critical Review of Concepts, Models, and Applications. *Journal of Danish Archaeology* 6:223–229.

Gellner, E.
1983 *Nations and Nationalism.* Basil Blackwell, Oxford.

George, D. R., and J. C. M. Bendremer
1995 Late Woodland Subsistence and the Origins of Maize Horticulture in New England.

Paper presented at the 60th annual meeting of the Society for American Archaeology, Minneapolis, Minnesota.

Gero, J., and M. W. Conkey (editors)
1991 *Engendering Archaeology: Women in Prehistory.* Basil Blackwell, Oxford.

Giddens, A.
1979 *Central Problems in Social Theory.* Cambridge University Press, Cambridge.
1984 *The Constitution of Society: Outline of the Theory of Structuration.* University of California Press, Berkeley and Los Angeles.
1991 Structuration Theory: Past, Present and Future. In *Giddens' Theory of Structuration: A Critical Appreciation,* edited by G. A. Bryant and D. Jary, pp. 201–221. Routledge, London and New York.

Gladwin, W., and H. S. Gladwin
1935 *The Eastern Range of the Red-on-Buff Culture.* Medallion Papers no. 36. Gila Pueblo, Globe, Arizona.

Gladwin, H. S., E. W. Haury, E. B. Sayles, and N. Gladwin
1937 *Excavations at Snaketown I: Material Culture.* Medallion Papers no. 25. Gila Pueblo, Globe, Arizona.

Glanzman, W. D., and S. J. Fleming
1985 Ceramic Technology at Prehistoric Ban Chiang, Thailand: Fabrication Methods. *MASCA Journal* 3:114–121.

Glassie, H.
1975 *Folk Housing in Middle Virginia: A Structural Analysis of Historic Artefacts.* University of Tennessee Press, Knoxville.

Goddard, I.
1978 Eastern Algonquian Languages. In *Handbook of North American Indians,* vol.15, *Northeast,* edited by B. G. Trigger, pp. 70–77. Smithsonian Institution, Washington, D.C.

Goodby, R. G.
1992 Diversity as a Typological Construct: Understanding Late Woodland Ceramics from Narragansett Bay. Paper presented at the 32nd Congress of the Northeastern Anthropological Association, Bridgewater, Massachusetts.
1994 *Style, Meaning, and History: A Contextual Study of 17th Century Native American Ceramics from Southeastern New England.* Ph.D. dissertation, Department of Anthropology, Brown University, Providence, Rhode Island. University Microfilms, Ann Arbor, Michigan.
1995 Native American Ceramics from the Rock's Road Site, Seabrook, New Hampshire. *The New Hampshire Archeologist* 35(1):46–60.

Goodenough, W.
1956 Residence Rules. *Southwestern Journal of Anthropology* 12:22–37.

Gookin, D.
1792 Historical Collections of Indians in New England. *Massachusetts Historical Society Collections* 1:141–227. Reprinted 1968, Johnson Reprint Corporation, New York.
1806 Historical Collections of the Indians in New England. *Collections of the Massachusetts Historical Society,* vol. 1. Massachusetts Historical Society, Boston.

Gosden, C.
1989   Prehistoric Social Landscapes of the Arawe Islands, West New Britain Province. *Archaeology in Oceania* 24(2):45–58.

Gosselain, O. P.
1992a Bonfire of the Enquiries. Pottery Firing Temperatures in Archaeology: What For? *Journal of Archaeological Science* 19(2):243–259.
1992b Technology and Style: Potters and Pottery among Bafia of Cameroon. *Man* (n.s.) 27(3):559–586.
1993   From Clay to Pottery, with Style: 1990–1992 Fieldwork in Cameroon. *Nyame Akuma* 39:2–7.
1994   Skimming through Potter's Agenda: An Ethnoarchaeological Study of Clay Selection Strategies in Cameroon. In *Society, Culture, and Technology in Africa*, edited by S. T. Childs, pp. 99–107. MASCA Research Papers in Science and Archaeology (vol. 11, supplement), Philadelphia, Pennsylvania.
1995   *Identités techniques: Le travail de la poterie au Cameroun méridional*. Ph.D. dissertation, Faculté de Philosophie et Lettres, University of Brussels.

Gosselain, O. P., and A. Livingstone Smith
1995   The Ceramic and Society Project: An Ethnographic and Experimental Approach to Technological Choices. In *The Aim of Laboratory Analyses of Ceramics in Archaeology*, edited by A. Lindhal and O. Stilborg, pp. 147–160. KVHAA Konferenser 34, Stockholm.

Gosselain, O. P., A. Livingstone Smith, H. Wallaert, G. W. Ewe, and M. Vander Linden
1996   Preliminary Results of the "Ceramic and Society Project" Fieldwork in Cameroon, December 1995–March 1996. *Nyame Akuma* 46.

Gosselain, O. P., and P. L. van Berg
1992   Style, individualité et taxonomie chez les potières Bafia du Cameroun. *Bulletin du Centre Genevois d'Anthropologie* 3:99–114.

Gould, R. A.
1980   *Living Archaeology*. Cambridge University Press, Cambridge.
1990   *Recovering the Past*. University of New Mexico Press, Albuquerque.

Graebner, F.
1909   Die melanesische Bogenkultur und ihre Verwandten. *Anthropos* 4:726–780.
1927   Kopfbanke. *Ethnologica* 3:1–13.

Graves, M. W.
1985   Ceramic Design Variation within a Kalinga Village: Temporal and Spatial Processes. In *Decoding Prehistoric Ceramics*, edited by B. A. Nelson, pp. 5–34. University of Southern Illinois Press, Carbondale.

Gregory, D. A.
1995   Prehistoric Settlement Patterns in the Eastern Tonto Basin. In *Roosevelt Community Development Study: New Perspectives on Tonto Basin Prehistory*, edited by M. D. Elson, M. T. Stark, and D. A. Gregory, pp. 127–184. Anthropological Papers no. 15. Center for Desert Archaeology, Tucson, Arizona.

Gregory, D. A., and G. Huckleberry
1994   *An Archaeological Survey in the Blackwater Area*, vol. 1, *The History of Human Settle-*

ment in the Blackwater Area. Cultural Resources Report no. 86. Archaeological Consulting Services, Tempe, Arizona.

Gregory, D. A., and J. Urry (editors)
1985 *Social Relations and Spatial Structures.* St. Martins, New York.

Groves, M.
1972 Hiri. In *Encyclopedia of Papua New Guinea,* edited by P. Ryan, pp. 523–527. Melbourne University Press in association with the University of Papua New Guinea, Melbourne.

Guenther, M. G.
1981 Bushman and Hunter-Gatherer Territoriality. *Zeitschrift fur Ethnologie* 106:109–120.

Guthrie, M.
1967–70 *Comparative Bantu: An Introduction to the Comparative Linguistic and Prehistory of the Bantu Language.* 4 vols. Gregg, Farnborough.

Haaland, R.
1978 Ethnographical Observations of Pottery-Making in Darfur, Western Sudan, with Some Reflections on Archaeological Interpretation. In *New Directions in Scandinavian Archaeology,* edited by K. Kristiansen and C. Poludan-Müller, pp. 47–61. The National Museum of Denmark, Copenhagen.

Haefeli, E., and K. Sweeney
1993 Revisiting *The Redeemed Captive:* New Perspectives on the 1704 Attack on Deerfield. *The William and Mary Quarterly* (3rd series) 52(1):3–46.

Hagstrum, M. B.
1988 Ceramic Production in the Central Andes, Peru: An Archaeological and Ethnographic Comparison. In *A Pot for All Reasons: Ceramic Ecology Revisited,* edited by C. C. Kolb and L. M. Lackey, pp. 127–145. Laboratory of Anthropology, Temple University, Philadelphia.

Hamilton, N., and D. Yesner
1985 Early, Middle and Late Woodland Ceramic Assemblages from Great Diamond Island, Casco Bay, Maine. In *Ceramic Analysis in the Northeast: Contributions to Methodology and Culture History,* edited by J. B. Petersen, pp. 39–72. Occasional Publications in Northeastern Anthropology no. 9. Rindge, New Hampshire.

Handsman, R.
1988 Algonkian Women Resist Colonialism. *Artifacts* 16(3–4):29–31.
1990 Corn and Culture, Pots and Politics: How to Listen to the Voices of Mohegan Women. Paper presented at the annual meeting of the Society for Historical Archaeology, Tucson, Arizona.

Hardin, M. A.
1984 Models of Decoration. In *The Many Dimensions of Pottery: Ceramics in Archaeology and Anthropology,* edited by A. C. Pritchard and S. E. van der Leeuw, pp. 573–614. University of Amsterdam Press, Netherlands.

Harding, T. G.
1967 *Voyagers of the Vitiaz Strait: A Study of a New Guinea Trade System.* American Ethnological Society Monograph 44. University of Washington Press, Seattle.

Härke, H.
1991 All Quiet on the Western Front? Paradigms, Methods and Approaches in West Ger-

man Archaeology. In *Archaeological Theory in Europe: The Last Three Decades*, edited by I. Hodder, pp. 187–222. Routledge, London.

Harries, P.
1991 Exclusion, Classification and Internal Colonialism: The Emergence of Ethnicity among the Tsonga-Speakers of South Africa. In *The Creation of Tribalism in Southern Africa*, edited by L. Vail, pp. 82–117. University of California Press, Berkeley.

Hasenstab, R. J.
1990 *Agriculture, Warfare, and Tribalization in the Iroquois Homeland of New York: A G.I.S. Analysis of Late Woodland Settlement*. Ph.D. dissertation, Department of Anthropology, University of Massachusetts, Amherst. University Microfilms, Ann Arbor, Michigan.

Haudricourt, A.-G.
1987 *La technologie, science humaine: Recherches d'histoire et d'ethnologie des techniques*. Éditions de la Maison des Sciences de l'Homme, Paris.

Hauptman, L.
1990 The Pequot War and Its Legacies. In *The Pequots of Southern New England*, edited by L. M. Hauptman and J. D. Wherry, pp. 69–80. University of Oklahoma Press, Norman.

Haury, E. W.
1932 *Roosevelt 9:6: A Hohokam Site of the Colonial Period*. Medallion Papers no. 1. Gila Pueblo, Globe, Arizona.
1945 *The Excavation of Los Muertos and Neighboring Ruins in the Salt River Valley, Southern Arizona*. Papers of the Peabody Museum of American Archaeology and Ethnology no. 24(1). Harvard University, Cambridge, Massachusetts.
1958 Evidence at Point of Pines for a Prehistoric Migration from Northern Arizona. In *Migrations in New World Culture History*, edited by R. Thompson, pp. 1–6. University of Arizona Bulletin no. 29. University of Arizona Press, Tucson.
1965 Stone: Palettes and Ornaments. In *Excavations at Snaketown: Material Culture*, edited by H. S. Gladwin, E. W. Haury, E. B. Sayles, and N. Gladwin, pp. 121–134. Reprinted. University of Arizona Press, Tucson. Originally published 1937, Medallion Papers no. 25, Gila Pueblo, Globe, Arizona.
1976 *The Hohokam: Desert Farmers and Craftsmen*. Excavations at Snaketown, 1964–1965. University of Arizona Press, Tucson.

Hayes, A., and J. A. Lancaster
1975 *Badger House Community, Mesa Verde National Park*. U.S. Department of the Interior, National Park Service, Washington, D.C.

Healan, D. M.
1974 *Residential Architecture and Household Patterning in Ancient Tula*. Ph.D. dissertation, University of Missouri, Columbia. University Microfilms, Ann Arbor.
1989 Synopsis of Structural Remains in the Canal Locality. In *Tula of the Toltecs, Excavations and Survey*, edited by D. M. Healan, pp. 54–67. University of Iowa Press, Iowa City.

Heath, D. (editor)
1963 *A Journal of the Pilgrims at Plymouth: Mourt's Relation* [1622]. Corinth Books, New York.

Heckenberger, M. J., J. B. Petersen, and N. Asch Sidell
1992   Early Evidence of Maize Agriculture in the Connecticut River Valley of Vermont. *Archaeology of Eastern North America* 20:125–149.

Hegmon, M.
1989   The Styles of Integration: Ceramic Style and Pueblo I Integrative Architecture in Southwestern Colorado. In *The Architecture of Social Integration in Prehistoric Pueblos*, edited by W. D. Lipe and M. Hegmon, pp. 125–142. Occasional Papers no. 1. Crow Canyon Archaeological Center, Cortez, Colorado.
1992   Archaeological Research on Style. *Annual Review of Anthropology* 21:517–536.
1995   *Style as a Social Strategy in the Early Puebloan Southwest.* Occasional Papers no. 5. Crow Canyon Archaeological Center, Cortez, Colorado.

Heidke, J. M.
1995   Overview of the Ceramic Collection. In *The Roosevelt Community Development Study*, vol. 2, *Ceramic Chronology, Technology, and Economics*, edited by J. M. Heidke and M. T. Stark, pp. 7–18. Anthropological Papers no. 14. Center for Desert Archaeology, Tucson, Arizona.

Heinz, H. J.
1972   Territoriality among the Bushmen in General and the !Ko in Particular. *Anthropos* 67:405–416.
1979   The Nexus Complex among the !Xo Bushmen of Botswana. *Anthropos* 74:465–480.

Held, G. J.
1947   *Papoea's van Waropen.* E. J. Brill, Leiden.
1957   *The Papuas of Waropen.* Koninklijk Instituut voor Taal-, Land- en Volkenkunde, translation series 2. Martinus Nijhoff, The Hague.

Herbich, I.
1981   *Luo Pottery: Socio-cultural Context and Archaeological Implications (An Interim Report).* Institute of African Studies, University of Nairobi Paper no. 155. Nairobi.
1987   Learning Patterns, Potter Interaction and Ceramic Style among the Luo of Kenya. *The African Archaeological Review* 5:193–204.

Herbich, I., and M. Dietler
1989   River-Lake Nilotic: Luo. In *Kenyan Pots and Potters,* edited by J. Barbour and S. Wandibba, pp. 27–40. Oxford University Press, Nairobi.
1991   Aspects of the Ceramic System of the Luo of Kenya. In *Töpferei- und Keramikforschung 2,* edited by H. Lüdtke and R. Vossen, pp. 105–135. Habelt, Bonn.
1993   Space, Time, and Symbolic Structure in the Luo Homestead: An Ethnoarchaeological Study of "Settlement Biography" in Africa. In *Actes du XIIe Congrès international des sciences préhistoriques et protohistoriques, Bratislava, Czechoslovakia, September 1–7, 1991,* vol. 1, edited by J. Pavúk, pp. 26–32. Archaeological Institute of the Slovak Academy of Sciences, Nitra.

Herron, M. K.
1986   A Formal and Functional Analysis of St. Johns Series Pottery from Two Sites in St. Augustine, Florida. In *Papers in Ceramic Analysis,* edited by P. M. Rice. *Ceramic Notes* 3:31–45.

Higgeson, Reverend
1629   New England's Plantation. In *Massachusetts Historical Society Collections* 1:117–124. Reprinted 1968, Johnson Reprint Corporation, New York.
Hill, J. N.
1970   *Broken K Pueblo: Prehistoric Social Interaction in the American Southwest.* Anthropological Papers of the University of Arizona no. 18, Tucson.
1985   Style: A Conceptual Evolutionary Framework. In *Decoding Prehistoric Ceramics,* edited by B. A. Nelson, pp. 128–153. Southern Illinois University Press, Carbondale.
Hillier, B., and J. Hanson
1984   *The Social Logic of Space.* Cambridge University Press, Cambridge.
Hindess, B., and P. Q. Hirst
1975   *Pre-Capitalist Modes of Production.* Routledge and Kegan Paul, London.
Hitchcock, R. K.
1978   *Kalahari Cattle Posts: A Regional Study of Hunter-Gatherers, Pastoralists, and Agriculturalists in the Western Sandveld Region, Botswana.* Botswana Government Printer, Gaborone, Botswana.
1980   Tradition, Social Justice, and Land Reform in Central Botswana. *Journal of African Law* 24(1):1–34.
1982   *The Ethnoarchaeology of Sedentism: Mobility Strategies and Site Structure among Foraging and Food Producing Populations in the Eastern Kalahari Desert, Botswana.* Ph.D. dissertation, Department of Anthropology, University of New Mexico, Albuquerque. University Microfilms, Ann Arbor, Michigan.
1987   Sedentism and Site Structure: Organizational Changes in Kalahari Basarwa Residential Locations. In *Method and Theory for Activity Area Research: An Ethnoarchaeological Approach,* edited by S. Kent, pp. 374–423. Columbia University Press, New York.
1988   *Monitoring, Research, and Development in the Remote Areas of Botswana.* Ministry of Local Government and Lands, Gaborone, Botswana.
1995   Centralization, Resource Depletion, and Coercive Conservation among the Tyua of the Northeastern Kalahari. *Human Ecology* 23(3):169–198.
Hitchcock, R. K., and P. Bleed
1997   Each According to Need and Fashion: Spear and Arrow Use among San Hunters of the Kalahari. In *Projectile Technology,* edited by H. Knecht, pp. 345–368. Plenum Press, New York.
Hitchcock, R. K., and J. D. Holm
1993   Bureaucratic Domination of Hunter-Gatherer Societies: A Study of the San in Botswana. *Development and Change* 24(2):305–338.
Hobsbawm, E.
1990   *Nations and Nationalism Since 1780: Programme, Myth, Reality.* Cambridge University Press, Cambridge.
Hodder, I.
1979a   Economic and Social Stress and Material Culture Patterning. *American Antiquity* 44:446–454.
1979b   Pottery Distributions: Service and Tribal Areas. In *Pottery and the Archaeologist,* edited

by M. Millett, pp. 7-24. Institute of Archaeology Occasional Publication no. 4. London.
1982 *Symbols in Action: Ethnoarchaeological Studies of Material Culture.* Cambridge University Press, Cambridge.
1985 Boundaries as Strategies: An Ethnoarchaeological Study. In *The Archaeology of Frontiers and Boundaries,* edited by S. W. Green and S. M. Perlman, pp. 141-159. Academic Press, New York.
1986 *Reading the Past: Current Approaches to Interpretation in Archaeology.* Cambridge University Press, Cambridge.
1990 Style as Historical Quality. In *The Uses of Style in Archaeology,* edited by M. Conkey and C. Hastorf, pp. 44-51. Cambridge University Press, New York.
1991 The Decoration of Containers: An Ethnographic and Historical Study. In *Ceramic Ethnoarchaeology,* edited by W. Longacre, pp. 71-94. University of Arizona Press, Tucson.
1992 *Reading the Past: Current Approaches to Interpretation in Archaeology.* 2nd edition. Cambridge University Press, Cambridge.

Hofmeyr, I.
1987 Building a Nation from Words: Afrikaans Language, Literature and 'Ethnic Identity,' 1902-1924. In *The Politics of Race, Class and Nationalism in Twentieth-Century South Africa,* edited by S. Marks and S. Trapido, pp. 95-123. Routledge, London.

Hosler, D.
1994 *The Sounds and Colors of Power.* The MIT Press, Cambridge, Massachusetts.

Hudson, D., R. Smith, and E. A. Smith
1978 Human Territoriality: An Ecological Reassessment. *American Anthropologist* 80(1):21-41.

Huysecom, E.
1992 Les percuteurs d'argile: Des outils de potières africaines utilisés de la préhistoire à nos jours. *Bulletin du Centre Genevois d'Anthropologie* 3:71-98.

Ingersoll, D. W., Jr.
1987 Introduction to Mirror and Metaphor. In *Mirror and Metaphor: Material and Social Constructions of Reality,* edited by D. W. Ingersoll and G. Bronitsky, pp. 1-16. University Press of America, Lanham, Maryland.

Ingersoll, D. W., Jr., and W. Macdonald
1977 Introduction. In *Experimental Archeology,* edited by D. W. Ingersoll, J. E. Yellen, and W. Macdonald, pp. xi-xvii. Columbia University Press, New York.

Ingersoll, D. W., Jr., J. E. Yellen, and W. MacDonald (editors)
1977 *Experimental Archeology.* Columbia University Press, New York.

Ingold, T.
1986 *The Appropriation of Nature: Essays on Human Ecology and Social Relations.* Manchester University Press, Manchester, England.
1988 Tools, Minds, and Machines: An Excursion into the Philosophy of Technology. *Techniques et Culture* (n.s.) 12:151-176.
1990 Society, Nature and the Concept of Technology. *Archaeological Review from Cambridge* 9(1):5-17.

1993a   Tools, Techniques and Technology. In *Tools, Language and Cognition in Human Evolution*, edited by K. R. Gibson and T. Ingold, pp.337–345. Cambridge University Press, Cambridge.

1993b   Tool-Use, Sociality and Intelligence. In *Tools, Language and Cognition in Human Evolution*, edited by K. R. Gibson and T. Ingold, pp. 429–445. Cambridge University Press, Cambridge.

Jameson, J. F.

1909   *Narratives of New Netherland, 1609–1664*. Barnes and Noble, New York.

Janssens, B.

1993   *Doubles réflexes consonantiques: Quatre études sur le Bantu de zone A*. Ph.D. dissertation, Faculté de Philosophie et Lettres, University of Brussels.

Johnson, D. A.

1992   Adobe Brick Architecture and Salado Ceramics at Fourmile Ruin. In *Proceedings of the Second Salado Conference, Globe, Arizona 1992*. Arizona Archaeological Society, Phoenix.

Johnson, E. S.

1993   *Some by Flatteries and Others by Threatening: Political Strategies in Seventeenth Century Native New England*. Ph.D. dissertation, Department of Anthropology, University of Massachusetts, Amherst. University Microfilms, Ann Arbor, Michigan.

Johnson, M.

1989   Conceptions of Agency in Archaeological Interpretation. *Journal of Anthropological Archaeology* 8:189–211.

1993   *Housing Culture: Traditional Architecture in an English Landscape*. Smithsonian Institution Press, Washington, D.C.

Jones, K., and M. Hegmon

1990   The Medium and the Message: A Survey of Information Conveyed by Material Culture in Middle Range Societies. Paper presented at the 56th annual meetings of the Society for American Archaeology, New Orleans.

Jordan, D.

1975   Factors Affecting New England Archaeology. *Man in the Northeast* 10:71–74.

Josselyn, J.

1988   *John Josselyn, Colonial Traveler: A Critical Edition of Two Voyages to New-England*, edited and introduced by P. J. Lindholdt. University Press of New England, Hanover, New Hampshire. Originally published 1674, G. Widdows, at the Green-Dragon in St. Paul's Churchyard, London.

Judd, N. M.

1916   *The Use of Adobe in Prehistoric Dwellings of the Southwest*. Holmes Anniversary Volume. Peabody Museum, Harvard University, Cambridge, Massachusetts.

1919   Archaeological Investigations at Paragonah, Utah. *Smithsonian Miscellaneous Collections* 70 (3).

1926   *Archaeological Observations North of the Rio Colorado*. Bureau of American Ethnology, Bulletin 82. Smithsonian Institution, Washington, D.C.

Kaiser, T.

1989   Steatite-Tempered Pottery from Selevac, Yugoslavia: A Neolithic Experiment in Ceramic Design. *Archaeomaterials* 3(1):1–10.

Keene, A. S., and E. S. Chilton
1995 Toward an Archaeology of the Pocumtuck Homeland. Paper presented at the 60th annual meeting of the Society for American Archaeology, Minneapolis.

Keener, R. T.
1965 The Phillips Site Excavation. *Bulletin of the Archaeological Society of Connecticut* 33:13–44.

Kelley, J. H.
1984 *The Archaeology of the Sierra Blanca Region of Southwestern New Mexico.* Anthropological Papers of the Museum of Anthropology no. 74. University of Michigan, Ann Arbor.
1995 Thoughts for the Durango Workshop on the Casas Grandes System. Paper presented at the Durango Conference, Durango, Colorado.

Kelly, R. C.
1976 *Etoro Social Structure.* University of Michigan Press, Ann Arbor.

Kelly, R. L.
1995 *The Foraging Spectrum: Diversity in Hunter-Gatherer Lifeways.* Smithsonian Institution Press, Washington, D.C.

Kent, S.
1990a Activity Areas and Architecture: An Interdisciplinary View of the Relationship between Use of Space and Domestic Built Environments. In *Domestic Architecture and the Use of Space,* edited by S. Kent, pp. 1–8. Cambridge University Press, Cambridge.
1990b *Domestic Architecture and the Use of Space: An Interdisciplinary Cross-Cultural Study.* Cambridge University Press, Cambridge.

Kent, S., and H. Vierich
1989 The Myth of Ecological Determinism-Anticipated Mobility and Site Spatial Organization. In *Farmers as Hunters: The Implications of Sedentism,* edited by S. Kent, pp. 96–130. Cambridge University Press, Cambridge and New York.

Kenyon, V. B.
1979 A New Approach to the Analysis of New England Prehistoric Pottery. *Man in the Northeast* 18:81–84.
1983 *River Valleys and Human Interaction: A Critical Evaluation of Middle Woodland Ceramics in the Merrimack River Valley.* Ph.D. dissertation, Department of Archaeology, Boston University. University Microfilms, Ann Arbor, Michigan.
1985 The Prehistoric Pottery of the Smyth Site. In *Ceramic Analysis in the Northeast: Contributions to Methodology and Culture History,* edited by J. B. Petersen, pp. 89–107. Occasional Publications in Northeastern Anthropology no. 9. Rindge, New Hampshire.

Kidder, A. V., H. S. Cosgrove, and C. B. Cosgrove
1949 The Pendleton Ruin, Hidalgo County, New Mexico. Carnegie Institution of Washington, Publication 585. *Contributions to American Anthropology and History* 50:107–152.

Kidder, A. V., and S. J. Guernsey
1919 *Archaeological Explorations in Northeastern Arizona.* Bureau of American Ethnology Bulletin 65. Smithsonian Institution, Washington, D.C.

Kingery, W. D.
1955 Factors Affecting Thermal Stress Resistance of Ceramic Materials. *Journal of the American Ceramic Society* 38(1):3–15.

Kleindeinst, M. R., and P. J. Watson
1956 "Action Archeology:" The Archeological Inventory of a Living Community. *Anthropology Tomorrow* 5(1):75–78.

Klemptner, L. J., and P. E. Johnson
1986 Technology and the Primitive Potter: Mississippian Pottery Development Seen Through the Eyes of a Ceramic Engineer. In *Ceramics and Civilization*, vol. 2, *Technology and Style*, edited by W. D. Kingery, pp. 251–271. The American Ceramic Society, Columbus, Ohio.

Knowles, R. L.
1974 *Energy and Form: An Ecological Approach to Urban Growth*. MIT Press, Cambridge, Massachusetts.

Kohler, T. A.
1993 News from the Northern American Southwest: Prehistory on the Edge of Chaos. *Journal of Archaeological Research* 1(4):267–321.

Kolb, C. C.
1989 The Current Status of Ceramic Studies. In *Ceramic Ecology, 1988: Current Research on Ceramic Materials*, edited by C. C. Kolb, pp. 377–421. BAR International Series 513, Oxford.

Kopytoff, I.
1986 The Cultural Biography of Things: Commoditization as Process. In *The Social Life of Things: Commodities in Cultural Perspective*, edited by A. Appadurai, pp. 64–91. Cambridge University Press, Cambridge.

Kraft, H. C.
1975 *The Archaeology of the Tocks Island Area*. Seton Hall, South Orange, New Jersey.

Kramer, C.
1979 Introduction. In *Ethnoarchaeology: Implications of Ethnography for Archaeology*, edited by C. Kramer, pp. 1–20. Columbia University Press, New York.
1982 *Village Ethnoarchaeology: Rural Iran in Archaeological Perspective*. Academic Press, New York.
1985 Ceramic Ethnoarchaeology. *Annual Review of Anthropology* 1:77–102.

Krause, R. A.
1985 *The Clay Sleeps: An Ethnoarchaeological Study of Three African Potters*. University of Alabama Press, Tuscaloosa.

Kristmanson, H., and M. Deal
1993 The Identification and Interpretation of Finishing Marks on Prehistoric Nova Scotian Ceramics. *Canadian Journal of Archaeology* 17.

Kroeber, A. L.
1939 *Cultural and Natural Areas of Native North America*. University of California Publications in American Archaeology and Ethnology, vol. 38. University of California, Berkeley.

Kuhn, R. D., and S. E. Bamann
1987   A Preliminary Report on the Attribute Analysis of Mohawk Ceramics. *Bulletin and Journal the New York State Archaeological Association* 94:40–46.
Kuhn, R. D., and R. E. Funk
1994   The Mohawk Klock and Smith Sites. Unpublished ms. in possession of authors.
Kumar, D., and S. Kadirgamar
1989   Introduction. In *Ethnicity: Identity, Conflict, and Crisis,* edited by D. Kumar and S. Kadirgamar, pp. vii–x. Arena Press, Hong Kong.
Kus, S., and V. Rasharijaona
1990   Domestic Space and the Tenacity of Tradition among Some Betsileo of Madagascar. In *Domestic Architecture and the Use of Space,* edited by S. Kent, pp. 21–33. Cambridge University Press, Cambridge.
LaFantasie, G.
1988   *The Correspondence of Roger Williams,* vols. 1 and 2. Brown University Press, University Press of New England, Hanover, New Hampshire.
Lambert, M. F.
1954   *Paa-ko, Archaeological Chronicle of an Indian Village in North Central New Mexico.* Monographs of the School of American Research no. 19. Santa Fe, New Mexico.
Larick, R.
1985   Spears, Style and Time among Maa-Speaking Pastoralists. *Journal of Anthropological Archaeology* 4:201–215.
1986   Age Grading and Ethnicity in the Style of Loikop (Samburu) Spears. *World Archaeology* 18(2):268–283.
1987   Men of Iron and Social Boundaries in Northern Kenya. In *Ethnicity and Culture,* edited by R. Auger, M. Glass, S. MacEachern, and P. McCartney, pp. 67–76. University of Calgary Press, Calgary.
1991   Warriors and Blacksmiths: Mediating Ethnicity in East African Spears. *Journal of Anthropological Archaeology* 10:299–331.
Lavin, L. F.
1980   Analysis of Ceramic Vessels from the Ben Hollister Site, Glastonbury, Connecticut. *Bulletin of the Archaeological Society of Connecticut* 43:3–46.
1986   Pottery Classification and Cultural Models in Southern New England Prehistory. *North American Archaeologist* 7(1):1–14.
1987   The Windsor Ceramic Tradition in Southern New England. *North American Archaeologist* 8(1):27–40.
1988   The Morgan Site, Rocky Hill, Connecticut: A Late Woodland Farming Community in the Connecticut River Valley. *Bulletin of the Archaeological Society of Connecticut* 5:7–22.
Lavin, L., F. Gudrian, and L. Miroff
1993   Pottery Production and Cultural Process: Prehistoric Ceramics from the Morgan Site. *Northeast Historical Archaeology* 21–22:44–63.
Lavin, L., and R. Kra
1994   Prehistoric Pottery Assemblages from Southern Connecticut: A Fresh Look at Ce-

ramic Classification in Southern New England. *Bulletin of the Archaeological Society of Connecticut* 57:35–51.

Lavin, L., and T. Mlroff

1992   Aboriginal Pottery from the Indian Ridge Site, New Milford, Connecticut. *Bulletin of the Archaeological Society of Connecticut* 55:39–51.

Lawry, S., and A. Thoma

1978   *A Spatial Development Plan for Remote Settlements in Northern Kgalagadi*. Kgalagadi District Council, Tsabong, and Hukuntsi, Botswana.

Leach, J. W., and E. Leach (editors)

1983   *The Kula: New Perspectives on Massim Exchange*. Cambridge University Press, Cambridge.

LeBlanc, S. A.

1982   Temporal Change in Mogollon Ceramics. In *Southwestern Ceramics: A Comparative Review*, edited by A. H. Schroeder, pp. 107–127. The Arizona Archaeologist no. 15. Arizona Archaeological Society, Phoenix.

1983   *The Mimbres People, Ancient Pueblo Painters of the American Southwest*. Thames and Hudson, London.

1989   Cultural Dynamics in the Southern Mogollon Area. In *Dynamics of Southwest Prehistory*, edited by L. S. Cordell and G. J. Gumerman, pp. 179–208. Smithsonian Institution Press, Washington, D.C.

Lechtman, H.

1977   Style in Technology: Some Early Thoughts. In *Material Culture: Style, Organization, and Dynamics of Technology*, edited by H. Lechtman and R. S. Merrill, pp. 3–20. West Publishing, New York and St. Paul, Minnesota.

1993   Technologies of Power: The Andean Case. In *Configurations of Power: Holistic Anthropology in Theory and Practice*, edited by J. S. Henderson and P. J. Netherly, pp. 244–280. Cornell University Press, Ithaca, New York.

Lechtman, H., and R. S. Merrill (editors)

1977   *Material Culture: Styles, Organization, and Dynamics of Technology*. 1975 Proceedings of the American Ethnological Society. West Publishers, New York and St. Paul, Minnesota.

Lechtman, H., and A. Steinberg

1979   The History of Technology: An Anthropological Point of View. In *The History and Philosophy of Technology*, edited by G. Bugliarello and D. B. Doner, pp. 135–160. University of Illinois Press, Champaign.

Lee, R. B.

1979   *The !Kung San: Men, Women, and Work in a Foraging Society*. Cambridge University Press, Cambridge and New York.

Lekson, S. H.

1986   *Great Pueblo Architecture of Chaco Canyon, New Mexico*. University of New Mexico Press, Albuquerque.

1988   The Idea of the Kiva in Anasazi Archaeology. *The Kiva* 53:213–234.

1989   Kivas? In *The Architecture of Social Integration in Prehistoric Pueblos*, edited by W. D. Lipe and M. Hegmon, pp. 161–167. Occasional Paper no. 1. Crow Canyon Archaeological Center, Cortez, Colorado.

1990   Prodigies of Prehistory: The Southwest's Remarkable Mimbres People. *Archaeology* 43(6):44–47.

1992   *Archaeological Overview of Southwestern New Mexico.* Prepared for the New Mexico State Historic Preservation Division, submitted by Human Systems Research, Inc., Las Cruces, New Mexico.

Lekson, S., and C. Cameron

1995   The Abandonment of Chaco Canyon, the Mesa Verde Migrations, and the Reorganization of the Pueblo World. *Journal of Anthropological Archaeology* 14:184–202.

Lekson, S., T. C. Windes, J. R. Stein, and W. J. Judge

1988   The Chaco Canyon Community. *Scientific American* 259(1):100–109.

Lemonnier, P.

1976   La description des *chaînes opératoires:* Contribution à l'analyse des systèmes techniques. *Techniques et Culture* 1:100–151.

1983   La description des systèmes techniques: Une urgence en technologie culturelle. *Techniques et Culture* 1 (n.s.):11–26.

1986   The Study of Material Culture Today: Towards an Anthropology of Technical Systems. *Journal of Anthropological Archaeology* 5:147–186.

1989   Bark Capes, Arrowheads and Concord: On Social Representations of Technology. In *The Meaning of Things: Material Culture and Symbolic Expression,* edited by I. Hodder, pp. 156–171. Unwin Hyman, London.

1990   Topsy Turvy Techniques: Remarks on the Social Representation of Techniques. *Archaeological Review from Cambridge* 9:27–37.

1992   *Elements for an Anthropology of Technology.* Anthropological Papers of the Museum of Anthropology, vol. 88. University of Michigan, Ann Arbor.

1993a   Introduction. In *Technical Choices: Transformation in Material Cultures since the Neolithic,* edited by P. Lemonnier, pp. 1–35. Routledge, London and New York.

1993b   *Technological Choices: Transformation in Material Cultures since the Neolithic.* Routledge, London and New York.

Lenig, D.

1965   The Oak Hill Horizon and Its Relation to the Development of Five Nations Iroquois Culture. *Research and Transactions of the New York State Archaeological Association* 15(1):1–114.

Leroi-Gourhan, A.

1943   *Évolution et techniques,* tome 1, *L'homme et la matière.* Albin Michel, Paris.

1964   *Le geste et la parole,* Tome 1, *Techniques et language.* Albin Michel, Paris.

1993   *Gesture and Speech (Le geste et la parole).* Translated from the French by Anna Bostock Berger. MIT Press, Cambridge, Massachusetts.

Levine, R. A., and D. T. Campbell

1972   *Ethnocentrism: Theories of Conflict, Ethnic Attitudes, and Group Behavior.* John Wiley and Sons, New York.

Lewis, M.

1991   Elusive Societies: A Regional-Cartographic Approach to the Study of Human Relatedness. *Annals of the Association of American Geographers* 81(4):605–626.

Lewis-Williams, J. D.
1982 The Economic and Social Context of Southern African San Rock Art. *Current Anthropology* 23:429–438.
1983 *Believing and Seeing: Symbolic Meanings in Southern San Rock Paintings*. Academic Press, London and New York.

Lightfoot, K. G.
1979 Food Redistribution among Prehistoric Pueblo Groups. *The Kiva* 44(4):319–339.

Lindauer, O.
1995 *The Archaeology of Schoolhouse Point Mesa, Roosevelt Platform Mound Study (1995 Draft)*. Roosevelt Monograph Series 8, Anthropological Field Studies 37. Office of Cultural Resource Management, Arizona State University, Tempe.

Lindsay, A. J., Jr.
1987 Anasazi Population Movements to Southern Arizona. *American Archaeology* 6(3):190–198.

Lipset, D. M.
1985 Seafaring Sepiks: Ecology, Warfare, and Prestige in Murik Trade. *Research in Economic Anthropology* 7:67–94.

Lizee, J. M.
1994 Cross-Mending Northeastern Ceramic Typologies. Paper presented at the 34th annual meeting of the Northeastern Anthropological Association. Geneseo, New York.

Lomnitz-Adler, C.
1991 Concepts for the Study of Regional Culture. *American Ethnologist* 18:195–214.

Longacre, W. A.
1970 *Archaeology as Anthropology: A Case Study*. Anthropological Papers of the University of Arizona no. 17. Tucson, Arizona.
1975 Material Culture Studies. *Reviews in Anthropology* 2(4):441–445.
1981 Kalinga Pottery: An Ethnoarchaeological Study. In *Pattern of the Past: Studies in Honor of David Clarke*, edited by I. Hodder, G. Isaac, and N. Hammond, pp. 49–66. Cambridge University Press, Cambridge.
1991a Ceramic Ethnoarchaeology: An Introduction. In *Ceramic Ethnoarchaeology*, edited by W. A. Longacre, pp.1–10. University of Arizona Press, Tucson.
1991b Sources of Ceramic Variability among the Kalinga of Northern Luzon. In *Ceramic Ethnoarchaeology*, edited by W. A. Longacre, pp. 95–111. University of Arizona Press, Tucson.

Longacre, W. A., and J. M. Skibo (editors)
1994 *Kalinga Ethnoarchaeology*. Smithsonian Institution Press, Washington, D.C.

Loose, A. A.
1974 *Archaeological Excavations Near Arroyo Hondo, Carson National Forest, New Mexico*. Report no. 4. USDA Forest Service, Albuquerque, New Mexico.

Luebben, R.
1968 Site TA 32: A Deep Pithouse and Surface Manifestation in North-Central New Mexico. *Fort Burgwin Research Center Papers on Taos Archaeology* 7:45–57.

Luedtke, B.
1980 The Calf Island Site and the Late Prehistoric Period in Boston Harbor. *Man in the Northeast* 20:1–23.
1986 Regional Variation in Massachusetts Ceramics. *North American Archaeologist* 7(2):113–135.

Lyneis, M. M.
1996   Pueblo II-Pueblo III Change in Southwestern Utah, the Arizona Strip, and Southern Nevada. In *The Prehistoric Pueblo World*, A.D. *1150–1350*, edited by M. A. Adler, pp. 11–28. University of Arizona Press, Tucson.

Lyons, P. J.
1987   Language and Style in the Peruvian. Montaña. In *Ethnicity and Culture*, edited by R. Auger, M. Glass, S. MacEachern, and P. McCartney, pp. 101–114. University of Calgary Press, Calgary.

Macdonald, W. K.
1990   Investigating Style, an Exploratory Analysis of Some Plains Burials. In *The Uses of Style in Archaeology*, edited by M. Conkey and C. Hastorf, pp. 52–60. Cambridge University Press, Cambridge.

MacEachern, A. S.
1990   Du Kunde: Processes of Montagnard Ethnogenesis in the Northern Mandara Mountains of Cameroon. Unpublished Ph.D. dissertation. Department of Archaeology, University of Calgary, Calgary.
1992a Defining Ethnicity: The Mandara Example. Paper presented at the annual conference of the Canadian Archaeological Association, London.
1992b Ethnicity and Ceramic Variation around Mayo Plata, Northern Cameroon. In *An African Commitment: Papers in Honour of Peter Lewis Shinnie*, edited by J. Sterner and N. David, pp. 211–230. University of Calgary Press, Calgary.
1993   Selling the Iron for Their Shackles: Wandala-Montagnard Interactions in Northern Cameroon. *Journal of African History* 33(2):241–270.
1994   'Symbolic Reservoirs' and Inter-Group Relations: West African Examples. *African Archaeological Review* 12:205–223.
1995   Iron Age Beginnings North of the Mandara Mountains, Cameroon and Nigeria. Paper presented at the 10th Pan-African Congress, Harare.
1997   Implications of Occupational Travel for West African Archaeology. In *Ancient Travelers: Proceedings of the Twenty-Seventh Chacmool Conference*, edited by J. Kahn. University of Calgary Archaeology Association, Calgary. In press.
1998   State Formation and Enslavement in Northern Cameroon and Northeastern Nigeria. In *West Africa during the Slave Trade: Archaeological and Historical Perspectives*, edited by C. DeCorse. Smithsonian Institution Press, Washington, D.C. In press.

MacNeish, R. S.
1952   Iroquois Pottery Types: A Technique for the Study of Iroquois Pottery. *National Museum of Canada Bulletin* 124.
1976   The *In Situ* Iroquois Revisited and Rethought. In *Culture, Change, and Continuity: Essays in Honor of James Bennett Griffin*, edited by C. E. Cleland, pp. 79–98. Academic Press, New York.

Madsen, D. B.
1989   *Exploring the Fremont*. University Museum of Natural History, University of Utah, Salt Lake City.

Magne, M., and R. G. Matson
1987   Projectile Point and Lithic Assemblage: Ethnicity in Interior British Columbia. In

Ethnicity and Culture, edited by R. Auger, M. F. Glass, S. MacEachern, and P. H. McCartney, pp. 227–242. Proceedings of the Eighteenth Annual Chacmool Conference, Archaeological Association of the University of Calgary, Calgary, Alberta.

Mahias, M.-C.
1993   Pottery Techniques in India: Technical Variants and Social Choice. In *Technological Choices: Transformation in Material Cultures Since the Neolithic,* edited by P. Lemonnier, pp. 157–180. Routledge, London and New York.

Malinowski, B.
1920a   Kula; the Circulating Exchange of Valuables in the Archipelagoes of Eastern New Guinea. *Man* 20:97–105.
1920b   War and Weapons among the Natives of the Trobriand Islanders. *Man* 20:10–12.
1921   The Primitive Economics of the Trobriand Islanders. *Economic Journal* 31:1–16.
1922   *Argonauts of the Western Pacific: An Account of Native Enterprise and Adventure in the Archipelago of Melanesian New Guinea.* Routledge and Kegan Paul, London.
1934   Stone Implements in Eastern New Guinea. In *Essays Presented to C. G. Seligman,* edited by E. E. Evans-Pritchard, R. Firth, B. Malinowski, and I. Shapera, pp. 189–196. Kegan Paul, Trench, Trubner and Company, London.

Maret, P. (editor)
1980   Ceux qui jouent avec le feu. La place du forgeron en Afrique centrale. *Africa* 50(3):263–279.

Marliac, A.
1991   *De la préhistoire à l'histoire au Cameroun septentrional.* Éditions de l'Orstom, Paris.

Marshall, L.
1976   *The !Kung of Nyae Nyae.* Harvard University Press, Cambridge, Massachusetts.

Marshall, M. P.
1982   Bis sa ani Pueblo: An Example of Late Bonito-Phase, Great-House Architecture. In *Bis sa ani: A Late Bonito Phase Community on the Escavada Wash, Northwest New Mexico,* vol. 2, pt. 1, edited by C. D. Breternitz, D. E. Doyel, and M. P. Marshall, pp. 169–358. Navajo Nation Papers in Anthropology no. 14. Window Rock, Arizona.

Marten, C.
1970   *The Wampanoags in the Seventeenth Century: An Ethnohistorical Survey.* Occasional Papers in Old Colony Studies. Plimoth Plantation, Plymouth, Massachusetts.

Matson, F. R.
1939   Further Technological Notes on the Pottery of the Young Site, Lapeer Country, Michigan. *Papers of the Michigan Academy of Science, Arts and Letters* 24(4):11–23.
1965   Ceramic Queries. In *Ceramic and Man,* edited by F. R. Matson, pp. 227–287. Aldine Publishing Company, Chicago.

Mauss, M.
1930   Les civilisations: Éléments et formes. In *Civilisation, le mot et l'idée,* pp. 456–479. Centre International de Synthèse, Paris.
1935   Les techniques du corps. *Journal de Psychologie* 32:271–293.
1941   Les techniques et la technologie. *Journal de Psychologie* 41:71–78.

May, P.
1977   Abelam. In *Pottery of Papua New Guinea: The National Collection,* edited by B. Egloff,

pp. 50–61. Trustees of the Papua New Guinea National Museum and Art Gallery, Port Moresby.

May, P., and M. Tuckson

1973   Coastal Pottery Villages, Wewak, New Guinea. *Pottery in Australia* 12:13–19.

1982   *The Traditional Pottery of Papua New Guinea*. Bay Books, Sydney, Australia.

McBride, K. A.

1984   *Prehistory of the Lower Connecticut River Valley*. Ph.D. dissertation, Department of Anthropology, University of Connecticut, Storrs. University Microfilms, Ann Arbor, Michigan.

1990   The Historical Archaeology of the Mashantucket Pequots, 1637–1900. In *The Pequots of Southern New England*, edited by L. M. Hauptman and J. D. Wherry, pp. 96–116. University of Oklahoma Press, Norman.

McBride, K. A., and R. E. Dewar

1987   Archaeology of the Mashantucket Pequots. In *The Pequots in Southern New England: The Fall and Rise of an American Indian Nation*, edited by L. M. Hauptman and J. D. Wherry, pp. 96–116. University of Oklahoma Press, Norman.

McCracken, G.

1990   *Culture and Consumption*. Indiana University Press, Bloomington.

McGahan, J.

1989   Vessel Lot Analysis: Indian Crossing Ceramics. Prepared for a course on Analysis of Material Culture, University of Massachusetts. Ms. on file, Department of Anthropology, University of Massachusetts, Amherst.

McGimsey, C. R., III

1980   *Mariana Mesa: Seven Prehistoric Settlements in West-Central New Mexico. A Report of the Upper Gila Expedition*. Peabody Museum, Harvard University, Cambridge, Massachusetts.

McGovern, P. E.

1986   Ancient Ceramic Technology and Stylistic Change: Contrasting Studies from Southwest and Southeast Asia. In *Ceramics and Civilization*, vol. 2, *Technology and Style*, edited by W. D. Kingery, pp. 33–52. The American Ceramic Society, Columbus, Ohio.

McGuire, R. H.

1982   The Study of Ethnicity in Historical Archaeology. *Journal of Anthropological Archaeology* 1:159–178.

McIntosh, R. J.

1989   Middle Niger Terracottas before the Symplegades Gateway. *African Arts* 22(2):74–83, 103–104.

McKenna, P. J., and M. L. Truell

1986   *Small Site Architecture of Chaco Canyon*. Reports of the Chaco Center, Publications in Archaeology 18D. National Park Service, Santa Fe, New Mexico.

McLaughlin, C.

1987   Style as a Social Boundary Marker: A Plains Indian Example. In *Ethnicity and Culture*, edited by R. Auger, M. F. Glass, S. MacEachern, and P. H. McCartney, pp. 55–66. Proceedings of the Eighteenth Annual Chacmool Conference, Archaeological Association of the University of Calgary, Calgary, Alberta.

McMullen, A.
1987   Looking for People in Woodsplint Basketry Decoration. In *A Key into the Language of Woodsplint Baskets,* edited by A. McMullen and R. Handsman, pp. 102–123. American Indian Archaeological Institute, Washington, Connecticut.

McNaughton, P.
1992   From Mande *Komo* to Jukun *Akuma*: Approaching the Difficult Problem of History. *African Arts* 25:76–85, 99–100.

Mead, M.
1938   The Mountain Arapesh: (1) An Importing Culture. *American Museum of Natural History Anthropological Papers* 36(3):139–349.

Meggitt, M. J.
1974   "Pigs Are Our Hearts!" The Te Exchange Cycle among the Mae Enga of New Guinea. *Oceania* 44:165–203.

Merrill, R. S.
1968   The Study of Technology. In *International Encyclopedia of the Social Sciences,* vol. 15, edited by D. L. Sills, pp. 576–589. Macmillan, New York.

Metcalfe, D., and K. M. Heath
1990   Microrefuse and Site Structure: The Hearths and Floors of the Heartbreak Hotel. *American Antiquity* 55(4):781–796.

Miedema, J.
1994   Trade, Migration, and Exchange: The Bird's Head Peninsula of Irian Jaya in a Comparative Perspective. In *Migration and Transformations: Regional Perspectives on New Guinea,* edited by A. J. Strathern and G. Sturzenhofecker, pp. 121–153. ASAO Monograph no. 15. University of Pittsburgh Press, Pittsburgh, Pennsylvania.

Miers, S., and M. Crowder
1988   The Politics of Slavery in Bechuanaland: Power Struggles and the Plight of the Basarwa in the Bamangwato Reserve, 1926–1940. In *The End of Slavery in Africa,* edited by S. Miers and R. Roberts, pp. 172–200. University of Wisconsin Press, Madison.

Miksa, E., and J. M. Heidke
1995   Drawing a Line in the Sands: Models of Ceramic Temper Provenance. In *The Roosevelt Community Development Study,* vol. 2, *Ceramic Chronology, Technology, and Economics,* edited by J. M. Heidke and M. T. Stark, pp. 133–205. Anthropological Papers no. 14. Center for Desert Archaeology, Tucson, Arizona.

Miller, B., and D. Boxberger
1994   Creating Chiefdoms: The Puget Sound Case. *Ethnohistory* 41(2):267–293.

Miller, D.
1985   *Artefacts as Categories: A Study of Ceramic Variability in Central India.* Cambridge University Press, Cambridge.
1987   *Material Culture and Mass Consumption.* Basil Blackwell, Oxford.

Mills, B. J.
1987   Ceramic Analysis. In *Archaeological Investigations at Eight Small Sites in West-Central*

Mintz, S.
1985   *Sweetness and Power: The Place of Sugar in Modern History.* Penguin, New York.

Moerman, M.
1965   Ethnic Classification in a Complex Civilization: Who Are the Lue? *American Anthropologist* 67:1215–1230.

Mohr-Chavez, K. L.
1992   The Organization of Production and Distribution of Traditional Pottery in South Highland Peru. In *Ceramic Production and Distribution,* edited by G. J. Bey and C. A. Pool, pp. 49–92. Westview Press, Boulder, Colorado.

Moore, C. C., and A. K. Romney
1994   Material Culture, Geographic Propinquity, and Linguistic Affiliation on the North Coast of New Guinea: A Reanalysis of Welsch, Terrell, and Nadolski (1992). *American Anthropologist* 96:370–392.
1995   Commentary on Welsch and Terrell's (1994) Reply to Moore and Romney (1994). *Journal of Quantitative Anthropology* 5:75–84.

Moore, H.
1986   *Space, Text, and Gender: An Anthropological Study of the Marakwet of Kenya.* Cambridge University Press, Cambridge.

Moquin, M.
1992   From *Bis sa ani* to Picuris: Early Pueblo Adobe Technology of New Mexico and the Southwest. *Traditions: The Adobe Journal* 8:10–27.

Morgan, L. H.
1901   *League of the Ho-de-no-sau-nee or Iroquois,* edited by H. M. Lloyd. Burt Franklin, New York.
1965   *Houses and House-Life of the American Aborigines.* University of Chicago Press, Chicago, Illinois.

Morris, E.
1915   The Excavation of a Ruin near Aztec, San Juan County, New Mexico. *American Anthropologist* 17:666–684.
1944   Adobe Bricks in a Pre-Spanish Wall near Aztec, New Mexico. *American Antiquity* 9(4):434–438.

Morton, T.
1969   *New English Canaan.* Da Capo Press, New York. Originally published in 1637.

Mrozowski, S.
1980   Aboriginal Ceramics. In *Burr's Hill, a 17th Century Wampanoag Burial Ground in Warren, Rhode Island,* edited by. S. Gibson, pp. 85–88. Haffenreffer Museum of Anthropology, Providence, Rhode Island.

Mulholland, M.
1988   Territoriality and Horticulture: A Perspective for Prehistoric Southern New England. In *Holocene Human Ecology in Northeastern North America,* edited by G. P. Nicholas, pp. 137–166. Plenum Press, New York.

Nabokov, P., and R. Easton
1989   *Native American Architecture*. Oxford University Press, Oxford.

Nash, M
1989   *The Cauldron of Ethnicity in the Modern World*. University of Chicago Press, Chicago, Illinois.

Neaher, N.
1979   Awka Who Travel: Itinerant Metal-Smiths of Southern Nigeria. *Africa* 49(4):352–366.

Neff, H.
1992   Ceramics and Evolution. In *Archaeological Method and Theory*, vol. 4, edited by M. B. Schiffer, pp. 141–193. University of Arizona Press, Tucson.
1993   Theory, Sampling, and Analytical Techniques in the Archaeological Study of Prehistoric Pottery. *American Antiquity* 58(1):23–44.

Nelson, B. A., and S. A. LeBlanc
1986   *Short-Term Sedentism in the American Southwest*. University of New Mexico Press, Albuquerque.

Nelson, M. C.
1991   The Study of Technological Organization. In *Archaeological Method and Theory*, vol. 3, edited by M. B. Schiffer, pp. 57–99. Academic Press, New York.

Netting, R. McC., R. R. Wilk, and E. J. Arnould (editors)
1984   *Households: Comparative and Historical Studies of the Domestic Group*. University of California Press, Berkeley.

Neuhauss, R.
1911   *Deutsch-Neu-Guinea*. 3 vols. Dietrich Reimer, Berlin.

Newton, D.
1989   Mother Cassowary's Bones: Daggers of the East Sepik Province, Papua New Guinea. *Metropolitan Museum Journal* 24:305–325.

Nicklin, K.
1971   Stability and Innovation in Pottery Manufacture. *World Archaeology* 3(1):13–48.
1979   The Location of Pottery Manufacture. *Man* (n.s.) 14(3):436–458.

Nieman, F. D.
1995   Stylistic Variation in Evolutionary Perspective: Inferences from Decorative Diversity and Interassemblage Distance in Illinois Woodland Ceramic Assemblages. *American Antiquity* 60(1):7–36.

Niemczycki, M. A. P.
1984   *The Origin and Development of the Seneca and Cayuga Tribes of New York State*. Rochester Museum and Science Center, Research Records 17. Rochester, New York.

O'Brien, M. J., and T. D. Holland
1992   The Role of Adaptation in Archaeological Explanation. *American Antiquity* 57(1):36–59.

O'Brien, M. J., T. D. Holland, R. J. Hoard, and G. L. Fox
1994   Evolutionary Implications of Design and Performance Characteristics of Prehistoric Pottery. *Journal of Archaeological Method and Theory* 1(3):259–304.

O'Hear, A.
1986   Pottery Making in Ilorin: A Study of the Decorated Water Cooler. *Africa* 56(2):175–192.

Ogot, B. A.
1967   *History of the Southern Luo.* East African Publishing House, Nairobi.
Ortner, S.
1984   Theory in Anthropology since the Sixties. *Comparative Studies in Society and History* 26:126–166.
Osborn, A. J.
1996   Cattle, Co-wives, Children, and Calabashes: Material Context for Symbol Use among the Il Chamus of West-Central Kenya. *Journal of Anthropological Archaeology* 15(1):107–136.
Otte, M.
1990   The Northwestern European Plain around 18,000 B.P. In *The World at 18,000 BP*, vol. 1, *High Latitudes,* edited by O. Sofer and C. Gamble, pp. 54–68. Unwin Hyman, London.
Otte, M., and L. Keeley
1990   The Impact of Regionalism on Paleolithic Studies. *Current Anthropology* 31(5):577–582.
Pacey, A.
1990   *Technology in World Civilization.* MIT Press, Cambridge, Massachusetts.
Parker, A. C.
1916   The Origin of the Iroquois as Suggested by Their Archaeology. *American Anthropologist* 18:479–507.
1968   *Parker on the Iroquois.* Syracuse University Press, Syracuse, New York.
Parkinson, R.
1900   Die Berlinhafen-Section: Ein Beiträg zur Ethnographie der Neu-Guinea-Küste. *Internationales Archiv für Ethnographie* 13:18–54.
1979   The Aitape Coast. Translated by J. J. Tschauder and P. Swadling. In *People of the West Sepik Coast,* edited by P. Swadling, pp. 35–107. National Museum and Art Gallery Record no. 7. National Museum and Art Gallery, Port Moresby.
Parsons, N.
1973   Khama III, the Bamangmato and the British, with Special Reference to 1895–1923. Ph.D. dissertation, University of Edinburgh, Edinburgh, United Kingdom.
Pedrick, K. E.
1992   Introduction. In *Developing Perspectives on Tonto Basin Prehistory,* edited by C. L. Redman, G. E. Rice, and K. E. Pedrick, pp. 1–4. Roosevelt Monograph Series 2. Anthropological Field Studies 26. Office of Cultural Resource Management, Arizona State University, Tempe.
Pendergast, J. F.
1973   *The Roebuck Prehistoric Village Site Rim Sherds: An Attribute Analysis.* Archaeological Survey of Canada Paper 8.
Perinbam, B.
1980   The Julas in Western Sudanese History: Long-Distance Traders and Developers of Resources. In *West African Culture Dynamics,* edited by B. Swartz and R. Dumett, pp. 455–475. Mouton, The Hague.
Perles, C.
1987   Strategies de gestion des outillages lithiques au néolithique. *Paléo* 2:257–283.

Petersen, J. B.
1980   *The Middle Woodland Ceramics of the Winooski Site, A.D. 1–1000*. Vermont Archaeological Society New Series, Monograph 1.
1985   Ceramic Analysis in the Northeast: Resume and Prospect. In *Ceramic Analysis in the Northeast: Contributions to Methodology and Culture History*, edited by J. B. Petersen, pp. 5–25. Occasional Publications in Northeastern Anthropology no. 9. Rindge, New Hampshire.

Petersen, J. B., and N. Hamilton
1984   Early Woodland Ceramic and Perishable Fiber Industries from the Northeast: A Summary and Interpretation. *Annals of the Carnegie Museum* 53(14):413–445.

Petersen, J. B., and M. Power
1985   Three Middle Woodland Ceramic Assemblages from the Winooski Site. In *Ceramic Analysis in the Northeast: Contributions to Methodology and Culture History*, edited by J. B. Petersen, pp. 109–159. Occasional Publications in Northeastern Anthropology no. 9. Rindge, New Hampshire.

Petersen, J. B., and D. Sanger
1991   An Aboriginal Ceramic Sequence for Maine and the Maritime Provinces. In *Prehistoric Archaeology in the Maritime Provinces: Past and Present Research*, edited by M. Deal and S. Blair, pp. 121–178. The Council of Maritime Premiers, Maritime Committee on Archaeological Cooperation, Fredericton, New Brunswick.

Pfaffenberger, B.
1988   Fetishized Objects and Humanized Nature: Toward an Anthropology of Technology. *Man* (n.s.) 23(2):236–252.
1992   Social Anthropology of Technology. *Annual Review of Anthropology* 21:491–516.

Phillips, D. A., Jr.
1989   Prehistory of Chihuahua and Sonora, Mexico. *Journal of World Prehistory* 3(4):373–401.

Pierret, A.
1995   Analyse technologique des céramiques archéologiques: Développements méthodologiques pour l'identification des techniques de façonnage. Ph.D. dissertation, University of Paris I, Panthéon-Sorbonne.

Pinçon, B., and D. Ngoie-Ngalla
1990   L'unité culturelle Kongo à la fin du XIXe siècle: L'apport des etudes céramologiques. *Cahiers d'Etudes Africaines* 118(30–32):157–178.

Plog, S.
1976   Measurement of Prehistoric Interaction between Communities. In *The Early Mesoamerican Village*, edited by K. V. Flannery, pp. 255–272. Academic Press, New York.
1978   Social Interaction and Stylistic Similarity: A Reanalysis. In *Advances in Archaeological Method and Theory*, vol. 1, edited by M. B. Schiffer, pp. 143–182. Academic Press, New York.
1980a   *Stylistic Variation in Prehistoric Ceramics: Design Analysis in the American Southwest*. Cambridge University Press, Cambridge.
1980b   Village Autonomy in the American Southwest: An Evaluation of the Evidence. In

Models and Methods in Regional Exchange, edited by R. E. Fry, pp. 135–146. SAA Papers no. 1. Society for American Archaeology, Washington, D.C.
1983 Analysis of Style in Artifacts. *Annual Review of Anthropology* 12:125–142.
1990 Sociopolitical Implications of Stylistic Variation in the American Southwest. In *The Uses of Style in Archaeology*, edited by M. Conkey and C. Hastorf, pp. 61–72. Cambridge University Press, Cambridge.
1995 Approaches to Style: Complements and Contrasts. In *Style, Society, and Person: Archaeological and Ethnological Perspectives*, edited by C. Carr and J. E. Neitzel, pp. 369–392. Plenum Press, New York and London.

Plog, S., and J. L. Hantman
1986 Multiple Regression Analysis as a Dating Method in the American Southwest. In *Spatial Organization and Exchange: Archaeological Survey on Northern Black Mesa*, edited by S. Plog, pp. 87–113. Southern Illinois University Press, Carbondale.

Pollock, S.
1983 Style and Information: An Analysis of Susiana Ceramics. *Journal of Anthropological Archaeology* 2:354–90.

Pope, G. D., Jr.
1953 The Pottery Types of Connecticut. *Bulletin of the Archaeological Society of Connecticut* 27:3–10.

Prezzano, S. C.
1992 Iroquois Community Patterns: Evidence of Long-Term Ethnic Identity at the Fringe of Iroquoia. Paper presented at the 57th annual meeting of the Society for American Archaeology, Pittsburgh, Pennsylvania.

Priddy, B.
1971 Some Modern Ghanaian Pottery. In *Papers Presented to the 4th Meeting of West African Archaeology: Jos, 1971*, edited by A. Fagg, pp. 72–81. Federal Department of Antiquities, Jos Museum, Jos, Nigeria.

Pryor, J.
1989 Market Forces in the Creation of Pomo Basketry Style and Pomo Ethnicity. *Research in Economic Anthropology* 11:181–216.

Rafferty, K.
1982 Hohokam Micaceous Schist Mining and Ceramic Craft Specialization: An Example from Gila Butte, Arizona. *Anthropology* 6(1–2):199–222.

Ramsden, P. G.
1977 *Refinement of Some Aspects of Huron Ceramic Analysis*. Archaeological Survey of Canada Paper 63.

Ranger, T.
1983 The Invention of Tradition in Colonial Africa. In *The Invention of Tradition*, edited by E. Hobsbawm and T. Ranger, pp. 211–262. Cambridge University Press, Cambridge.
1991 Missionaries, Migrants and the Manyika: The Invention of Ethnicity in Zimbabwe. In *The Creation of Tribalism in Southern Africa*, edited by L. Vail, pp. 118–150. University of California Press, Berkeley.

Rapoport, A.
1969   House Form and Culture. Prentice-Hall, Inc., Englewood Cliffs, New Jersey.
1990   Systems of Activities and Systems of Settings. In *Domestic Architecture and the Use of Space,* edited by S. Kent, pp. 9–20. Cambridge University Press, Cambridge.

Redman, C. L.
1993   *People of the Tonto Ruin: Archaeological Discovery in Prehistoric Arizona.* Smithsonian Institution, Washington, D.C.

Reid, J. J.
1973   *Growth and Response to Stress at Grasshopper Pueblo, Arizona.* Ph.D. dissertation, Department of Anthropology, University of Arizona. University Microfilms, Ann Arbor, Michigan.
1989   A Grasshopper Perspective on the Mogollon of the Arizona Mountains. In *Dynamics of Southwestern Prehistory,* edited by L. S. Cordell and G. J. Gumerman, pp. 65–97. Smithsonian Institution Press, Washington, D.C.

Reina, R. E., and R. M. Hill
1978   *The Traditional Pottery of Guatemala.* University of Texas Press, Austin.

Renfrew, C.
1974   Space, Time and Polity. In *The Evolution of Social Systems,* edited by M. Rowlands and J. Friedman, pp. 89–114. Duckworth, London.

Renfrew, C., and P. Bahn
1991   *Archaeology.* Thames and Hudson, London.

Rice, G. E.
1985   *Studies in the Hohokam and Salado of the Tonto Basin.* Office of Cultural Resource Management Report no. 63. Arizona State University, Tempe, Arizona.

Rice, P. M.
1981   Evolution of Specialized Pottery Production: A Trial Model. *Current Anthropology* 22:219–240.
1984   Change and Conservatism in Pottery Producing Systems. In *The Many Dimensions of Pottery,* edited by S. E. van der Leeuw and A. C. Pritchard, pp. 231–293. University of Amsterdam, Amsterdam.
1987   *Pottery Analysis: A Sourcebook.* University of Chicago Press, Chicago, Illinois.
1991   Specialization, Standardization, and Diversity: A Retrospective. In *The Ceramic Legacy of Anna O. Shepard,* edited by R. L. Bishop and F. W. Lange, pp. 257–279. University Press of Colorado, Niwot.
1996   Recent Ceramic Analysis: 1. Function, Style, and Origins. *Journal of Archaeological Research* 4(2):133–163.

Richardson, M. (editor)
1974   *The Human Mirror: Material and Spatial Images of Man.* Louisiana State University Press, Baton Rouge.

Richard, M.
1977   *Traditions et coutumes matrimoniales chez les Mada et les Mouyeng, Nord-Cameroun.* Anthropos-Institut, Haus Volker und Kulturen, St. Augustin.

Riesenfeld, A.
1946   Rattan Cuirasses and Gourd Penis-Cases in New Guinea. *Man* 46:31–36.

Ritchie, W. A.
1944   *The Pre-Iroquoian Occupation of New York State*. Rochester Museum of Arts and Science, Memoir 1. Rochester, New York.
1958   *An Introduction to Hudson Valley Prehistory*. New York State Museum and Science Service Bulletin 367. Albany.
1980   *The Archaeology of New York State*. Revised edition. Harbor Hill, Harrison, New York. Originally published 1965.
Ritchie, W. A., and R. S. MacNeish
1949   Pre-Iroquoian Pottery of New York State. *American Antiquity* 15:97–124.
Robbins, M.
1959   Some Indian Burials from Southeastern Massachusetts, Part 2: The Wapanucket Burials. *Bulletin of the Massachusetts Archaeological Society* 20:61–67.
Robertson, I.
1992   Hoes and Metal Templates in Northern Cameroon. In *An African Commitment: Papers in Honour of Peter Lewis Shinnie*, edited by J. Sterner and N. David, pp. 231–240. University of Calgary Press, Calgary.
Robinson, P.
1990   *The Struggle Within: The Indian Debate in Seventeenth Century Narragansett Country*. Ph.D. dissertation, Department of Anthropology, State University of New York at Binghamton. University Microfilms, Ann Arbor, Michigan.
Robinson, P., and G. Gustafson
1982   A Partially Disturbed 17th Century Indian Burial Ground in Rhode Island: Recovery, Preliminary Analysis, and Protection. *Bulletin of the Connecticut Archaeological Society* 45:41–50.
Robinson, P., M. Kelly, and P. Rubertone
1985   Preliminary Biocultural Interpretations from a Seventeenth-Century Narragansett Indian Cemetery in Rhode Island. In *Cultures in Contact: The Impact of European Contacts on Native American Cultural Institutions* A.D. *1000–1800*, edited by W. Fitzhugh, pp. 107–130. Smithsonian Institution Press, Washington, D.C.
Robinson, W. J., J. W. Hannah, and B. G. Harrill
1972   *Tree-Ring Dates from New Mexico I, O, U, Central Rio Grande Area*. Laboratory of Tree-Ring Research, University of Arizona, Tucson.
Rodatz, H.
1908   Aus dem neuen Bezirk Eitape. *Deutsches Kolonialblatt* 19:16–17.
1909   Eine Expedition im Norden von Kaiser-Wilhelmsland. *Deutsches Kolonialblatt* 20: 174–176.
Roseberry, W.
1989   *Anthropologies and Histories: Essays in Culture, History, and Political Economy*. Rutgers University Press, New Brunswick, New Jersey.
Rouse, I.
1945   Styles of Pottery in Connecticut. *Bulletin of the Massachusetts Archaeological Society* 7(1):108.
1947   Ceramic Traditions and Sequences in Connecticut. *Bulletin of the Archaeological Society of Connecticut* 21:10–25.

1960 The Classification of Artifacts in Archaeology. *American Antiquity* 25:313–323.
1964 *Prehistory in Haiti: A Study in Method.* Reprinted. Yale University Publication in Anthropology 21, New Haven, Connecticut. Originally published 1939.

Roux, V.
1994 La technique du tournage: Définition et reconnaissance par les macrotraces. In *Terre cuite et société: La céramique, document technique, économique, culturel,* edited by F. Audouze and D. Binder, pp. 45–58. Éditions APDCA, Juan-les-Pins.

Roux, V., B. Bril, and G. Dietrich
1995 Skills and Learning Difficulties Involved in Stone Knapping: The Case of Stone-Bead Knapping in Khambhat, India. *World Archaeology* 27(1):63–87.

Rubertone, P.
1989 Archaeology, Colonialism, and 17th Century Native America: Towards an Alternative Interpretation. In *Conflict in the Archaeology of Living Traditions,* edited by R. Layton, pp. 32–45. Unwin Hyman Ltd., London.
1993 Grave Remembrances: Enduring Traditions among the Narragansett. Unpublished ms. in possession of author.

Rye, O. S.
1976 Keeping Your Temper under Control: Materials and the Manufacture of Papuan Pottery. *Archaeology and Physical Anthropology in Oceania* 11(2):106–137.
1981 *Pottery Technology: Principles and Reconstruction.* Manuals on Archaeology No. 4. Taraxacum, Washington, D.C.

Rye, O. S., and C. Evans
1976 *Traditional Pottery Techniques of Pakistan.* Smithsonian Contributions to Anthropology 21, Washington D.C.

Sackett, J. R.
1973 Style, Function, and Artifact Variability in Paleolithic Assemblages. In *Explanation of Culture Change,* edited by C. Renfrew, pp. 317–25. Duckworth, London.
1977 The Meaning of Style in Archaeology: A General Model. *American Antiquity* 43(3):369–380.
1982 Approaches to Style in Lithic Archaeology. *Journal of Anthropological Archaeology* 1(1):59–112.
1985a Style and Ethnicity in the Kalahari: A Reply to Wiessner. *American Antiquity* 50(1):154–159.
1985b Style, Ethnicity and Stone Tools. In *Status, Structure and Stratification: Current Archaeological Reconstructions,* edited by M. Thompson, M.-T. Garcia, and F. Kense, pp. 277–282. Chacmool. Calgary, Alberta.
1986 Isochrestism and Style: A Clarification. *Journal of Anthropological Archaeology* 5:266–277.
1990 Style and Ethnicity in Archaeology: The Case for Isochrestism. In *The Uses of Style in Archaeology,* edited by M. W. Conkey and C. A. Hastorf, pp. 32–43. Cambridge University Press, Cambridge.

Sagard, Father G.
1968 *The Long Journey to the Country of the Hurons,* edited by G. M. Wrong. Translated by H. H. Langton. Reprinted, originally published in 1939. Greenwood Press, New York.

Sahlins, M. D.
1972 *Stone Age Economics*. Aldine, Chicago, Illinois.
1976 *Culture and Practical Reason*. University of Chicago Press, Illinois.
1981 *Historical Metaphors and Mythical Realities: Structure in the Early History of the Sandwich Islands Kingdom*. University of Michigan Press, Ann Arbor.
1985 *Islands of History*. University of Chicago Press, Illinois.
1994 Cosmologies of Capitalism: The Trans-Pacific Sector of "the World System." In *Culture/Power/History: A Reader in Contemporary Social Theory*, edited by N. B. Dirks, G. Eley, and S. B. Ortner, pp. 412–455. Princeton University Press, Princeton, New Jersey.

Salisbury, N.
1990 Indians and Colonists in Southern New England after the Pequot War. In *The Pequots of Southern New England*, edited by L. M. Hauptman and J. D. Wherry, pp. 81–95. University of Oklahoma Press, Norman.
1993 Facing the Eastern Door: New England Algonquians and the Iroquois. Paper presented at the 2nd Mashantucket Pequot History Conference, Mystic, Connecticut.

Salwen, B.
1978 Indians of Southern New England and Long Island. In *Handbook of North American Indians*, vol. 15, *Northeast*, edited by B. G. Trigger, pp. 160–176. Smithsonian Institution, Washington, D.C.

Salwen, B., and A. Ottesen
1972 Radiocarbon Dates for a Windsor Occupation at the Shantok Cove Site. *Man in the Northeast* 3:8–19.

Samarin, W.
1984 Bondjo Ethnicity and the Colonial Imagination. *Canadian Journal of African Studies/Revue Canadien des Études Africaines* 18(2):345–365.

Sampson, C. G.
1988 *Stylistic Boundaries among Mobile Hunter-Foragers*. Smithsonian Institution Press, Washington, D.C.

Saraswati, B., and N. K. Behura
1966 *Pottery Techniques of Peasant India*. Anthropological Survey of India, Memoir 13, Calcutta.

Schaafsma, C. F.
1979 The "El Paso Phase" and Its Relationship to the "Casas Grandes Phenomenon." In *Jornada Mogollon Archaeology: Proceedings of the First Jornada Conference*, edited by P. H. Beckett and R. N. Wiseman. Historic Preservation Bureau, State Planning Division, Santa Fe, New Mexico.

Schaafsma, P.
1992 *Rock Art in New Mexico*. Museum of New Mexico Press, Santa Fe.

Schaafsma, P., and C. F. Schaafsma
1974 Evidence for the Origin of the Pueblo Katchina Cult as Suggested by Southwestern Rock Art. *American Antiquity* 39:535–545.

Schapera, I.
1930 *The Khoisan Peoples of South Africa: Bushmen and Hottentots*. Routledge and Kegan Paul, London.
1938 *A Handbook of Tswana Law and Custom*. Frank Cass, London.

1943 *Native Land Tenure in the Bechuanaland Protectorate.* Lovedale Press, Alice, South Africa.

Schiffer, M. B.

1976 *Behavioral Archaeology.* Academic Press, New York.

1978 Methodological Issues in Ethnoarchaeology. In *Explorations in Ethnoarchaeology*, edited by R. Gould, pp. 229–247. University of New Mexico Press, Albuquerque.

1990 The Influence of Surface Treatment on Heating Effectiveness of Ceramic Vessels. *Journal of Archaeological Science* 17(4):373–381.

Schiffer, M. B., and J. M. Skibo

1987 Theory and Experiment in the Study of Technological Change. *Current Anthropology* 28(5):595–622.

Schiffer, M. B., J. M. Skibo, T. C. Boelke, M. A. Neupert, and M. Aronson

1994 New Perspectives on Experimental Archaeology: Surface Treatments and Thermal Response of the Clay Cooking Pot. *American Antiquity* 59(2):197–217.

Schlanger, N.

1991 Le fait technique total: La raison pratique et les raisons de la pratique dans l'oeuvre de Marcel Mauss. *Terrain* 16:114–130.

Schlereth, T. J. (editor)

1982 *Material Culture Studies in America.* American Association for State and Local History, Nashville, Tennessee.

Schortman, E. M.

1989 Interregional Interaction in Prehistory. *American Antiquity* 54:52–65.

Searle, A. B., and R. W. Grimshaw

1959 *The Chemistry and Physics of Clays and Other Ceramic Materials.* 3rd edition. Interscience, New York.

Sebastian, L., and F. Levine

1989 The Protohistoric and Spanish Colonial Periods. In *Living on the Land: 11,000 Years of Human Adaptation in the Southeastern New Mexico*, edited by L. Sebastian and S. Larralde, pp. 92–104. Cultural Resources Series no. 6. Bureau of Land Management, Safford, Arizona.

Sellet, F.

1993 Chaîne Opératoire: The Concept and Its Applications. *Lithic Technology* 18(1–2):106–112.

Shanks, M., and C. Tilley

1987 *Re-constructing Archaeology.* Cambridge University Press, Cambridge.

Sheller, P.

1977 *The People of the Central Kalahari Game Reserve: A Report on the Reconnaissance of the Reserve, July–September, 1976.* Report to the Government of Botswana. Gaborone, Botswana.

Shennan, S. J.

1989 Introduction: Archaeological Approaches to Cultural Identity. In *Archaeological Approaches to Cultural Identity*, edited by S. J. Shennan, pp. 1–32. Unwin Hyman, London.

Shepard, A. O.

1936 Technology of Pecos Pottery. In *The Pottery of Pecos*, vol. 2, edited by A. V. Kidder and A. O. Shepard, pp. 389–587. Papers of the Phillips Academy, Southwestern Expedition 7. Andover, Massachusetts.

Shutler, R., Jr.
1961 *Lost City: Pueblo Grande de Nevada*. Nevada State Museum Anthropological Paper 5. Carson City, Nevada.

Sigaut, F.
1985 More (and Enough) on Technology! *History and Technology* 2:115–132.
1987 Un couteau ne sert pas à couper, mais en coupant: Structure, fonctionnement et fonction dans l'analyse des objets. In *25 Ans d'études technologiques en préhistoire*, pp. 21–34. Éditions APDCA, Juan-les-Pins.
1991 Aperçus sur l'histoire de la technologie en tant que science humaine. *Actes et Communications* 6:67–82.

Silberbauer, G. B.
1965 *Report to the Government of Bechuanaland on the Bushman Survey*. Government Printer, Gaberones, Botswana.
1972 The G/wi Bushmen. In *Hunters and Gatherers Today*, edited by M. G. Bicchieri, pp. 271–326. Holt, Rinehart, and Winston, New York.
1981a *Hunter and Habitat in the Central Kalahari Desert*. Cambridge University Press, Cambridge and New York.
1981b Hunter/Gatherers of the Central Kalahari. In *Omnivorous Primates: Gathering and Hunting in Human Evolution*, edited by R. S. O. Harding and G. Teleki, pp. 455–498. Columbia University Press, New York.

Silver, A.
1980 Comment on Maize Cultivation in Coastal New York. *North American Archaeologist* 2(2):117–130.

Simmons, W.
1970 *Cautantowwit's House: An Indian Burial Ground on the Island of Conanicut in Narragansett Bay*. Brown University Press, Providence, Rhode Island.
1978 Narragansett. In *Handbook of North American Indians*, vol. 15, *Northeast*, edited by B. G. Trigger, pp. 190–197. Smithsonian Institution, Washington, D.C.
1989 *The Narragansett*. Chelsea House Publishers, New York.

Simmons, W., and G. Aubin
1975 Narragansett Kinship. *Man in the Northeast* 9:21–31.

Sinopoli, C. M.
1991 *Approaches to Archaeological Ceramics*. Plenum Press, New York.

Sires, E. W., Jr.
1984 Hohokam Architecture and Site Structure. In *Hohokam Archaeology along the Salt-Gila Aqueduct, Central Arizona Project*, vol. 9, *Synthesis and Conclusions*, edited by L. S. Teague and P. L. Crown, pp. 115–140. Archaeological Series no. 150. Arizona State Museum, University of Arizona, Tucson.
1987 Hohokam Architectural Variability and Site Structure during the Sedentary-Classic Transition. In *The Hohokam Village: Site Structure and Organization*, edited by D. E. Doyel, pp. 171–182. Southwestern and Rocky Mountain Division of the American Association for the Advancement of Science, Glenwood Springs, Colorado.

Skibo, J. M.
1992   Pottery Function: A Use-Alteration Perspective. Plenum Press, New York.
Skibo, J. M., M. B. Schiffer, and N. Kowalski
1989a   Ceramic Style Analysis in Archaeology and Ethnoarchaeology: Bridging the Analytical Gap. *Journal of Anthropological Archaeology* 8:388–409.
Skibo, J. M., M. B. Schiffer, and K. D. Reid
1989b   Organic-Tempered Pottery: An Experimental Study. *American Antiquity* 54(1):122–146.
Sklenár, K.
1983   *Archaeology in Central Europe: The First 500 Years*. St. Martin's Press, New York.
Smiley, T. L., S. A. Stubbs, and B. Bannister
1953   *A Foundation for Dating of Some Late Archaeological Sites in the Rio Grande Area, New Mexico; Based on Studies in Tree-Ring Methods and Pottery Analyses*. University of Arizona Bulletin 24(3), Laboratory of Tree-Ring Research Bulletin 6. Tucson.
Smith, A. D.
1986   *The Ethnic Origins of Nations*. Basil Blackwell, Oxford.
Smith, C.
1944   Clues to the Chronology of Coastal New York. *American Antiquity* 1:87–98.
1947   An Outline of the Archaeology of Coastal New York. *Bulletin of the Archaeological Society of Connecticut* 21:3–9.
1950   The Archaeology of Coastal New York. *Anthropological Papers of the American Museum of Natural History* 43(2):94–200.
Smith, F. T.
1989   Earth, Vessels, and Harmony among the Gurensi. *African Arts* 22(2):60–65.
Snow, D. R.
1980   *The Archaeology of New England*. Academic Press, New York.
1994   *The Iroquois*. Blackwell, Cambridge, Massachusetts.
1995   *Mohawk Valley Archaeology: The Sites*. The Institute for Archaeological Studies, University at Albany, State University of New York, Albany.
Solecki, R.
1950   The Archaeological Position of Historic Fort Corchaug, L. I., and Its Relation to Contemporary Forts. *Bulletin of the Archaeological Society of Connecticut* 24:3–40.
1957   Shantok Influence on Eastern Long Island. *American Antiquity* 23:171–173.
Solway, J. S.
1987   Commercialization and Social Differentiation in a Kalahari Village, Botswana. Unpublished Ph.D. dissertation, University of Toronto, Ontario.
Solway, J. S., and R. B. Lee
1990   Foragers, Genuine or Spurious? Situating the Kalahari San in Prehistory. *Current Anthropology* 31(2):109–146.
Southall, A.
1952   *Lineage Formation among the Luo*. International African Institute Memorandum 26. London.

1970 The Illusion of Tribe. In *The Passing of Tribal Man in Africa*, edited by P. Gutkind, pp. 28–51. Brill, Leiden.

Spector, J. D.

1993 *What This Awl Means: Feminist Archaeology at a Wahpeton Village*. Minnesota Historical Society Press, St. Paul.

Sperber, D.

1975 *Rethinking Symbolism*. Cambridge University Press, Cambridge.

Speth, J.

1988 Do We Need Concepts Like "Mogollon," "Anasazi," and "Hohokam" Today? A Cultural Anthropological Perspective. *The Kiva* 53:201–204.

Spriggs, M., and D. Miller

1979 Ambon-Lease: On Study of Contemporary Pottery Making and Its Archaeological Relevance. In *Pottery and the Archaeologist*, edited by M. Millet, pp. 25–34. Institute of Archaeology Occasional Publication 4. London.

Stahl, A.

1994 Innovation, Diffusion and Culture Contact: The Holocene Archaeology of Ghana. *Journal of World Prehistory* 8(1):51–110.

Stanislawski, M. B.

1978 If Pots Were Mortal. In *Explorations in Ethnoarchaeology*, edited by R. Gould, pp. 201–228. University of New Mexico Press, Albuquerque.

Stanislawski, M. B., and B. B. Stanislawski

1978 Hopi and Hopi-Tewa Ceramic Tradition Networks. In *The Spatial Organisation of Culture*, edited by I. Hodder, pp. 61–76. University of Pittsburgh Press, Pittsburgh, Pennsylvania.

Stark, M. T.

1995a Commodities and Interaction in the Prehistoric Tonto Basin. In *The Roosevelt Community Development Study: New Perspectives on Tonto Basin Prehistory*, edited by M. D. Elson, M. T. Stark, and D. A. Gregory, pp. 307–342. Anthropological Papers no. 15. Center for Desert Archaeology, Tucson, Arizona.

1995b Economic Intensification and Ceramic Specialization in the Philippines: A View from Kalinga. *Research in Economic Anthropology* 16:179–226.

1995c Cultural Identity in the Archeaological Record: The Utility of Utilitarian Ceramics. In *The Roosevelt Community Development Study*, vol. 2, *Ceramic Chronology, Technology, and Economics*, edited by J. M. Heidke and M. T. Stark, pp. 331–362. Anthropological Papers no. 14. Center for Desert Archaeology, Tucson, Arizona.

Stark, M. T., J. J. Clark, and M. D. Elson

1995 Causes and Consequences of Migration in the 13th Century Tonto Basin. *Journal of Anthropological Archaeology* 14:212–246.

Stark, M. T., and J. M. Heidke

1995 Early Classic Period Variability in Utilitarian Ceramic Production and Distribution. In *The Roosevelt Community Development Study*, vol. 2, *Ceramic Chronology, Technology, and Economics*, edited by J. M. Heidke and M. T. Stark, pp. 363–394. Anthropological Papers no. 14. Center for Desert Archaeology, Tucson, Arizona.

Stark, M. T., J. M. Vint, and J. M. Heidke
1995   Compositional Variability in Utilitarian Ceramics at a Colonial Period Site. In *The Roosevelt Community Development Study*, vol. 2, *Ceramic Chronology, Technology, and Economics*, edited by J. M. Heidke and M. T. Stark, pp. 273–296. Anthropological Papers no. 14. Center for Desert Archaeology, Tucson, Arizona.

Starna, W.
1990   The Pequots in the Early Seventeenth Century. In *The Pequots of Southern New England*, edited by L. M. Hauptman and J. D. Wherry, pp. 33–47. University of Oklahoma Press, Norman.

Steen, C. R.
1965   Excavations in Compound A, Casa Grande National Monument, 1963. *The Kiva* 31(2):59–82.
1983   The San Juan Anasazi in the 13th Century. In *Proceedings of the Anasazi Symposium, 1981*, compiled and edited by J. E. Smith, pp. 167–174. Mesa Verde National Park/Mesa Verde Museum Association, Inc., Cortez, Colorado.

Stein, J. R., and P. J. McKenna
1988   *An Archaeological Reconnaissance of a Late Bonito Phase Occupation near Aztec Ruins National Monument, New Mexico*. Division of Anthropology, Branch of Cultural Resources Management, Southwest Cultural Resources Center, National Park Service, Santa Fe, New Mexico.

Steinberg, A.
1977   Technology and Culture: Technological Styles in the Bronzes of Shang China, Phrygia and Urnfield Central Europe. In *Material Culture: Style, Organization, and Dynamics of Technology*, edited by H. Lechtman and R. S. Merrill, pp. 53–86. West, St. Paul, Minnesota.

Steponaitis, V.
1983   *Ceramics, Chronology and Community Patterns: An Archaeological Study at Moundville*. Academic Press, New York.
1984   Technological Studies of Prehistoric Pottery from Alabama: Physical Properties and Vessel Function. In *The Many Dimensions of Pottery*, edited by S. E. van der Leeuw and A. C. Pritchard, pp. 81–127. University of Amsterdam, Amsterdam.

Sterner, J.
1989   Who is Signaling Whom? Ceramic Style, Ethnicity, and Taphonomy among the Sirak Bulahay. *Antiquity* 63:451–459.

Sterner, J., and N. David
1993   Transformers Transformed: Aspects of Caste and Iron Technology in the Mandara Mountains. Paper presented at the workshop "Transformations, Technology and Gender in African Metallurgy," Oxford.

Stimmell, C., R. B. Heiman, and R. G. V. Hancock
1982   Indian Pottery from the Mississippi Valley: Coping with Bad Raw Materials. In *Archaeological Ceramics*, edited by J. S. Olin and A. D. Franklin, pp. 219–228. Smithsonian Institution Press, Washington D.C.

Stimmell, C., and R. L. Stromberg
1986   A Reassessment of Thule Eskimo Ceramic Technology. In *Ceramics and Civilization*, vol. 2, *Technology and Style*, edited by W. D. Kingery, pp. 237–250. The American Ceramic Society, Columbus, Ohio.

Stothers, D. M.
1977   *The Princess Point Complex*. Archaeological Survey of Canada Paper 58. Ottawa.

Stott, M. A., and B. Reynolds
1987   Material Anthropology: Contemporary Approaches to Material Culture. In *Material Anthropology: Contemporary Approaches to Material Culture*, edited by B. Reynolds and M. A. Stott, pp. 1–12. University Press of America, Lanham and New York.

Strathern, A. J.
1971   *The Rope of Moka: Big-Men and Ceremonial Exchange in Mount Hagen, New Guinea*. Cambridge University Press, Cambridge.
1979   *Ongka: A Self-Account by a New Guinea Big-Man*. Duckworth, London.

Stubbs, S. A., and W. S. Stallings
1953   *The Excavation of Pindi Pueblo*. New Mexico Monographs of the School of American Research and the Laboratory of Anthropology 18. Santa Fe.

Stuiver, M., and G. W. Pearson
1993   High-Precision Bidecadal Calibration of the Radiocarbon Time Scale, A.D. 1950–500 B.C. and 2500–6000 BC. *Radiocarbon* 35(1):1–23.

Stuiver, M., and P. J. Reimer
1993   Extended 14C Data Base and Revised CALIB 3.0 14C Age Calibration Program. *Radiocarbon* 35(1):215–230.

Sturtevant, W. C.
1975   Two 1761 Wigwams at Niantic, Connecticut. *American Antiquity* 40(4):437–444.

Suttles, W.
1987   Cultural Diversity within the Coast Salish Continuum. In *Ethnicity and Culture*, edited by R. Auger, M. F. Glass, S. MacEachern, and P. H. McCartney, pp. 243–250. Proceedings of the Eighteenth Annual Chacmool Conference, Archaeological Association of the University of Calgary, Calgary, Alberta.

Swihart, S.
1992   A Challenge to MacNeish's Owasco-Iroquois Seriational Continuum and the *In Situ* Hypothesis of Iroquois Origins. Paper presented at the 32nd Congress of the Northeastern Anthropological Association, Bridgewater, Massachusetts.

Tanaka, J.
1976   Subsistence Ecology of Central Kalahari San. In *Kalahari Hunter-Gatherers: Studies of the !Kung San and Their Neighbors*, edited by R. B. Lee and I. DeVore, pp. 98–119. Harvard University Press, Cambridge, Massachusetts.
1980   *The San, Hunter-Gatherers of the Kalahari: A Study in Ecological Anthropology*. University of Tokyo Press, Japan.

Taylor, J. S.
1983   *Commonsense Architecture*. W. W. Norton and Son, New York.

Taylor, R. B.
1988   *Human Territorial Functioning: An Empirical Evolutionary Perspective on Individual and Small Group Cognitions, Behaviors, and Consequences.* Cambridge University Press, Cambridge and New York.

Terrell, J.
1993   Regional Studies in Anthropology: A Melanesian Prospectus. *Current Anthropology* 34:177–179.

Terrell, J., and R. L. Welsch
1990a   Return to New Guinea: In the Field. *The Bulletin of the Field Museum of Natural History* 61(5):1, 10–11.
1990b   Trade Networks, Areal Integration, and Diversity along the North Coast of New Guinea. *Asian Perspectives* 29:156–165.

Terry, R. D., and G. V. Chilingar
1955   Data Sheet 6. *Geotimes.* Available from the American Geological Institute, Washington, D.C. Reprinted from the *Journal of Sedimentary Petrology* 25:229–234.

Thomas, D. S. G., and P. A. Shaw
1991   *The Kalahari Environment.* Cambridge University Press, Cambridge and New York.

Thomas, P. A.
1979   *In the Maelstrom of Change: The Indian Trade and Cultural Process in the Middle Connecticut River Valley: 1635–1665.* Ph.D. dissertation, Department of Anthropology, University of Massachusetts, Amherst. University Microfilms, Ann Arbor, Michigan.

Thompson, E. P.
1991   *Customs in Common.* The Merlin Press, London.

Thomsen, H. H.
1960   Occurrence of Fired Bricks in Pre-Conquest Mexico. *Southwestern Journal of Anthropology* 16:428–441.

Thorbahn, P. F.
1988   Where are the Late Woodland Villages in Southern New England? *Bulletin of the Massachusetts Archaeological Society* 49(2):46–57.

Tiesler, F.
1969–70 Die intertribalen Beziehungen an der Nordkuste Neuguineas im Gebiet der Kleinen Schouten-Inseln. *Abhandlungen und Berichte des Staatlichen Museums für Volkerkunde Dresden* 30:1–122, 31:111–195; Plates C1–C14.

Tilley, C.
1989   Interpreting Material Culture. In *The Meaning of Things: Material Culture and Symbolic Expression,* edited by I. Hodder, pp. 185–194. Cambridge University Press, Cambridge.

Topic, J.
1994   Ethnogenesis in Huamachuco. *Andean Past* 5 (Latin American Studies Program, Cornell University, Ithaca, New York).

Tozzer, A. M.
1921   *Excavation of a Site at Santiago Ahuitzotla, D.F., Mexico.* Bureau of American Ethnology Bulletin 74. Washington, D.C.

Trigger, B.
1978　The Development of the Archaeological Culture in Europe and America. In *Time and Traditions: Essays in Archaeological Interpretation,* edited by B. Trigger, pp. 75–95. Edinburgh University Press, Edinburgh, Great Britain.
1989　*A History of Archaeological Thought.* Cambridge University Press, Cambridge.

Tuck, J. A.
1971　*Onondaga Iroquois Prehistory: A Study in Settlement Archaeology.* Syracuse University Press, New York.
1978　Northern Iroquoian Prehistory. In *Handbook of the North American Indians,* vol. 15, *The Northeast,* edited by B. G. Trigger, pp. 322–333. Smithsonian Institution, Washington, D.C.

Tuckson, M.
1977　Tumleo Island. In *Pottery of Papua New Guinea: The National Collection,* edited by B. Egloff, pp. 76–77. Trustees of the Papua New Guinea National Museum and Art Gallery, Port Moresby.

Tuckson, M., and P. May
1975　Pots and Potters in Papua New Guinea. *Australian Natural History* 18:168–173.

Tuggle, H. D.
1982　Q Ranch Ceramic and Site Chronology. In *Cholla Project Archaeology,* vol. 3, *The Q Ranch Region,* edited by J. J. Reid, pp. 23–32. Archaeological Series no. 161. Arizona State Museum, University of Arizona, Tucson.

Turner, T.
1969　Tchirkin, a Central Brazilian Tribe and its Symbolic Language of Bodily Adornment. *Natural History* 78:50–59.
1980　The Social Skin. In *Not Work Alone: A Cross-Cultural View of Activities Superfluous to Survival,* edited by J. Cherfas and R. Lewin, pp. 112–140. Sage Publications, Beverly Hills, California.

Upham, S., P. L. Crown, and S. Plog
1994　Alliance Formation and Cultural Identity in the American Southwest. In *Themes in Southwestern Prehistory,* edited by G. J. Gumerman, pp. 183–210. School of American Research Press, Santa Fe, New Mexico.

Vail, L.
1991　Introduction: Ethnicity in Southern African History. In *The Creation of Tribalism in Southern Africa,* edited by L. Vail, pp. 1–19. University of California Press, Berkeley.

Valiente-Noailles, C.
1993　*The Kua: Life and Soul of the Central Kalahari Bushmen.* A. A. Balkema, Rotterdam, The Netherlands.

van As, A.
1984　Reconstructing the Potter's Craft. In *The Many Dimensions of Pottery,* edited by S. E. van der Leeuw and A. C. Pritchard, pp. 131–164. University of Amsterdam, Amsterdam.

van Beek, W.
1987　*The Kapsiki of the Mandara Hills.* Waveland Press, Prospect Heights.

van Grunderbeeck, M.
1983　*La premier âge du fer au Rwanda et au Burundi: Archéologie et environnement.* Brussels.

van der Leeuw, S. E.
1984   Dust to Dust: A Transformational View of the Ceramic Cycle. In *The Many Dimensions of Pottery: Ceramics in Ethnoarchaeology and Anthropology*, edited by S. E. van der Leeuw and A. C. Pritchard, pp. 707–778. University of Amsterdam, Amsterdam.
1993   Giving the Potter a Choice: Conceptual Aspects of Pottery Techniques. In *Technological Choices: Transformation in Material Cultures since the Neolithic*, edited by P. Lemonnier, pp. 238–288. Routledge, London and New York.

van der Leeuw, S. E., D. A. Papousek, and A. Coudart
1991   Technical Traditions and Unquestioned Assumptions: The Case of Pottery in Michoacan. *Techniques et Culture* (n.s.) 17–18: 145–173.

Vandiver, P. B., W. A. Ellingson, T. K. Robinson, J. J. Lobick, and F. H. Seguin
1991   New Applications of X-Radiographic Imaging Technologies for Archaeological Ceramics. *Archaeomaterials* 5(2):185–207.

Vansina, J.
1990   *Paths in the Rainforest: Toward a History of Political Tradition in Equatorial Africa*. University of Wisconsin Press, Madison.

Veit, U.
1989   Ethnic Concepts in German Prehistory: A Case Study on the Relationship between Cultural Identity and Archaeological Objectivity. In *Archaeological Approaches to Cultural Identity*, edited by S. Shennan, pp. 35–56. Unwin Hyman, London.

Vernon-Jackson, H.
1960   Craft Work in Bida. *Africa* 25(1):51–73.

Vieira Powers, K.
1995   The Battle for Bodies and Souls in the Colonial North Andes: Intraecclesiastical Struggles and the Politics of Migration. *Hispanic American Historical Review* 75(1):31–56.

Vierich, H.
1981   The Kua of the Southeastern Kalahari: A Study in the Socioecology of Dependency. Unpublished Ph.D. dissertation, Department of Anthropology, University of Toronto, Ontario.

Vierich, H., and R. Hitchcock
1996   Kua: Farmer/Foragers of the Eastern Kalahari, Botswana. In *Cultural Diversity among Twentieth Century Foragers: An African Perspective*, edited by S. Kent, pp. 108–124. Cambridge University Press, Cambridge and New York.

Vincent, J.
1974   The Structuring of Ethnicity. *Human Organization* 33:375–379.

Virot, C.
1994   L'association argile. In *Terre cuite et société: La céramique, document technique, economique, culturel*, edited by F. Audouze and D. Binder, pp. 351–362. Éditions APDCA, Juan-les-Pins.

Wade, J.
1993   The Wife of the Village: Understanding Caste in the Mandara Mountains. Ms. on file with the author.

Wagner, P. L.
1960   *The Human Use of the Earth*. The Free Press, New York.

Wagner, R.
1989 Conclusion: The Exchange Context of the Kula. In *Death Rituals and Life in the Societies of the Kula Ring,* edited by F. H. Damon and R. Wagner, pp. 254–274. Northern Illinois University Press, DeKalb.

Wahlman, M.
1972 Yoruba Pottery Making Techniques. *Baessler-Archiv* (n.f.) 20:313–346.

Wahome, E.
1989 Ceramics and History in the Iron Age of Northern Cameroon. Unpublished M.A. thesis. University of Calgary, Calgary.

Walker, N.
1995 The Archaeology of the San: The Late Stone Age of Botswana. In *Speaking for the Bushmen,* edited by A. J. G. M. Sanders, pp. 54–87. The Botswana Society, Gaborone, Botswana.

Washburn, D.
1977 *A Symmetry Analysis of Upper Gila Area Ceramic Design.* Harvard University Press, Cambridge, Massachusetts.

Watson, J. B.
1970 Society as Organized Flow: The Tairora Case. *Southwestern Journal of Anthropology* 26(2):107–124.

1977 Pigs, Fodder, and the Jones Effect in Postipomoean New Guinea. *Ethnology* 16: 57–70.

1981 The Exchange Strategies of Crowded Partners. In *Persistence and Exchange: Papers from a Symposium on Ecological Problems of the Traditional Societies of the Pacific Region,* XIV Pacific Science Congress, Khabarovsk, U.S.S.R., August–September, 1979, edited by R. W. Force and B. Bishop, pp. 151–153. Pacific Science Association, Honolulu.

1983 *Tairora Culture: Contingency and Pragmatism.* Anthropological Studies in the Eastern Highlands of New Guinea, vol. 5. University of Washington Press, Seattle.

1990 Other People Do Other Things: Lamarckian Identities in Kainantu Subdistrict, Papua New Guinea. In *Cultural Identity and Ethnicity in the Pacific,* edited by J. Linnekin and L. Poyer, pp. 17–41. University of Hawaii Press, Honolulu.

Weinstein-Farson, L.
1991 Land Politics and Power: The Mohegan Indians in the Seventeenth and Eighteenth Centuries. *Man in the Northeast* 42:9–16.

Welsch, R. L.
1994 Pig Feasts and Expanding Networks of Cultural Influence in the Upper Fly-Digul Plain. In *Migrations and Transformations: Regional Perspectives in New Guinea,* edited by A. J. Strathern and G. Sturtzenhofecker, pp. 85–119. University of Pittsburgh Press, Pittsburgh, Pennsylvania.

1996a Language, Culture, and Data on the North Coast of New Guinea. *Journal of Quantitative Anthropology* 6(4):209–234.

1996b Collaborative Regional Anthropology in New Guinea: From the New Guinea Micro-Evolution Project to the A. B. Lewis and Beyond. *Pacific Studies* 19(3).

1998 *An American Anthropologist in Melanesia: A. B. Lewis and the Joseph N. Field South Pacific Expedition, 1909–1913.* University of Hawaii Press, Honolulu. In press.

Welsch, R. L., and J. Terrell
1991   Continuity and Change in Economic Relations along the Aitape Coast of Papua New Guinea, 1909–1990. *Pacific Studies* 14(4):113–128.
1994   Reply to Moore and Romney. *American Anthropologist* 96(2):392–396.
Welsch, R. L., J. Terrell, and J. A. Nadolski
1992   Language and Culture on the North Coast of New Guinea. *American Anthropologist* 94(3):568–600.
Wendorf, F., and E. K. Reed
1955   An Alternative Reconstruction of Northern Rio Grande Prehistory. *El Palacio* 62(5–6):131–173.
Whallon, R., Jr.
1968   Investigations of Late Prehistoric Social Organization in New York State. In *New Perspectives in Archaeology*, edited by S. R. Binford and L. R. Binford, pp. 223–244. Aldine, Chicago, Illinois.
Wheat, J. B.
1955   Mogollon Culture Prior to A.D. 1000. Memoir no. 82 (Occasional Paper). *American Anthropologist* 57(3).
White, R.
1993   Introduction. In *Gesture and Speech,* by A. Leroi-Gourhan, pp. xiii–xxii. MIT Press, Cambridge, Massachusetts.
Whittlesey, S. M., and J. J. Reid
1982   Cholla Project Perspectives on Salado. In *Cholla Project Archaeology*, vol. 1, *Introduction and Special Studies*, edited by J. J. Reid, pp. 63–80. Archaeological Series no. 161. Arizona State Museum, University of Arizona, Tucson.
Wiessner, P.
1977   *Hxaro: A Regional System for Reducing Risk among the !Kung San.* Ph.D. dissertation, Department of Anthropology, University of Michigan, Ann Arbor. University Microfilms, Ann Arbor, Michigan.
1983   Style and Social Information in Kalahari San Projectile Points. *American Antiquity* 48(2):253–276.
1984   Reconsidering the Behavioral Basis for Style: A Case Study among the Kalahari San. *Journal of Anthropological Archaeology* 3(3):190–234.
1985   Style or Isochrestic Variation? A Reply to Sackett. *American Antiquity* 50(1):160–166.
1990   Is There a Unity to Style? In *The Uses of Style in Archaeology*, edited by M. W. Conkey and C. A. Hastorf, pp. 105–112. Cambridge University Press, Cambridge.
Wilcox, D. R.
1991   Hohokam Social Complexity. In *Chaco and Hohokam: Prehistoric Regional Systems in the American Southwest,* edited by P. L. Crown and W. J. Judge, pp. 253–276. School of American Research Press, Santa Fe, New Mexico.
1994   Macroregional Systems in the North American Southwest and Their Relationships. In *Great Towns and Regional Polities,* edited by J. Neitzel. University of New Mexico Press, Albuquerque.

Wilcox, D. R., and L. O. Shenk
1977   *The Architecture of the Casa Grande and Its Interpretation.* Arizona State Museum Archaeological Series no. 115. University of Arizona, Tucson.

Wilcox, D. R., and C. Sternberg
1981   *Additional Studies of the Architecture of the Casa Grande and Its Interpretation.* Arizona State Museum Archaeological Series no. 146. University of Arizona, Tucson.

Willey, G. R., and D. Lathrap
1956   An Archaeological Classification of Culture Contact Situations. In *Seminars in Archaeology: 1955,* edited by R. Wauchope, pp. 3–30. Society for American Archaeology, Washington D.C.

Willey, G. R., and P. Phillips
1958   *Method and Theory in American Archaeology.* University of Chicago Press, Illinois.

Willey, G. R., and J. A. Sabloff
1993   *A History of American Archaeology.* 3rd edition. W. H. Freeman and Company, New York.

Williams, F. E.
1932–33   Trading Voyages from the Gulf of Papua. *Oceania* 3(2):139–166.

Williams, L.
1972   *Ft. Shantok and Ft. Corchaug: A Comparative Study of Seventeenth Century Culture Contact in the Long Island Sound Area.* Ph.D. dissertation, Department of Anthropology, New York University. University Microfilms, Ann Arbor, Michigan.

Williams, R.
1866   *A Key into the Language of America.* Rhode Island Historical Society, Providence.
1963   *The Complete Writings of Roger Williams,* vol. 1. Russell and Russell, New York. Originally published 1643, Gregory Dexter, London.
1973   *A Key into the Language of America.* Wayne State University Press, Detroit, Michigan.

Wills, W. H., and R. D. Leonard
1994   *The Ancient Southwestern Community: Models and Methods for the Study of Prehistoric Social Organization.* University of New Mexico Press, Albuquerque.

Wilmsen, E. N.
1989   *Land Filled with Flies: A Political Economy of the Kalahari.* University of Chicago Press, Illinois.

Wimberly, M., and P. Eidenbach
1980   *Reconnaissance Study of the Archaeological and Related Resources of the Lower Puerco and Salado Drainages, Central New Mexico.* Human Systems Research, Inc., Tularosa, New Mexico.

Wissler, C.
1923   *Man and Culture.* Thomas Crowell, New York.

Wobst, H. M.
1977   Stylistic Behavior and Information Exchange. In *For the Director: Research Essays in Honor of James B. Griffin,* edited by C. E. Cleland, pp. 317–342. University of Michigan, Museum of Anthropology, Ann Arbor.
1978   The Archaeo-Ethnology of Hunter-Gatherers or the Tyranny of the Ethnographic Record in Archaeology. *American Antiquity* 43(2):303–309.

Woichom, J. W.
1979   Canoes to Boats: Ali Exchange Tradition in Transition on the Siau Coast of the Northwest Papua New Guinea since 1896. Paper presented to the 49th ANZAAS Congress.

Wolf, E.
1982   *Europe and the People without History.* University of California Press, Berkeley.

Wood, J. S.
1987   *Checklist of Pottery Types for the Tonto National Forest: An Introduction to the Archaeological Ceramics of Central Arizona.* The Arizona Archaeologist 21. Phoenix.

Wood, W.
1968   *New Englands Prospect.* De Capo Press, Amsterdam.
1977   *New England's Prospect,* edited by Alden T. Vaughan. University of Massachusetts Press, Amherst. Originally published 1634, Thomas Cotes, London.

Woods, A. J.
1985   An Introductory Note on the Use of Tangential Thin Sections for Distinguishing between Wheel-Thrown and Coil/Ring-Built Vessels. *Bulletin of the Experimental Firing Group* 3:100–114.
1986   Form, Fabric, and Function: Some Observations on the Cooking Pot in Antiquity. In *Ceramics and Civilization,* vol. 2, *Technology and Style,* edited by W. D. Kingery, pp. 157–172. The American Ceramic Society, Columbus, Ohio.

Wormington, H. M.
1955   *A Reappraisal of the Fremont Culture.* Denver Museum of Natural History, Proceedings 1. Denver, Colorado.

Worsley, P.
1984   *The Three Worlds: Culture and World Development.* University of Chicago Press, Illinois.

Wotzka, H.-P.
1993   Zum traditionellen Kulturbegriff in der prähistorischen Archäologie. *Paideuma* 39:25–43.

Wright, J. V.
1980   The Role of Attribute Analysis in the Study of Iroquoian Culture Prehistory. In *Proceedings of the 1979 Iroquois Pottery Conferences,* edited by C. F. Hayes III, pp. 21–26. Rochester Museum and Science Center, Research Records 13. Rochester, New York.

Wright, R. P.
1985   Technology and Style in Ancient Ceramics. In *Ceramics and Civilization: Ancient Technology to Modern Science,* vol. 2, edited by W. D. Kingery, pp. 1–20. The American Ceramic Society, Inc., Columbus, Ohio.

Yampolsky, M., and C. Sayer
1993   *The Traditional Architecture of Mexico.* Thames and Hudson Ltd., London.

Yellen, J. E.
1977   *Archaeological Approaches to the Present: Models for Reconstructing the Past.* Academic Press, New York.

Yellen, J. E., and H. Harpending
1972   Hunter-Gatherer Populations and Archaeological Inference. *World Archaeology* 4(2):244–253.

Yengoyan, A.
1985 Digging for Symbols: The Archaeology of Everyday Material Life. *Proceedings of the Prehistoric Society* 51:329–334.

Young, D., and R. Bonnichsen
1984 *Understanding Stone Tools: A Cognitive Approach.* Peopling of the Americas Process Series 1. University of Maine, Orono.

Young, L. C., and T. Stone
1990 The Thermal Properties of Textured Ceramics: An Experimental Study. *Journal of Field Archaeology* 17(2):195–203.

Young, W. R. (editor)
1969 *An Introduction to the Archaeology and History of the Connecticut Valley Indian.* Springfield Museum of Science, Massachusetts.

Index

Acacia species, 15, 101
Acculturation, resistance to, 177–178
Acephalous societies, 115, 118
Acorns, 139
Action archaeology, 4
Activity areas
　archaeological, 39, 40
　contemporary, 38, 40–47, 91
　and design variation, 42
　material items in, 40, 43–44
Actualistic research, 4
Adamawa-Ubangi language group, 83
Adams, E. Charles, 203–204
Adams, R. McC., 47
Adirondack Mountains, 149
Adobe, coursed, architecture
　brick-making, 187
　bricks, 187, 190, 205
　course dimensions, 188
　distribution, 189, 190, 201–202, 204–205.
　　See also Katchina religion
　materials, 187, 188, 191, 204
　multistoried, 188, 201
　origin, 190, 203
　post-reinforced, 199
　in Rio Grande region, 185, 194–199
　style, 191–192
　style determinants of, 191–204

　techniques, 189, 193, 201
　use, temporal and spatial, 183, 185
Adzes, 38
Africa, 275
　archaeology, 113, 115
　colonialism, 111, 115
　social fields, 10
　sociopolitical systems, 22
　See also Cameroonian potters; Kalahari Desert region; Luo; Manadara Mountains region
Age, 101
Age regiments, Tswana, 44
Agnatic groupings, 55, 56, 57
　tensions, 57
Agricultural areas, 16, 17(fig.), 18
Agropastoralists (in Kalahari), 12, 17, 18, 19–29, 32
　technology, 40–46, 65
　territorial units, 22, 31
Algonquian (language), 137, 163
Algonquian Indians
　ceramics, 143, 144(table), 147(table), 150, 153, 156(fig.), 158(table)
　and Iroquois, ceramic differences, 148–157, 159
　-Iroquois interaction, 142–143, 159
　Late Woodland sites, 135, 136–137, 138
　settlement patterns, 139–141, 142
　social organization, 142

［*continues*］

341

342 *Index*

See also Narragansett tribe; Pequot-Mohegan tribe; Wampanoag tribe
Allen, K.M. S., 157
Alliances, 10, 1921
  band, 31–32
  for defense, 55
  and language, 32
  marriage, 73
  trade, 32, 46
  See also Friendships
Aluminum, 85
Ammunition, 18
Analytic instrumentation, 5
Anasazi, 184(fig.), 185
  ceramics, 185, 228
  stone or adobe pueblos, 185, 190
Ancestors, 30, 31, 53
Andean metallurgy, 267
Anga, 268
Anglo-American archaeological tradition, 1, 3–5
Angola, 46
Appadurai, A., 233
Appliqués, 119, 144(table)
Archaeology
  and ethnicity, 111–113, 232–233
  and ethnography, 73–74, 109, 110, 121, 239
  New World-American, 112
  North American, 109, 132
  Old World-West European, 112, 232
  synthesis, 1–2, 10–11
  traditions, 1, 3–7, 30, 79
Archaeometry, 79
Architecture
  activities and purposes, 186–187
  avoidance and encounter patterns, 187
  change, 201, 256–260
  multistoried, 186
  See also Adobe, coursed, architecture; Houses; Room contiguity index; Luo, houses; *under* Tonto Basin
Arnold, Dean E., 133
Arnold, P. J., 100, 101
Arrows, 17, 38, 276
  exchanges of, 44, 51, 59
  as made everywhere, 65, 66
  point sharpeners, 39
  repair kit, 17

shafts, 43
shaft straighteners, 38
Arroyo Hondo Pueblo, 196(fig.)
Artifacts. *See* Material culture; *individual items*
Ash, 16
Assemblage, 107, 109, 146, 147(table)
Assimilation, 186
Atlatl, 40
Attribute analysis, 146–157, 168–169(tables)
Attribute value, 146
Austronesian language family, 62, 63, 64
  non-, 62, 63, 64
Autochthony, 116
Awls, 38
Axes, 17
Aztec site (N.M.), 184(fig.), 192

Bafia, 91
Bags, 51, 59, 65–66
Bakgalagadi, 19, 20, 21, 27
Bakhurutshe, 19
Balala, 20(fig.)
Bali, 270
Ballcourts, 199, 205
Bamangwato, 17
  administrative structure, 27
  cattle identification, 42
  chiefs, 22, 44
  class divisions, 27
  kinship, 22
  land tenure system, 19, 21, 22, 35
  lineages, 18, 22
  material items, 40–41, 42, 45
  natural resources, 22
  polygynous, 42
  status, 19
  and Tyua, 44
Bamboo, 58
  borers (tool), 56
Bamileke Fe fe, 84
  potters, 85
Bamum, 84
  potters, 85
Bands (sociopolitical), 22, 26, 31–32
Banen, 92, 93(fig.)
Bantu, 18, 45
Bantu (language), 83, 92
Baobab trees (*Adansonia digitata*), 32

Bark, 86(fig.), 89, 139
Barter, 70, 84, 85
Barth, Frederik, 112–113, 271, 272
Bartram, Laurence E., Jr., 7, 9, 16
Basarwa. *See* San
Baskets, 18, 33, 36, 40, 43
  as exchange goods, 51, 54, 59, 61, 65, 84
  exchanged rarely, 66
  for sale, 84
  tradition maintained, 178
Batswana (people of larger Tswana tribal
    polity), 44
Bayei, 20(fig.)
Beads, 14, 16, 18, 71, 117, 276
  -baking implements, 18
Beans, 35, 139
Bear, 139
Bechuanaland Protectorate (now Botswana), 36
Behavioral archaeologists, 79
Behura, N. K., 100
Berries, 35, 139
Betel nut, 54
Bi: (Coccinia rehmannii), 34
Binford, Lewis R., 3, 4, 38, 78, 233
Biś sa ani site, 184(fig.), 189(fig.)
  architecture, 192, 203
  style determinants, 192–194
Black market
  in animal resources, 27
Black palm bows, 54, 58, 61, 64–65
  decorations, 65
Bleed, P., 17
Boas, Franz, 3
Body decoration, 43, 242
Bone
  beads, 16
  containers, 38
  daggers, 66, 67
  needles, 38
  projectile points, 43
Boreholes, 16, 20–21, 26, 29
Botswana, Republic of (formerly Bechuanaland
    Protectorate), 14, 36
  currency (pula), 35
  land boards (1970), 28
  land reform, 28–29
  sociopolitical class system, 20(fig.)
Bourdieu, P., 8, 9, 235. *See also* Practice theory

Bowls, 18, 41, 54, 59
Bows, 17, 38, 64–65, 68
  as exchange goods, 51, 54, 58, 61
Bracelets, 41
Branch, D. P., 191
Braun, D. P., 100
Breternitz, C. D., 190
Brick, 187
Bridelia ferruginea, 86(fig.)
Bride-price system, 53, 64, 67, 71
Brooms, 41
Builders, 186
Bureau of Reclamation, 213
Burials, and broken vessels, 101
Burnished red slip, 120
Bushmen. *See* San
Byers, D. S., 136

Calabashes, 41
  decorated, 41–42
Cameron, Catherine M., 7, 9
Cameroon. *See* Cameroonian potters; Mandara
    Mountains region
Cameroonian potters
  ceremonial pottery, 85
  dail-use pottery, 85
  decline of craft, 85
  diffusion of technology, 94–97, 98(figs.), 275
  and ethnicity, 97, 275
  flexibility, 99
  income, 85
  languages, 83, 84(fig.), 93(fig.)
  learning period and sites, 83, 84(fig.), 91,
    93(fig.), 94, 96- -97(figs.), 103
  mobility, 97(figs.), 99
  population, 83
  postfiring materials, 86(fig.), 87
  pottery bartered, exchanged, or sold, 84–85
  shaping techniques, 133–134
  spatial distribution, 95, 97, 98(figs.)
  subsistence agriculture, 84
  and symbolic system, 91, 101
  techniques, 86(fig.), 87–91, 92
  variations, 87–89, 101–103
Canoes, 54, 61, 63–64
  as exchange goods, 51, 63
Canyon heads, 195
Carr, C., 277

Carrying bag, 37
  men's, 17
  women's, 17–18
Casa Grande site, 184(fig.), 189
Casas Grandes site (Mexico), 184(fig.), 189, 190
Cash crops, 84
Cash economy, 59
Cash-for-work programs (Botswana), 29
Cassava, 84
Cassowary, 66
Castes. *See* Technical specialists
Cattle, 28, 35
  brands, 42
  disease, 18, 35
  dung, 41
  faunal remains, 47
  fences, 28
  identification criteria, 42
  inherited, 42
  loan system, 19
  ranches, 28, 29
  for rituals, 19
  as wealth, 19
  and water points, 26
Cattle posts, 16, 17, 18, 19
  and mobility, 35, 36
  1960s, 1970s, 1990s, 28, 29
Caves, 46
Ceci, L., 141
Cement, 256
Cemeteries, 16
Cenchrus ciliaris, 16
Central Arizona Project, 213
Central District (Botswana), 14(fig.), 16, 28
Ceramic ecology, 4, 79, 82, 134
Ceramics, 16, 43, 84, 144(table), 145(fig.)
  aging, 101
  assemblages, 146
  attribute analysis, 146–157, 168–169(tables), 172–176
  attribute value, 146
  for beer, 119
  broken, 101, 104, 125
  constraints (perceived). *See* Ceramic Ecology
  daily use, 85, 118–119, 157
  and dry season, 84, 100, 105n3
  durability, 69, 151, 152
  efficiency, 81
  as exchange goods, 18, 51, 54, 58, 61–63, 84, 85, 119. *See also* Trade
  firing, 86(fig.), 87, 100, 143, 151, 215(table)
  for food storage and preparation, 119
  and gender, 84, 85, 118, 125, 143, 250
  household level production, 157
  learning, 250, 253. *See also under* Cameroonian potters
  lip form, 147
  manufacture and use, 79, 80, 85, 143, 157
  New England (17th century), 171–176
  plain ware, 218, 227
  postfiring treatment, 86(fig.), 87
  raw materials, 80, 86(fig.), 87, 89, 150, 217–218, 215(table)
  rims, 146, 147(table), 168–169(tables)
  shapes, 134, 143, 158(table)
  shaping techniques, 133–134, 228, 278
  sherd location, 147
  sherds, 101, 127, 137, 137, 146, 173
  smudging, 215(table)
  as specialized craft, 250
  surface treatment, 144(table), 147, 152–157, 158(table), 167, 168(table)
  symbolic use, 101, 171
  temper, 144(table), 146, 147(table), 148, 149–151, 168(table), 171, 217, 227, 253–254
  thermal expansion coefficients, 149, 151–152
  thickness, 146, 147(table)
  tools for, 144(table)
  as tourist item, 85
  traditions, 123–124, 125, 158(table), 167, 169, 171–172
  transporting, 152
  utilitarian, 210, 212. *See also subhead* daily use
  variation, adjuncts and instrumental, forms, 159
  wall thickness, 151–152
  waterproofing, 89
  See also Cameroonian potters; Clay; Decoration; Earthenware; Late Woodland; Manadara Mountains region; Material culture; Pots; Tonto Basin
Ceramic sociology, 266
Ceremonial vessels, 85

Chaco region, 183, 184(fig.), 192–194
Chad, Lake, Basin, 124
Chadic (language), 115, 125
Chaînes opératoires, 2–3, 5–6, 9, 79, 82, 85, 87, 92, 94, 101, 236, 238, 244, 262n3
  choice strategies, 248, 253–254, 255
  process, 238, 246
  and variations, 101–103
Chert tools, 142, 148(fig.)
Chestnuts, 139
Chickens, 84
Chiefs, 20(fig.), 27, 85, 249
  and labor, 22
  and land, 22
  places, 16, 44
Chihuahuan Desert, 184(fig.), 202
Childe, V. Gordon, 239, 266
Childs, S. T., 275
Chilton, Elizabeth S., 7, 9, 136, 275
Cibola region, 228
Cibola White Wares, 216
Circumcision, 101
Circumstantionalists, 112
Clams, 56
Clan motifs, as decoration, 64, 65
Clark, G. A., 112, 124
Clark, Jeffery J., 8, 9, 10
Clarke, D. L., 233
Clay, 16, 56, 58, 80
  analysis, 87, 89, 105n8
  beads, 16
  fashioning, 86(fig.), 92, 93(fig.), 96(fig.), 102
  figurines, 85, 117
  firing, 86(fig.), 87, 100
  firing for use as temper, 253
  firing fuel, 87
  paste behavior, 87
  pipes, 85
  plasticity, 87
  postfiring, 86(fig.), 89–90, 101
  processing, 86(fig.), 87
  selection, 90–91
  thermal expansion coefficient, 149
  See also Adobe, coursed, architecture; Ceramics; Earthenware pots
Click languages, 44
Clothing, 241–242

Cocoa, 84
Coffee, 84
Coiling, 86(fig.), 87, 88(fig.), 92, 93(fig.), 167
  spiral-, 61–62, 63, 69, 86(fig.), 87, 93(fig.)
Collared pots. *See* Decoration, collared
Colonization, 21, 111, 162, 166, 275. *See also* European contact
Comaroff, John, 113
Commoners, 20(fig.), 27
Communal land, 19, 21
Communication. *See under* Material culture
Communities of culture, 10, 69
Community
  characteristics, 57–58, 60, 69, 70, 75n5
  differences, 71
  specializations, 59, 60, 61–65
Conkey, M. W., 175
Connecticut, 135, 138, 142, 144(table)
Conquest, 22. *See also* Colonization; European contact
Consumption, anthropology of, 235, 244, 246, 254, 255(fig.)
Containers, 17, 40, 41, 43, 51
  contemporary, 85
  decorated, 42
  for poison, 38, 40
Contextual analysis, 173
Cooking area, 42
Cord, 17, 45
Cordell, L. S., 186
Cordon fences, 28, 29
Corn, 84. *See also* Maize
Corporate groups, 55, 56, 109
Cosmologies in the Making (Barth), 112–113, 272
Cowgill, G. L., 146
Craft markets, 85
Crafts, 17, 51, 84
  specialization, 114
Cranberries, 139
Croes, D. R., 275
Cross-cultural investigation, 111, 208
Crows, 47
Cultigens, 139, 142
Cultural anthropologists, 3
Cultural areas, 3, 10, 60, 185, 208
  and movement of people, 185–186
  Nevada-Utah, 189, 190

346   Index

"Cultural Contexts of Ethnic Differences"
    (Eriksen), 273
Cultural economy, 235
Culture, as ". . . means of adaptation," 78
Culture change, 200, 220
Culture histories, 132, 232
Cushing, Frank Hamilton, 188

Daga houses, 15
Daggers, 66, 67
Darwinian archaeologists, 4, 81
David, N., 276
Death, and site abandonment, 33
DeBoer, W. R., 102
Decoration (ceramics)
    Algonquian-Iroquois, differences, 152–157
    Algonquian-Iroquois similarities, 138(table),
        142, 144(table), 145(fig.)
    brushed surface, 144(table), 168(tables)
    castellations, 143, 144(table), 145(fig.), 171–172
    and chronology, 132, 158(table)
    collared, 142, 143, 144(table), 145(fig.),
        147(table), 153, 158(table), 167, 169(table), 171
    collars, lobate, 171
    color, 144(table), 147, 167, 171
    cord impressed, 144(table), 153, 169(table)
    corrugated, 227–228, 229(fig.)
    dentate stamped, 153, 167, 169(table)
    effigies, 167, 169(table), 172, 173
    fabric impressed, 153
    forming tasks, 215(table)
    impressed, 144(table), 153, 154–155(figs.)
    innovation, 254
    incised, 144(table), 147(table), 153,
        154–156(figs.), 158(table), 167, 169(table),
        170(fig.)
    linear punctated, 153
    lip form, 169(table)
    for maker identity, 46, 65–66, 132, 191, 253
    Narragansett and Wampanoag attributes,
        169(table)
    notched, 147(table), 153, 169(table), 212
    punctate, 153, 155(fig.), 169(table)
    rocker stamped, 153
    scraped, 147(table), 155
    shell impressed, 144(table), 153
    shell stamped, 169(table)
    simple, 277

    social function, 237
    stamped surface, 144(table), 153, 169(table),
        150, 154, 158(table)
    and technical choice, 176, 216
    indicating unity, 177–178
    use, 153, 154, 158(table)
    for user identity, 41–42, 61, 64, 65, 67, 68, 69,
        156, 212
    wiped, 147(table), 152
    See also Body decoration; Style; Technical
        choices
Dens, 46
Desert Archaeology, Inc., 213
Design variation
    on cattle brands, 42
    on daggers, 67
    by household kinship type, 42
    by lineage, 66
    on string bags, 65–66
    See also Decoration
Dibbles, 91
Dietler, Michael, 8, 9, 10, 270, 271, 278
Diffusion
    cultural, 111
    language, 79
    technology, 66, 94–97, 98(figs.), 190
Digging sticks, 15, 17, 40
DiPeso, C. C., 188, 190, 203
Displacement settlements, Kalahari, 28, 29, 48
Divorce, 123
Dobres, M.-A., 269, 270
Dogs, 36, 47
Domestic servants, 18
Donkeys, 18, 28, 36
Draper, P., 39
Drawing a lump pottery technique, 86(fig.), 87,
    92, 93(fig.)
Drawing a ring of clay pottery technique,
    86(fig.), 87, 93(fig.)
Drills (tools), 18
Drought, and territorial boundary protection, 23
Dry season, 55
    and pottery, 84, 100, 105n3
Dung, 41
Dyadic relationships, 10

Eagle Ridge site, 219(fig.)
Earrings, 41

Earthenware pots
　language associations, 62–63
　production, 61–63
　production sequence, 214–216
　shapes, 61
　as specialized exchange item, 51, 54, 58, 61–63
Eastern Highlanders (New Guinea), 53
Eastern/Hoa, 32
Eastern Niantic, 165
Effigies. *See under* Decoration
Eggshell. *See* Ostrich eggshell
Elephant hunting, 27
Elk, 139
El Paso-Casas Grandes area, 184(fig.), 201, 204, 205
Elson, Mark D., 8, 9, 10
Emblemic style, 191, 192–193, 203–205, 206, 242
Endogamy, 83
Engelbrecht, W., 143
English Cob technique, 188–190
Environmental conditions
　Africa, 126
　Kalahari, 47
　Southwest, 191, 192, 194, 199, 200, 203, 205–206, 216
Eriksen, T. H., 272, 273
Ethnic boundaries, 9–10, 13, 28, 29–30, 44–45, 48, 97, 110, 120, 272, 273
　fluidity, 112–113, 114
　signalling, 242–243
Ethnic Groups and Boundaries: The Social Organization of Culture Difference (Barth), 271
Ethnicity, 111, 112, 115, 232, 272–274. *See also under* Archaeology; Material culture
Ethnoarchaeology, 4, 5
　and contemporary traditional societies, 4–5, 9, 12–13, 16–17, 46, 73–74, 110, 118, 129, 139, 162, 234, 235, 266
　and faunal material, 46–47
　field methods, 263n8
　and material culture, 13, 108–109, 213
　and patterning, 11
　and social boundaries, 8, 12–13, 162–163
Ethnogenesis, 124
Ethnography, 4–5, 9, 73–74, 79, 109, 110–112
Ethnohistory, 9
Ethnolinguistic groups, 50, 51
Ethnonym, 107, 108, 117

European archaeological tradition, 1, 2–3, 5–7. *See also* Technology, and culture
European contact (North American), 138–139, 142, 161, 171, 176, 178, 275
Exchange labor, 18, 19, 21, 30, 35–36
Exchange of goods and services, 10, 18, 19, 23, 30, 33, 45–46, 51, 54, 84, 119
　made everywhere, 60, 65–66
　most desired, 54
　rare, 60, 66–68
　silent, 70, 72(fig.)
　and specialized products, 55, 56, 58–59
Exogamous chiefdoms (Cameroon), 83
Exogamy, 55, 83, 114, 122
Experimental archaeology, 4, 79

Facilities, 40–41
Family fields (Tawana village), 17(fig.)
Faunal remains, 14, 16, 46–47
Feldspar, 149
Feminist archaeologists, 133Feathers, 18
Fences, 16, 28, 29, 41
Fence wire, 43
Field Museum, 67
Figurines, 85
Files (tools), 41
Fish, 54, 58
Fishing, 32, 84, 139
Fishing communities, 56
Fissioning and fusion, seasonal, 142, 143
Fitch, J. M., 191
Floors, 15, 16, 41
Food
　collecting, 139
　gifts, 51, 54
　preparation, 39, 41, 119
　storage, 39, 119
Food programs, 29
Foragers, 12, 15, 17, 32, 36
　bands, 22, 26
　carrying bags, 17–18
　and land area size, 28, 29, 48
　mobility strategies, 34–35
　and social boundaries, 29
　and storage, 37
　technology, 40
　territorial unit, 22
　and yields, 34

348   Index

Formal variation, 1
   in finished product, 2
   during manufacturing sequence. *See Chaînes opératoires*
   See also under Material culture
Fort Shantok site, 170–172
Fossiles directeurs, 107
Four Corners area, 192
Fourmile-style pottery, 204
France, 1
Fried, M. H., 273
Friendships, 10, 32, 73
   and gender, 55, 75n4
   and gifts and hospitality, 54–55, 59, 60, 70
   and land, 22
   and language, 58
Fruits, 139
Frying pans, 61
Function
   and material culture variability, 4
   v. style, 7, 81, 237–238
   and use, 79, 80. *See also* Ceramic Ecology
Fungi, 139
Funnels, 41

Game scouts, 45, 47
G//ana, 17, 23, 26, 29
   mobility, 33
   projectile points, 43
Gbaya, 99
Geometric design, 142
Germany, 54
Ghana, 101
Giant clam shells (*Tridacna*), 56, 64
Giddens, A., 269, 270
Gila Plain ware, 217
Gila River, 217, 218
Glass
   beads, 14, 16, 43, 46, 64
   beads for bride-price, 71
   containers, 85
   jewelry, 41
Global systems, 52
Globular bodied pots, 143, 145(fig.), 171
Goats, 19, 28, 84
Goodby, Robert G., 7, 9, 10, 271, 275
Gookin, D., 143

Gosselain, Olivier P., 7, 9, 130, 133, 275, 277, 278
Gourds, 41–42, 66
Grain, 33, 41
Grain bins
   contemporary, 41, 257(fig.)
   Iron Age (Kalahari), 15
Granulometric data, 87, 89, 90(fig.), 105n8
Grasses, 15, 16, 22
Grazing
   areas, and hunting, 43
   land, 17(fig.), 19–20, 22, 28, 29
   land, and San, 28
   over-, 21
   Tswana areas and districts, 17(fig.), 19–20, 22, 27, 29
   year-around, 27
"Great sand face," 26
Grewia flava, 45
Griffin Wash, 224(fig.), 226(fig.), 227
Grinding stones, 38, 91
Grit temper, 144(table), 147(table), 168(table)
Grog, 101
Guatemala, 215–216
Guida Farm site (Mass.), 135, 136, 148–157
Guides, 18
Guns, 18, 27
Gurensi, 101
G/wi, 14(fig.), 17, 23, 26, 29
   mobility, 33, 35
   perceived as dangerous, 44
   projectile points, 43, 44
   and Tyua, 44–45

Habitus concept, 9, 245, 246, 247, 248, 260, 263n9, 269, 270, 277
   and choices, 259
   and ethnicity, 273
   personal transformation of, 253
Hair baskets, 66, 67
Hamlets, 55, 57, 69
Hanson, J., 216
Hardin, M. A., 240
Hardveldt, 15, 16, 19
Headbands, 41
Head pads, 122
Headrests (pillows), 66, 67

Healers, 18, 43
Hedge Apple site, 217, 218(fig.)
Hegmon, Michelle, 8, 9, 10
Heinz, H. J., 31
Herbich, Ingrid, 8, 9, 10, 270, 271, 278
Herding. *See* Agropastoralists
Herero, 19, 20(fig.), 21
    cattle identification, 42
    containers, 42
    as immigrants, 28
    material items, 40–41, 45
    polygynous, 42
    spears, 44
Hides. *See* Skins
Higgeson (reverend), 139
Hill, J. N., 133
Hillier, B., 216
Hilltop sites, 15–16
Historical evidence, 9
Hitchcock, Robert K., 7, 9, 16
Hodder, I., 175, 242–243
Hoes, 91, 127
Hoffman, C. R., 269, 270
Hohokam region, 184, 185, 199–200, 210(fig.)
    ceramics, 185, 220
    Classic period of culture change, 200
    pit houses, 185, 199
    religion, 205
    site layout, 199
    in Tonto Basin, 217, 220, 228
Hoof-and-mouth disease, 35
Hornblende, 149
Horn container, 17, 38, 40
Horses, 18
Horticulture, 138, 139
Hosler, D., 190
Households
    family types, 42
    production, 157
Houses, 40
    Algonquian, 139–141
    form, 256, 259
    Iron Age (Kalahari), 15, 16
    Iroquois, 140(fig.), 141
    Kua, 43
    land for, 19
    on long-term residential sites, 41

Luo, 256–260
    in pits. *See* Pit structures
    Tswana, 16
    See also Architecture
Hovenweep National Monument, 197 (caption)
Hudson Valley, 138, 140
Hunter-gatherers, 24–25(table), 143. *See also* Foragers
Hunting, 17, 18, 27, 37, 84, 139
    with animals, 36
    areas (Tswana), 17(fig.), 20
    for black market, 27
    blinds, 39, 45
    with bows and arrows, 43, 44, 68
    illegal, 43, 44, 47
    with spears, 68
    technology, 38
Huron, 143
    technology and prey, 44, 45
Hyena, brown (*Hyaena brunnea*), 46

Iberomaurusian, 113
Ideology, 30, 204, 206
Il Chamus, 42
Illness, 43
Immigrants, 20(fig.), 27
Implements, defined, 40
Incision, 119
India, 100, 134
Information exchange model, 4, 81–82, 156, 191, 209–210, 240, 241, 263n7
Ingold, T., 269
Inheritance, 23, 42, 259–260
Insect (*Diamphidia nigro-ornata*) larvae, as poison, 38
Interaction zones, 10, 12–13
Irian Jaya, 67
Iron Age sites
    Kalahari, 15–16
    Mandara Mountains region, 117, 124, 125–126, 127
Iron hoes, 127
Iron ore, 16
Iron pots, 18
Iroquoian language family, 137

Iroquois Indians
    and Algonquian ceramic differences, 148–157, 159
    -Algonquian interaction, 142–143, 159
    ceramics, 136, 143, 144(table), 145(fig.), 150, 151, 153, 155--156, 158(table)
    diet, 139
    Late Woodland site, 137–138
    long house, 140(fig.), 141
    maize, 139
    matrilocal-matrilineal groups, 142
    population, 142
    settlement patterns, 141, 142, 151
    sociopolitical organization, 141–142
    vessel size and communal dining, 151
Islamic states (Africa), 115
Isochrestic variation, 240, 268
    defined, 267
    See also Technical choice
Ivory, 18, 27, 41

Jacal, 187, 188, 198
Jewelry, 41, 117. *See also* Beads, Necklaces
Johnson, E. S., 142
Joking, 55
"Jones effect," 52–53
Josselyn, John, 139, 141
Ju/'hoansi, 22, 23, 32, 46, 47
Jumperbars, 91

Kalahari Desert region, 13, 14–15
    archaeological surveys, 15–16
    contemporary sites, 16
    continuous occupation period, 16
    ethnic groups in research area, 16–17, 18, 32
    ethnoarchaeological research region, 14(fig.), 16–17
    game animal loss, 18, 28
    government settlements, 28, 29
    house and kraal sites, 15, 16
    hunter-gatherer populations, 24–25(table)
    material culture (archaeological), 16
    material culture (contemporary,subsistence), 17–18, 47
    mobility strategies, 33–37
    population (research area), 16
    population density, 24–25(table)
    population displacement, 28, 29
    ranches, 28
    technology (archaeological), 45
    technology (contemporary), 17–18, 45
    territorial boundaries, 22–27
Kalanga, 19, 20(fig.), 21
Karosses, 37
Katchina religion, 203–205
Kayenta region, 183, 184(fig.), 228
Keene, Arthur S., 136
Kelley, J. H., 204
Kenya, 103, 234, 242
Kettles, brass, 171
Khama I (Bamangwato chief), 18
Khoi koi, 20(fig.)
King Philip's War (1675), 165, 178
Kinship
    -based villages, 55–56
    and design variation, 42, 69
    fictive, 32
    land land, 21, 22, 23, 25–26
    lineages, 55–56
    networks, 72–73
    nexuses, 31–32
    and obligations, 56
    and resource allocation, 42
Kintampo, 113
Kivas, 195, 204
Klock site (N.Y.), 135–136, 137, 148–156
Knecht, Ueli, 85
Koma-Ndera, 101
Kossina, , 232
Kraals, 15, 16
Kua, 17–19
    band territories, 26, 31–32
    dogs and warthog bones, 47
    and government programs, 29
    hunting, 37
    land, 21, 23, 29, 30–31
    livestock, 28
    mobility strategies, 33–34, 35
    plant gathering, 36
    projectile points, 43–44
    research area, 14(fig.)
    status, 18, 20–21
    technology, 40, 43
    tools, 39

and Tyua, 44–45
work, 19, 21
Kula, 50, 52, 53
Kulturkreis school, 111, 232
!Kung area, 14(fig.)

Labor
  for chiefs, 22
  for grazing district oversser, 27
  investment and land rights, 21–22
  See also Exchange labor
Ladder motif, 142
Land boards (Botswana), 28
Landless, 20, 27
Land management system. *See* Land tenure system; Land use; *under* Tswana
Land reform, 28
Land tenure system, 48
  Bamangwato, 19
  and chiefs, 22, 28
  and colonization, 21
  and communal land, 21
  and government, 28, 29
  and immigration, 28
  and kinship, 21
  and natural resources, 22, 27, 29
  and population density, 22
  Tswana, 19–20, 27, 28, 29
Land use, 21, 23, 28–29
  interpretations, 30
  and mobility, 34
Language, 244
  diffusion, 79
  group alliance, 32
Lanick, R., 275
Late Archaic, 142
Late Woodland, 132, 275
  ceramics, 136, 137, 143, 144(table), 167, 169–170, 174
  ceramics, 19th century, 174
  culture, 137–146
  sites, 135–136, 138
Lavin, L. F., 145–146
Leather bags, decorated, 43
Lechtman, Heather, 266, 267–268
Lemonnier, Pierre, 6, 12, 30, 66, 68, 99–100, 108, 114, 115, 133, 175, 209, 268, 274

Leroi-Gourhan, A., 5, 6, 114
Letlhakane River, 15
Levine, F., 204
Lewis, A. B., 61, 62, 65, 67, 72
Lichens, 139
Linden Corrugated ceramics, 228, 229(fig.)
Lineages, 18, 22, 55–56, 69. *See also* Kinship; Territorial lineage groups
Linguistically distinct co-resident populations, 13, 28, 30, 123
Linguisitic evidence, 9
Little Colorado area, 184(fig.), 204–205
Livestock, 84
  density, 36
  development program (Botswana), 28
  disease, 18, 35
  loan system, 19
  ranches, 28, 29
Longhouse, 140(fig.)
Los Muertos site, 188
Luo, 103, 234
  agriculturalists, 250
  ceramic distribution, 254–256
  ceramic microstyles, 250, 253, 254, 275
  ceramic raw materials, 253–254
  ceramics, 250, 251–252(figs.), 263n7
  and change, 256–258, 260, 278
  homestead pattern, 257–260
  houses, 256–260, 278
  kinship, 257–258, 259–260
  land shortage, 257–258
  language, 248
  potter communities, 250, 253
  settlement pattern, 248, 249(fig.)
  sociopolitical organization, 249, 257
  women, 250, 253
Luo(language), 248

Macaranga spinosa, 86(fig.)
MacEachern, Scott, 7, 10, 50, 73, 74, 266, 271, 276
Magic, 18
Maize, 41, 138, 139, 142
  cooking pots, 149, 151–152
Makgadikgadi Pans, 14(fig.), 15, 32
Malinowski, B., 50, 53
Mandara Archaeological Project, 115

352   *Index*

Mandara Mountains region, 276
  blacksmiths, 127
  ceramics, 117, 118-119
  ceramics, functional catergories, 119, 122
  ceramic traditions, 120-121, 126
  ethnic groups, 116, 117(fig.), 118, 120, 128-130
  Iron Age sites, 117, 124, 125-126, 127
  languages, 115, 123
  map, 116(fig.)
  montegnard communities, 119, 120
  pot decoration, 119, 120
  pot handles, 119, 120(fig.)
  pot use, 119, 122
  pot variation, 119-122, 123-124
  sociopolitical groups, 115, 116-117, 118
  territorial lineage groups, 116, 118-120, 122-123
Marriage
  as alliance, 71
  interethnic, 123
  and land rights, 21, 22
  and lineage, 123
  patterns, 10, 30, 31-32, 52, 53, 64, 67
  and residence, 55, 103
Marshall, M. P., 190
MAS. *See* Massachusetts Archaeology Society
Masonry compounds, 220, 221(fig.), 222, 225(fig.)
Massachusetts, 135
Massachusetts Archaeology Society (MAS), 136
Massachusetts Bay Colony, 165
Massachusetts Bay region, 138-139
Massasoit (sachem), 164
Material culture
  and acculturation, 178
  and communication, 243, 244, 277
  context of design, manufacture, and distribution, 130
  continuity, 124-125, 178
  and decorative design, 240, 244
  differences and ethnicity and ideology, 30, 68, 69, 109, 113, 114, 267
  distribution through different groups, 50-52, 66-67, 68, 108, 122, 128, 177, 185-186, 206, 216, 235, 270
  and economic reductionism, 241, 242
  elaboration indicating resistance to change, 177-178, 179

  and ethnicity, 45, 69, 83, 125, 275
  and formal variation, 2, 3-7, 9, 47-48, 265, 277. *See also* Style; Technology
  general distribution of, 60, 65-66, 113, 124
  and interpretation, 14, 30, 45, 110
  and language association, 62, 64, 76n14, 124
  and meaning, 243
  and nonmaterial cultural elements, 109, 115, 234
  and production center location, 52
  production and consumption, 244, 246
  and sedentary and mobile groups, 39, 40-41, 46, 48, 150, 156- -157
  similarity among ethnic groups, 45, 69, 108, 114, 173, 185-186
  and social networks, 51
  and spatial and temporal extent, 107-108, 113, 124, 185, 208
  specialized item, 60, 61-65
  and structure, 51, 239, 245
  and techniques and culture, 1, 2, 5-6, 9, 10, 30, 73, 183, 217. *See also* Practice theory; Techniques
  technology and important items of, 38
  and technology studies, 3-5
  and tool types, 40
  variability, 5, 39, 43, 44, 47, 69, 108, 122, 133, 159, 184
  See also Architecture; Ceramics; Social boundaries, and material culture patterning; Style; Technical systems; *under* Kalahari Desert region; Sepik Coast
Matrilocal-matrilineal group, 142
Mauss, M., 5, 236, 238
McBride, K. A., 141, 172
McDonald Corrugated ceramics, 228, 229(fig.)
Mead, Margaret, 73
Meat, 18, 27, 33
  hunt distribution, 37
Meddler Point site, 217, 224-225(figs.), 227
Melons, 139
  Citrullus vulgaris, C. naudinianus, 34, 35
Melpa, 53
Men's cult
  house ornaments, 61
  ritual, 53
Merrill, R. S., 266

Mesoamerica, 190
Metal
 axes, 17
 hoes, 127
 jewelry, 41
 projectile points, 43
 tools, 10, 18, 33, 41, 43, 127
Metallurgy, 190, 267, 277
Mexico, 190
Miantonomi (sachem), 166, 176–177
Micaceous temper, 144(table), 217–218
Microstyles, 250, 253, 254, 275
Middens, 15, 16
Middlemen trade, 70–71, 72(fig.), 85
"Middle-range" research, 4, 277
Middle Woodland, 153
Migratory movements, 79, 114, 124, 183–184, 185–186, 197, 200–201, 206, 217, 273–274
 individual/family, 125, 197–198
Milk, 36
Miller, D., 134
Millet, 41
Mimbres, 184, 200–201, 204
Missiles, 40
Mobility patterns, 33–37
 factors, 33, 34
 interpretations, 30
 logistical, 33–34, 35
 and material items, 39–40
 residential, 33, 34, 35, 36
 and season, 34, 35
 and technical style, 79, 123–124
Mogollon, 184(fig.), 185, 228
Mogollon Rim, 187, 228
Mohawk Indians, 176
Mohawk Valley, 135, 142
Mohegan. *See* Pequot-Mohegan tribe
Moka, 53
Mongongo (*Ricinodendron rautanenii*), 38
Montauk Indians, 176
Moore, C. C., 67
Mora (Wandala), 115, 116(fig.), 118
Moravian folk costumes, 241
Morgan, Lewis Henry, 141, 151
Mortars, 40, 41, 91
Mosquito nets, 67
Motor habits, 102

Mountain Arapesh, 73
Mud. *See* Adobe, coursed, architecture
Multilingualism,
Murik, 70–71
Murik baskets, 59, 61, 64–65
Murik Lakes villages, 70
Mussels, 139
Mystic (Conn.). *See* Pequot War

Name relationship, 23, 32
Namibia, 22
Narragansett Bay, 139
Narragansett tribe, 163
 cemeteries, 171
 ceramics, 167, 168(table), 169–170, 171, 172–176, 178–179
 and the English, 163–164, 165, 166, 176
 European brass kettles, 171
 language, 163
 marriage, 163, 165
 and Pequot-Mohegan, 163, 165–166, 174
 sachems, 164–165, 166, 176
 socioeconomic structure, 163–164
 territory, 163, 164(fig.), 166
 and Wampanoag, 164–165, 173–174
Nationalism, 112
Natural resources
 conservation, 23
 management, 31–32
 and mobility strategies, 33–37
 use of, 22, 27, 30–31, 33, 38
Necklaces, 41
Needles, 38
Nelson, M. C., 12
Neolithic, 113
Nevada, 190, 203
New Archaeology, 3–4, 209, 293
New England, 135, 138, 142, 143–146, 163, 273, 275
 technological patterning, 161, 176–181
New Guinea
 material culture, 50, 52–53
 See also Sepik Coast; Social fields
New York, 135, 144(table)
Nexus, 31–32
Ngwato Land Board, 15
Niantic. *See* Eastern Niantic

354 *Index*

Nigeria. *See* Mandara Mountains region
Nilotic language, 248
Northern Khoe. *See* Tyua
Northwest Coast (U.S.), 275
Nsei, 85
Nutcracking stones, 38
Nuts, 139

Obligation, 19
Obliteration technique, 228
Ochers, 58
Operational sequences. *See Chaînes opératoires*
Ornaments
  as exchange goods, 51, 58
  on houses, 61
Osborn, A. J., 42
Ostrich (*Struthio camelus*), 45
Ostrich eggshell
  beads, 18, 43, 45–46
  collecting ban, 46
  designs, 43, 45, 46
  necklaces, 18, 46
  as trade item, 45–46
Ostrich feathers, 18
Outrigger canoes, 54, 61, 63–64
Overgrazing, 21
Oxen, 36

Paddle-and-anvil pottery technique, 61, 62, 69, 228
Paddles, as exchange goods, 51
Palm oil, 85
Palm trees, 33, 84
Pans (depressions), 15
  and residential locations, 31, 36
Papua New Guinea, 268
Paquimé. *See* Casas Grandes
Parker, A. C., 139
Parsons, N., 19
Patrilineage, territorial, 116, 118, 122
Patrilocality, 103, 240, 250, 253
Patrivirilocality, 55
Patron-client relationship, 18, 27, 30, 44
  and land, 21
Payson Basin, 224
Peanuts, 84
Penis gourds, 66, 67

Pequot-Mohegan tribe tribe, 163
  ceramics (17th century sites), 170–172, 177, 178 179
  and the English, 163–164, 165, 166, 177
  European brass kettles, 171
  language, 163
  marriage, 163, 165
  and Narragansett, 163, 165–166, 177
  sachems, 164, 165
  socioeconomic structure, 163
  territory, 163, 164(fig.), 166
Pequot War (1636–1637), 165, 177
Pern, 190, 203
Pestles, 38, 41, 91
Peterson, J. B., 146
Pfaffenberger, B., 270
Phillips, P., 73, 109
Phoenix Basin, 217, 220, 227, 228
Pigments, 58
Pigs, 84
  feast, 53
  fodder, 52
Pinching pottery technique, 86(fig.), 87, 88(fig.), 93(fig.)
Pindi Pueblo, 188
Pine Hill site (Mass.), 135, 136–137, 148–156
  attribute analysis, 147(table)
Pipes, 85
Pit structures, 185, 198, 199, 201, 217, 218–219(figs.)
Plant resources, 22, 27, 84
  for food, 34, 35, 36
  for pottery, 84, 86(fig.), 89, 90
  symbolic use, 101
  water-bearing, 33, 34
Plastering, 193
Plastic, 41, 85
Plates, 41
Platform mounds, 199, 205, 214(fig.), 216, 224(fig.), 226, 227
Plazas, 195–196(figs.), 204, 217
Pleistocene, 113
Pliers, 41
Plows, 41
Plumes, 54
Plymouth Colony, 164
Pocket knives, 106
Pocumtucks, 135

Poison, 38, 44
  container, 17
Pole fences, 41
Pole houses, 15
Political alliances, 10
Polygyny, 42
Population density, 130
  of hunter-gatherers (Kalahari), 24–25(table)
  Iroquois, 142, 157
  and land conflicts, 22
  and water, 36
Population mixing, 79
Porcupines, 47
Porridge stirrers, 41
Pots, 40
  broken, 101, 104, 124
  carrying posture with, 122
  collared, 142, 143
  constricted neck, 143
  cooking, 61, 119, 125, 149, 151–152, 171, 250
  as exchange goods, 51, 54, 58, 61–63, 84–85, 119
  head pads, 122
  iron, 18
  porridge, 119
  production techniques, 61–63, 69, 86(fig.), 87, 92, 93(fig.), 119, 123, 151, 214–216
  sauce, 119
  shapes, 61, 119, 142–146, 215–216
  size, 171
  stands, 41, 119
  for trade, 33, 119, 120
  tripod, 125
  watercarrying, 119, 122
  water storage, 250, 251–252(figs.)
  See also Ceramics; Decoration; Earthenware pots
Potter communities, 250
Pottery. *See* Ceramics; Decoration; Earthenware pots; Pots
Pottery-making communities, 56
Practice theory, 8, 9, 235, 239, 243, 244, 245, 260, 269–270, 273
  doxa, 247
Pre-modern societies, 112
Primordialists, 112
Processual archaeology, 10
  post-, 30
Procyanidins, 89

Projectile points, 14
  materials, 43, 178
  and social boundaries, 212
  variation, 43
Projet Maya-Wandala, 115
Psychology, 113
Public meeting areas, 16
Puddled technique, 188
Puebloan architecture, 186, 201
Pueblo culture, 187
Pula (Botswana currency), 35

Quartz, 149, 150
Quill projectile points, 43
Quivers, 38

Rainmakers, 18
Rakes, 41
Ranches, 28, 29
Raw materials, as gifts, 51, 58
RCD. *See* Roosevelt Community Development Study
Redman, C. L., 224
Refugees, 20(fig.), 27
Regional systems, 52, 58
Religion. *See* Katchina religion
Resettlement. *See* Displacement settlements
Residential sites
  long-term, 41
  material items, 40, 41, 43–44
  material items, similarity among ethnic groups, 45
  shifts, 57. *See also* Southwest
Resin, 86(fig.)
Resource field, 70–73
Rhinoceros horn, 27
Rinderpest, 18
Rings, 56
Rio Grande region, northern, 183–184, 204
  architecture, 183–184, 188, 189, 196(fig.), 197, 203
  ceramics, 183
  and San Juan region, differences, 183–184, 195, 197, 198–199
  and San Juan region, similarities, 183, 195
  site layout, 195
  technological style, 194–195, 197–199
  Valdez phase, 189, 198
Ritchie, William A., 138, 143

Rituals, 55, 101, 116
Ritual specialists, 18
River Bushmen. *See* Tyua
River valleys (fossil), 15
Rock art, 14, 43, 204
Rock shelters, 195
Roles, 12, 30, 242
Romney, A. K., 67
Roofs, 22, 41, 58, 256, 259
Room block form, 220, 221(fig.), 222–223, 224, 226–227
Room contiguity index, 220–224
Roosevelt Community Development Study (RCD), 209, 213, 214(fig.)
Roosevelt Lake, 213, 214(fig.)
Roots, 34, 35, 139
Rope, 17
Rouletting, 119
Rouse, I., 136, 145, 146, 172
Rowe Ruin, 198(fig.)
Royal family, 20(fig.)
Rubbing stone, 18
Rumens, 33–34

Sachems, 163, 164, 165
Sackett, James R., 104, 108, 159, 179, 191, 237, 240, 267, 268
Sacred sites, 22, 31, 39
Sagard, G., 143
Sago, 54, 55
  distribution, 58
  importance, 56, 57
  leaf, 58
  -producing communities, 56, 58
  pancakes, 61
  pudding pots, 61
  in silent trade, 70
Saguaro Muerto site, 224(fig.), 227
Saharan (language), 115
Sahlins, M. D., 39, 239
Salado culture, 200, 209. *See also* Tonto Basin
Salado Polychromes, 216, 230
Salt, 33
Salt River, 213
San (Bushmen, Basarwa), 13, 17, 18, 19
  arrows, 276
  and government programs, 29
  grazing land denied, 28
  land, 23, 25–26, 28, 29
  ornaments, 46, 276
  in sociopolitical class system, 20, 27
  territorial unit, 22–23, 29
Sanaga, 92, 93(fig.)
Sand Canyon Pueblo, 195(fig.)
Sandstone, 184, 199
Sanger, D., 146
San Juan region, northern, 183–184
  architecture, 183–184, 195–196, 199
  ceramics, 183
  and Rio Grande region, differences, 183–184, 195, 197, 198–199
  and Rio Grande region, similarities, 183, 195
  site layout, 195
Saraswati, B., 100
Savannah, 15
Scale, 10, 12, 42
  ethnographic, 111
  spatial and temporal, 153
Scallop shell, 153
Schaafsma, Polly, 204
Schapera, I., 18
Schiffer, M. B., 6, 101
Schoolhouse Point site, 224(fig.), 227
Schortman, E. M., 272
Seasonal encampment, 136
Sebastian, L., 204
Sedge, 65, 67
Self-identification, 116, 242, 276. *See also* Social identity
Sepik Coast (New Guinea), 50, 51
  behavior expectations, 53, 55, 60
  bush communities, 59, 65, 66, 70, 72–73
  cash economy, 59
  cooperation, 53, 56
  exchange goods, 51, 54, 56
  German control (1900s), 54
  languages, 51, 58, 59, 62–63, 68
  map, 63(fig.)
  markets, 59
  material culture, distribution and specialization, 60–68, 70- -73, 128
  social networks, 51, 52, 55, 57–58
  trade relationships, 70–73
  villages, 55–59
  warfare, 54
Serfs, 20(fig.), 27

Seroka shrub (*Commiphora pyracanthoides*), 38
Serra Hills (Sepik Coast), 71, 72(fig.)
Setswana (language), 44
Shamans, 43
Shantock Tradition, 171–172, 174, 175
Sheep, 19, 28
Shell
  ornaments, 58
  rings, 56, 58, 61, 63, 64
  as temper, 144(table), 168(table)
  tools, 58, 144(table), 153, 169(table)
Shenk, L. O., 188
Sherds. *See under* Ceramics
Shrubs, 15
  insect larvae, 38
Shua, 32
Sigaut, F., 238
Signification, 244
Silent trade, 70
Sinew, 45
Sister exchange, 64, 67, 71
Site damage, 138
Site functions, 39, 40–41
Skins, 18, 27, 37, 40
Sko language family, 62, 63, 64, 67
Slab-building pottery technique, 62, 63, 69, 86(fig.), 93(fig.)
Slag, 16
Slave-raiding, 124
Sleeping bags, 66, 67
Slipping, 120, 121
Smith, C., 143, 145
Smith, F. T., 101
Smoked fish, 54, 58, 70
Social alliances, 30
  and gifts, 51
  and land, 21, 22
Social boundaries
  as abstract and ideological, 271
  changes, 28–29, 274, 275. *See also* Colonization; European contact
  difficulties in archaeological record, 73–74, 104, 162, 185–186. *See also* Ethnoarchaeology
  as dynamic and unstable, 73, 108, 181
  and ethnicity, 9–10, 28, 29–30, 44–45, 48, 97, 272–274
  identification and analysis possibility, 8, 9

  and language, 30, 44, 83, 84(fig.), 92, 93(fig.), 103
  and local traditions, 209. *See also* Tonto Basin
  and material culture patterning, 1–2, 3, 8, 9, 10, 12, 51, 92, 108, 109, 175, 179–180, 187, 208, 233–235, 236, 238, 242, 245, 246–247, 260, 270
  particularist approach, 10
  and perception, 44
  and scale, 10, 47, 48, 109, 208–209
  synthesis approach, 11
  and technological patterning, 161, 162, 180–181, 211–213, 216, 218, 271, 274
  See also Social fields; Style; Technology; Territorial boundaries
Social change, 248, 256–258, 260
Social, economic, and political relations. *See* Social fields
Social Evolution (Childe), 266
Social facts, 238, 240, 266, 272
Social fields, 10, 70
  as bounded social formations, 51, 52, 69, 71, 72(fig.), 73
  defined, 52–53
  discontinuity, 71, 73
  extent, 59–60
  as playing fields, 53, 59–60
  as unbounded friendship field, 60, 69, 72(fig.), 73
Social groups, 1–2, 208, 233
  in archaeological record, 8, 185
  in Botswana, 20(fig.)
  See also Ethnoarchaeology
Social identity
  and architecture, 186, 187
  and social boundaries, 161, 169–170, 177–178
  and technical choices, 79, 82, 83, 92, 93(fig.), 101, 156
  and technical systems, 12, 48
  and territorial boundaries, 37
Socializing, 55
Social linkages, 10, 32
Social networks, 51, 52, 58
  despite differences, 58, 60, 68
  extent of, 59, 67
  See also Social fields
Social space, 13

Social status, 242
Social units, 13
Sociopolitical systems
  centralized power, 83, 142
  class, 20(fig.), 27
  land basis, 22, 26, 29
  tribal, 141–142
Sociotechnical system, 270–271
Socorro region, 184(fig.), 192, 193, 194
Sonoran Desert, 184(fig.), 202
Sorghum, 41, 84, 119
Southern Adobe Belt, 202, 203
Southwest (American), 183
  architecture, 183–185, 186
  ceramics, 183, 185
  deserts, 202
  migration, 184, 273–274
  See also Tonto Basin
Spatial analysis, 2, 3
Spatial discontinuity, 208, 211
Spatial patterning, 5, 6
  and scale, 10
Spatial variability, 1, 4, 13, 47, 103, 105, 108, 113, 185, 213, 216, 227
Spears, 17, 38, 44, 66, 68, 275
Spear throwers, 40, 66, 67
Spiral-coiling pottery technique, 61–62, 63, 69, 86(fig.), 87, 93(fig.)
Spoons, 41
Springfield Museum (Mass.), 136
Squash, 139, 142
Stallings, W. S., 188
Stanislawski, M. B., 175
Stark, Miriam T., 8, 9, 10
Statuettes, 117
Status, 12, 18, 242
Steel, 41
Sterner, J., 275
Stick fences, 41
Stirrers, 41
Stone, 38
  flaked, 178
  masonry, 185, 198(fig.), 199, 206, 221(fig.), 222, 225(fig.)
  spalls, 187
  tools, 178, 212
Stone Age sites, Kalahari, 15

Storage, 37, 38, 39, 41
  pots, 61, 119
Strathern, A. J., 53
Strawberries, 139
String bags, 51, 59
  as made everywhere, 65–66
Structuralists, 30, 239, 245
Stubbs, S. A., 188
Style
  of action, 236, 237, 240, 253
  as choice, 82, 109, 132–134, 159, 160, 191, 253
  as communication, 82, 240–241, 265–266.
    See also Emblemic style
  constraints on, 237
  and consumption, 254
  copying, 132
  decorative, 133
  defined, 133, 265
  differences, 30, 69, 102–103, 108–109, 114–115
  distributions, 128, 254–256
  vs. function, 7, 79, 80, 81, 237, 241, 264, 278
  interpretations, 79, 80, 175
  learning context, 276
  material, 236, 238, 243, 244, 246
  and material culture variability, 4–5, 104, 113, 114–115, 161, 175, 185, 222, 237, 250, 251–252(figs.)
  as method of study, 2, 4, 8, 82–83, 104, 114–115
  as polythetic phenomenon, 104, 253
  shared, 275
  similarities, 142–143
  and social action, 238, 245
  and social boundaries, 8–9, 79, 104, 107, 114, 134, 162
  typologies, 132
  See also Technical/technological style
Subsistence technology, 17–18. *See also* Agropastoralists
Supernatural, 258–259
Surface water, 20
Sweet potatoes, 52
Symbols, 244
Systems theory, 4

Taphonomic effect, 114, 121
Taro, 52, 84
Tattoos, 43

Technical choices, 2, 79, 82, 83, 92, 93(fig.), 134, 159–160, 161, 176, 179, 191, 211, 271, 278
  attribute analysis, 146, 148–160, 174–175
Technical specialists, 127, 134
Technical/technological style, 83, 133
  and ceramic decoration, 176
  cultural conservatism in, 212
  as determinants of choice, 191–192, 194, 198–199, 200, 201– 202, 205, 206
  and social identity, 82, 103–104, 113, 134, 176, 178–179, 183, 191, 211–213, 216, 228, 230, 274–276
  stability, 191
  See also Technology, and choice
Technical systems
  and mobility strategies, 33–37
  and resource management, 31–32, 45, 47
  roles, 12, 30
  as signifying system, 30, 31, 43, 191. *See also* Emblemic style
  and social boundary maintenance, 30–33
  variability, 47, 48, 61
Techniques, 235, 236, 261
  and habitus, 246
  and practice, 239, 247–248, 260
Technological patterning. *See under* Social bounderies
Technologie. See Technology, choice and cognition
Technology
  as adaptive stategy, 79
  analysis, 146
  anthropology of, 1, 35, 211
  as behavior system, 6, 9, 78, 79, 83, 211
  and choice, 5, 6, 8, 10, 12, 30, 47, 48, 68–69, 82, 83, 85– 91, 99–104, 109, 134, 159–160, 179, 209, 267
  choice and cognition, 2, 39, 78, 79, 90–91, 102, 235, 245– 246, 260, 261n2, 262n3, 268
  complexity, 277
  and culture, 1, 2, 5–7, 9, 12
  curated, 39
  diffusion, 66, 94–97, 98(figs.), 190, 275–276
  efficiency, 38
  expedient, 39
  and gender, 56
  as maintainable system, 17
  and material culture and variability, 4, 9, 39, 42, 47, 78, 87, 102
  and production of goods, 56, 58–59, 60, 61–63, 102–104, 134, 209, 216, 237
  as raw materials and production steps, 4, 6, 78, 79, 80, 85, 86(fig.), 87–91, 99, 150
  as reliable system, 17
  as research interest, 5
  and social boundaries, 271, 278
  and society, 267, 268–269
  and style, 264, 266–267
  and transformation processes association, 91
  See also Social identity; Technical/technological style; Technical systems
Teeth, first, 101
Terracing, 15
Terrell, John Edward, 7, 9, 10, 128, 130
Territorial boundaries
  administration, 27
  between (no-man's-land), 28
  and communications, 31
  and drought, 23, 31
  economic importance, 30
  of foragers and agropastoralists, 22–27, 29, 30–31, 35–37, 47
  functions, 23, 31
  interpretations, 30
  marking of, 31, 32
  overlapping, 28
  and sedentary groups, 37
  size, 31
  social and symbolic importance, 30–31, 37
Territorial lineage groups, 116, 118, 122–123
Teti, 32
TGLP. *See* Tribal Grazing Land Policy
Thatched roofs, 22, 41, 58
Thermal efficiency, 186
Threshing floors, 41
Titicut site (Mass.), 170(fig.), 174
Tobacco, 142
  as exchange item, 18, 33, 54
Tonto Basin
  Archaic period, 217
  architecture, 216, 217, 220–227, 230
  Ceramic period, 213, 217

[continues]

ceramics, 214, 217–220, 227–228, 229(fig.), 230
Classic period (A.D. 1250), 209, 214, 216, 220–228
Colonial period (A.D. 750), 209, 213–214, 217, 220, 230
cultural groups, 213, 214, 230
Hohokam migration to, 217, 220
local systems, 10, 209, 210(fig.), 213, 230
migration, 218, 224
settlement morphology, 224–227
sites, 214(fig.)
social boundaries, 209
technical choices, 209
and trade, 218
Tonto Corrugated ceramics, 228, 229(fig.)
Tools, 18, 33, 56, 237, 238
  in archaeological contexts, 38, 39, 45, 142
  defined, 40
  as exchange goods, 58
  as facilities, implements, and missiles, 40
  hidden, 46
  multipurpose, 40, 91
  storage/caching, 40
  substitutions, 178
  technological organization for, 38, 39–40
  and technological style, 212
  variability and purpose, 43, 44, 45, 91
  worn out, 39
Totemic exogamous groups, 32
Towns, Tswana, 16, 19
Trade, 32, 33, 45–46, 114, 132, 142, 218. *See also* Exchange of goods and services
Transport, 18, 33, 36
Trapping, 139
Traps, 41
Trash heaps, 16, 224–226(figs.)
Travel time, 219, 231
Trees, 15, 32, 33, 38, 86(fig.), 89
Tribes. *See* Colonization, and ethnic groups
Tribal Grazing Land Policy (TGLP) (Botswana), 28
Tribute payment, 18, 22, 27
Trigger, B., 233
Trobriand Islanders, 53
Trunks, 41
Tsassi, 32
Tshua, 32
Tshwa. *See* Tyua

Tswana, 18
  as agropastoralists, 17(fig.), 19
  chiefs, 19, 20(fig.), 22, 29
  contemporary sites, 16
  customary law, 20
  on G/wi, 44
  kinship, 22
  land management system, 17(fig.), 19–20, 22, 27, 28, 29
  spears, 44
  towns, 16, 19
  tribal territory boundary, 17(fig.), 29
  tribes, 19. *See also* Bamangwato
  villages, 16, 17(fig.), 19
  and Tyua, 44
Tubers, 34, 35, 54
Tumleo Island, 56
Turkey, 139
"Turtle-backs," 187, 188
Twine, 45
Twin Towers (Hovenweep), 197(fig.)
Typological analysis, 146
Typologies
  inference from, 132
Tyua, 17, 21
  and Bamangwato, 44
  on cattle posts as nonterritorial, 37
  headman/headwoman, 32
  herders, 35–36
  hunting, 36, 37, 44, 45
  and Kua and G/wi, 44–45
  land rights, 28
  language, 32, 44
  material items, 40–41, 45
  mobility, 35, 36–37
  plant gathering, 36
  socioeconomic system, 32, 44
  spears, 44, 45
  territorial boundaries, 23, 32, 37
  tools, 39
  and trade, 36
  and Tswana, 44

Uncas (sachem), 165, 177
Undecorated artifacts, 6, 218, 227
United Colonies of New England, 165
University of Massachusetts Archaeology Field School, 136

University of New Mexico Kalahar Project, 15
Upper Paleolithic (Europe), 113
Urewe, 113
Utah, 190, 203
Utilitarian goods, 6, 101, 210, 211, 212
 decorated, 41–42, 43
Utilitarian purpose, 4, 211

Vessel lot, 173
Veterinary cordon fences, 28, 29
Victoria, Lake, 248, 249(fig.)

Villages
 agricultural, 18, 19
 establishment of, 57
 Iroquoian, 141
 kinship based, 55–56, 57–58, 69, 141
 Sepik Coast, 51, 54, 55–56, 57–59, 69, 70
 and technology choice, 69
 Tswana extended family, 16
 See also Puebloan architecture
Vultures, 47

Wage workers, 12, 17, 28, 35
Wagner, P. L., 40
Wagons, 18
Walls, 15, 20, 41, 41, 187
 cobbled, 188, 201
 courses, 188–190
 fabric. *See* Adobe, coursed, architecture, materials
 mud, 187, 188, 201
 plastered, 193
 puddled/poured, 188, 190
 retaining, 15
 Wattle and daub, 256
 See also Room contiguity index
Walnuts, 139
Wampanoag tribe, 163
 cemeteries, 171
 ceramics, 167, 168(table), 169–170, 171, 172–176, 178–179
 and the English, 163–164, 165
 European brass kettles, 171
 language, 163
 marriage, 163
 and Narragansett, 164–165, 173–174
 sachems, 164

 socioeconomic structure, 163
 territory, 163, 164(fig.)
Wandala state (Africa), 115
Warthog bones, 47
Washburn, D., 240
Water
 access to, 20, 22, 26–27, 33
 -bearing plants, 33
 -carrying pots, 119, 122
 control system, 270
 storage, 41
 and territoriality, 23, 26
 See also Bore holes; Wells
Watering troughs, 41
Watson, J. B., 52, 53
Wattle and daub, 256
Weaving, 84
 pattern, 65, 66
Wells, 16, 22, 36
Welsch, Robert L., 7, 9, 10, 52, 125, 130
Wequamugs (Mohegan), 165
West Atlantic languages (Africa), 115
Western Sandveld Region (Central District), 16, 17, 24–25(table)
Wet season, 56, 100
White, R., 7
White Mountain Redwares, 216
Wiessner, P. 191, 242, 243
Wigwams, 139–141
Wilcox, D. R., 188
Wildlife, 22, 27
 and cordon fences, 28
Willey, G. R., 73, 109
Williams, Roger, 139, 141, 143, 165, 178
Windsor Tradition, 144(table), 145(fig.), 158(table), 167, 169
Witchcraft protection, 43
Wire, 43
Wobst, H. M., 4, 81, 156, 191, 240, 241
Women
 burials, and broken vessels, 101
 carrying bags, 17–18
 cultivators, 85, 250
 as gatherers, 36
 post-marital resocialization, 250, 253
 potters, 84, 85, 103, 118, 123, 124, 143, 176, 178, 250, 262n5
 and travel, 75n4

Wood, W., 138
Wooden:
  bowls, 14, 19, 64
  containers, 18, 41
  pestles and mortars, 91
  projectile points, 43
  spoons, 41
  tools, 38, 40

Woronocos, 135
Wright, R. P., 275

Yamba, 84
  potters, 85
Yams, 54, 56
Yellen, J. E., 47
Young, William, 136